How to Ace Statics with Jeff Hanson

HOW TO ACE STATICS WITH JEFF HANSON

JEFF HANSON, ASHLEY BERNAL, and JAMES PITARRESI

McGraw Hill

NEW YORK CHICAGO SAN FRANCISCO ATHENS LONDON
MADRID MEXICO CITY MILAN NEW DELHI
SINGAPORE SYDNEY TORONTO

How to ACE Statics with Jeff Hanson

Library of Congress Control Number: 2022037611

1 2 3 4 5 6 7 8 9 LWI 26 25 24 23 22

ISBN 978-1-264-27830-5
MHID 1-264-27830-6

Sponsoring Editors
Bob Argentieri, Wendy Rinaldi

Production Supervisor
Lynn M. Messina

Acquisitions Coordinator
Elizabeth M. Houde

Project Manager
Patricia Wallenburg, TypeWriting

Copy Editor
Alison Shurtz

Art Director, Cover
Jeff Weeks

Illustrator
Jacob Hanson

Composition
TypeWriting

This “unbook” is dedicated to you, the student, whose desire to learn has driven you to seek other methods and sources of knowledge for your success. Students just like you have made the Jeff Hanson YouTube channel one of the most popular engineering education channels on the internet. Through the overwhelming outpouring of success stories and comments from students around the world, the authors were compelled to do even more to further improve student success. We thank you for all of your support and wish you the highest level of encouragement to reach your dreams and make a difference in the world.

About the Authors

Jeff Hanson is a Lecturer in the Department of Mechanical Engineering at Texas Tech University, Lubbock, Texas. He received his PhD in Systems and Engineering Management from Texas Tech. His popular YouTube channel with instructional engineering course videos has hundreds of thousands of followers worldwide.

Ashley Bernal is an Associate Professor of Mechanical Engineering at Rose-Hulman Institute of Technology, Terre Haute, Indiana. She received her PhD in Mechanical Engineering from Georgia Tech and previously worked at Boeing as a Subsystems Engineer on Joint Unmanned Combat Air Systems.

James Pitarresi is a Distinguished Teaching Professor of Mechanical Engineering and Vice Provost and Executive Director, Center for Learning and Teaching Binghamton University, State University of New York. He received his PhD in Civil Engineering at the University of Buffalo.

Contents

Introduction

Years ago, Jeff Hanson began making instructional videos to help his students understand the complexities of statics in relatable and practical ways. These videos grew into an entire YouTube channel course used by hundreds of thousands of students around the world.

This "unbook" is the next step in supporting student learning. It's the result of over 50 years of teaching statics and finding the best ways to explain the concepts and important aspects of this foundational course. We have extensive experience in seeing commonly made mistakes and the most difficult concepts for students to understand, so we not only cover the concepts but also help you avoid the common problems and give you tools to remember the difficult concepts.

We call this an "unbook" because we've created a book that looks like the most organized set of class notes ever written for Statics. Every page is user-friendly, and at the end of every chapter, you'll find summaries to help you review. This is a workbook with space for you to draw diagrams and work out solutions in the book. Throughout the book you'll find many examples of how the problems presented in the book relate to real-world systems.

The Structure of the Book

The book is broken into Levels, and as you gain experience and level up, you will be equipped with the skills you need for more and more complex problems. The levels of the book are ordered as you would typically see them in a Statics class, and for each level, you will find:

- A **Calculator Problem**, which is a printed, worked out, step-by-step solution
- A **Video Example Problem** with a completely worked out solution
- A **Test Yourself Problem** (with the solution in the back of the book) that will make sure you can solve the problem on your own

You'll also find:

- **Pro Tips** from *Ref Jeff*, which are helpful tips to avoid mistakes and memory aids to help you remember concepts and application methods
- **Pitfalls** from *Hard Hat Jeff*, which are the most common mistakes made by students, so you can avoid them
- **Recipes** from *Chef Jeff* that show the step-by-step approach to that type of problem
- **Sample Exams** (two complete sets!): One set has complete video solutions, and the other set has fully worked out solutions in the back of the book

How to Use the Videos

All of the videos that accompany the book can be accessed on the Jeff Hanson Statics channel on YouTube using the following QR code:

as well as this link: **https://tinyurl.com/5n8vvamu**.

The proper way to use the video solutions is to not just watch the video through because that will make it look easy. Instead, start the video, press pause, and try to solve the problem on your own. Press play if you get stuck and need help in a tough spot. If you watch Tiger Woods play golf on Sunday, you're not going to be playing in the PGA on Monday; you would have to practice—a lot! The same goes for these videos: watching Jeff solve these problems will not guarantee you will be able to solve them. You have to practice working them yourself.

Video solutions to the first practice exam can also be found on Jeff's channel.

Ultimately, your success in this course depends on the amount of practice you put in. The more problems you work, the more comfortable you'll become with this material and the faster you'll be able to solve the problems. So practice these methods, and in no time, you'll be a pro too!

HOW TO ACE STATICS WITH JEFF HANSON

Level 1

Introduction to Vector Mechanics

WATCH VIDEO **STATICS LESSON 1**
Introduction and Newton's Laws, Scalars, and Vectors

What Is Statics?

Study of mechanics concerned with the analysis of forces and moments acting on physical systems which do not experience acceleration ($\vec{a} = 0$).

Statics will allow you to analyze structures such as those below:

Statics Is Based on Newton's First and Third Laws

First Law

- An object at rest remains at rest or an object in motion remains in motion (constant velocity) unless acted upon by an outside force.
- $\Sigma\vec{F} = \vec{0}$ (sum all forces and set equal to zero)

Second Law

- Force on an object is equal to its mass multiplied by its acceleration
- $\Sigma\vec{F} = m\vec{a}$ (dynamics)

Third Law

- When bodies interact, for every action, there is an equal and opposite reaction
- Allows construction of FBDs (free body diagrams)

Scalars and Vectors

Scalars

A quantity described by magnitude alone.

Examples: t (time), m (mass), v (volume), ρ (density), T (temperature)

Vectors

A quantity described by magnitude as well as direction.

Examples: $\vec{F}$ (force), $\vec{a}$ (acceleration), $\vec{x}$ (displacement), $\vec{v}$ (velocity)

- Arrow (represents direction)
- Line length (represents magnitude)
- Can be written in polar form (magnitude and angle) or Cartesian form ($\hat{\imath}, \hat{\jmath}, \hat{k}$)

Line of Action (LOA)

A line on which a vector lies continuing in both directions indefinitely.

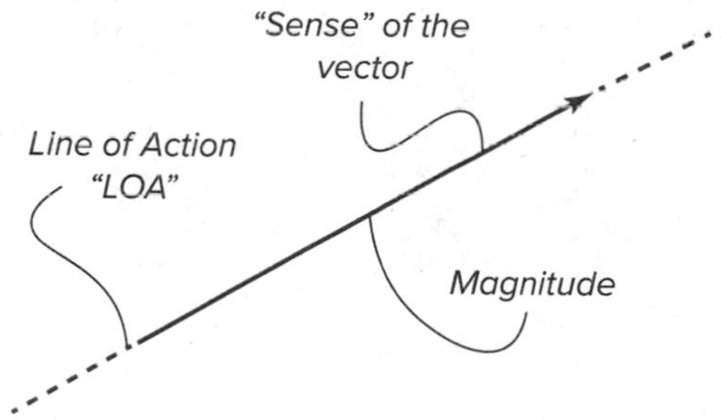

Note: When given two points that lie on a vector, you can find its position vector. For instance, you can find $\overrightarrow{AB}$ by taking point B in ($\hat{\imath}, \hat{\jmath}, \hat{k}$) and subtracting point A ($\hat{\imath}, \hat{\jmath}, \hat{k}$). This is referred to as the *position vector*. Hansonism for position vector: "How do you get to grandma's house?"

PITFALL

Don't forget for position vector $\overrightarrow{AB}$ you take $B - A$ and for position vector $\overrightarrow{BA}$ you take $A - B$.

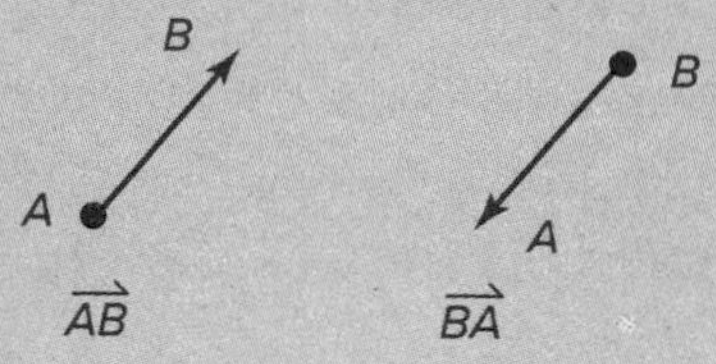

EXAMPLE: POSITION VECTOR

Find the position vector $\overrightarrow{BA}$.

Remember a "position vector" simply tells you how to get from one point to another.

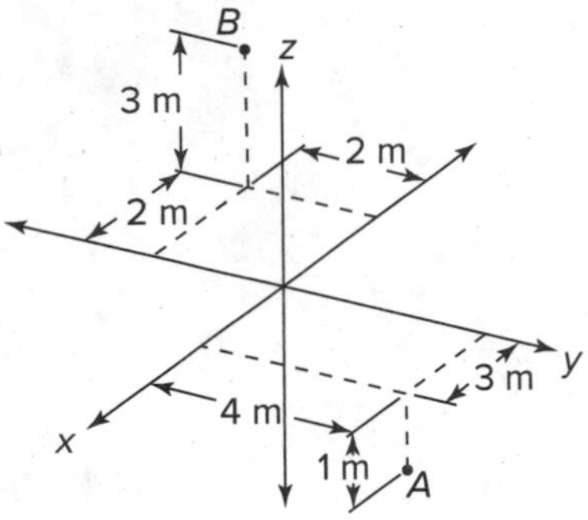

STEP 1: Identify the coordinates of the two points:

$A = (3, 4, -1)$

$B = (-2, -2, 3)$

STEP 2: Identify the direction of the vector (which is the start point and which is the end point). $\overrightarrow{BA}$ should read as vector $A - B$, going from B to point A.

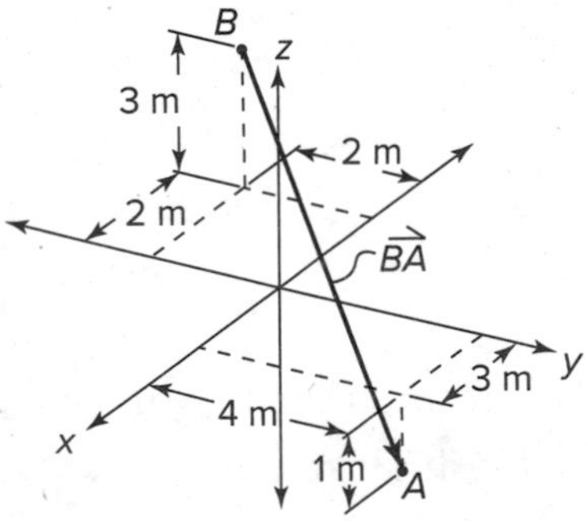

STEP 3: Subtract the ending point from the starting point.

Note: Be careful with your signs!

$(3\hat{\imath}, 4\hat{\jmath}, -1\hat{k}) - (-2\hat{\imath}, -2\hat{\jmath}, 3\hat{k}) = (5\hat{\imath} + 6\hat{\jmath} - 4\hat{k})$

So $(5\hat{\imath} + 6\hat{\jmath} - 4\hat{k})$ is your position vector which tells you how to get from one point to another, or in other words the Hansonism "how to get to grandma's house."

Note: Now see if you can find the answer for $\overrightarrow{AB}$. ***Hint:*** The answer is the same except all the signs are flipped.

✓ TEST YOURSELF 1.1

SOLUTION TO TEST YOURSELF: Introduction to Vector Mechanics

For the given vectors, find the position vector for $\vec{r}_{AB}$, $\vec{r}_{AC}$, $\vec{r}_{AD}$, and $\vec{r}_{AE}$.

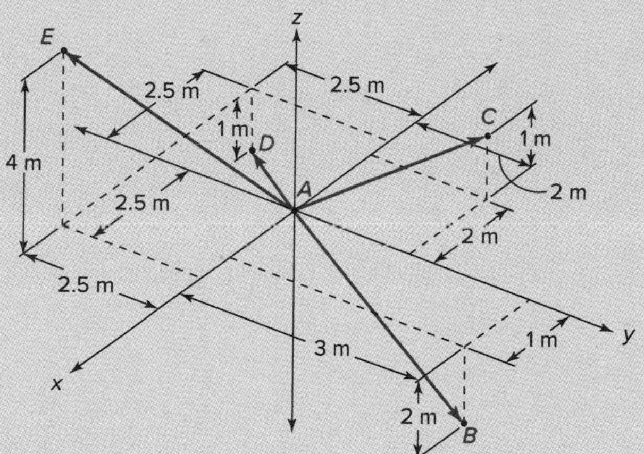

As a challenge, can you find position vector $\vec{r}_{BC}$, $\vec{r}_{BE}$, $\vec{r}_{DE}$?

ANSWERS

$\vec{r}_{AB} = (1\hat{i} + 3\hat{j} - 2\hat{k})$ m

$\vec{r}_{AC} = (-2\hat{i} + 2\hat{j} + 1\hat{k})$ m

$\vec{r}_{AD} = (-2.5\hat{i} - 2.5\hat{j} - 1\hat{k})$ m

$\vec{r}_{AE} = (2.5\hat{i} - 2.5\hat{j} + 4\hat{k})$ m

$\vec{r}_{BC} = (-3\hat{i} - 1\hat{j} + 3\hat{k})$ m

$\vec{r}_{BE} = (1.5\hat{i} - 5.5\hat{j} + 6\hat{k})$ m

$\vec{r}_{DE} = (5\hat{i} + 0\hat{j} + 5\hat{k})$ m

WATCH VIDEO

STATICS LESSON 2
Vector Language, Introduction to Vector Addition

CONCURRENT VECTOR FORCES

Two or more vector forces passing through the same point.

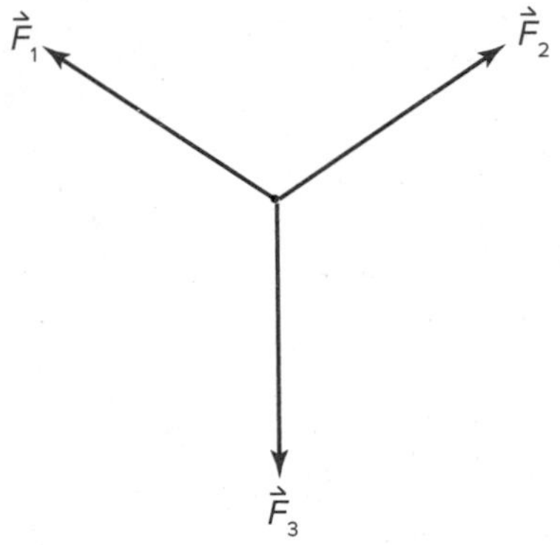

Concurrent and Coplanar Vectors

Note: We will learn how to solve this in the next level.

COPLANAR VECTOR FORCES

Two or more vectors contained in the same plane.

CHALLENGE QUESTION

If you have two concurrent vector forces, will they always be in the same plane? What about three concurrent vector forces?

Resultant of Vector Forces

A single equivalent resultant vector replacing two or more vector forces.

$\vec{r} = \vec{a} + \vec{b}$

where $\vec{r}$ is the resultant of vector $\vec{a}$ plus vector $\vec{b}$.

ANSWER

Any two vectors that are concurrent (that is, they meet at a point), will, by definition, form a plane. However, if there are three or more concurrent vectors, then generally, they do not form a plane except in special cases.

Let's Learn to Add Vectors!

For all of the following four cases, calculate the resultant vector ($\vec{r}$).

ASSUME FOR EACH CASE:

$|\vec{a}| = 5$ lbs (The magnitude of $\vec{a}$ is 5 lbs.)

$|\vec{b}| = 7$ lbs (The magnitude of $\vec{b}$ is 7 lbs.)

CASE 1

- Both vectors are along the same LOA
- Sense (direction) is the same
- Simply add together

$\vec{r} = 5 \text{ lbs} + 7 \text{ lbs}$

$= 12 \text{ lbs}$ . . . but this is not complete

PRO TIP

Remember if you have a vector on one side of the equal sign, you must have a vector on the other.

POLAR COORDINATE SYSTEM REVIEW

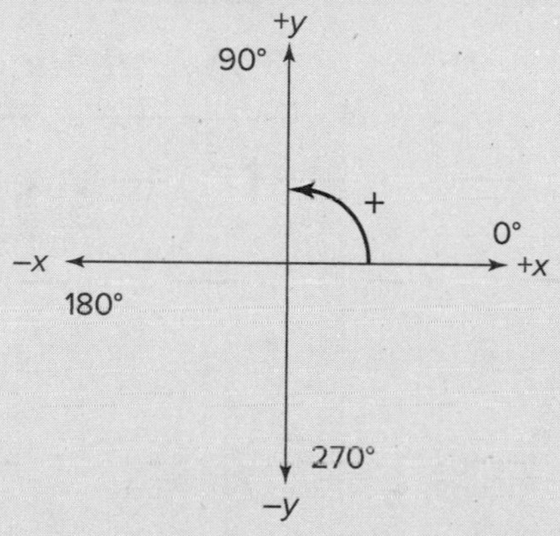

The resultant vector is in the positive x direction so: $\vec{r} = 12 \text{ lbs} \angle 0°$ (called the polar form of the vector).

CASE 2

- Both vectors are on the same LOA
- Vector $\vec{b}$ in this case is opposite direction to $\vec{a}$
- Simply subtract

  $\vec{r} = 5 \text{ lbs} - 7 \text{ lbs} = -2 \text{ lbs} \angle 0°$

  OR

  $\vec{r} = 5 \text{ lbs} - 7 \text{ lbs} = 2 \text{ lbs} \angle 180°$

WATCH VIDEO

STATICS LESSON 3
The Triangle Rule for Adding Vectors to Find a Resultant

TIME OUT

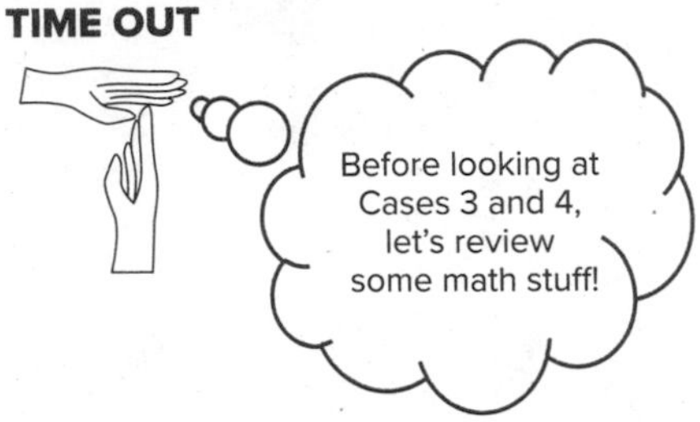
Before looking at Cases 3 and 4, let's review some math stuff!

Recall how to add two vectors together:

Triangle Rule
Parallelogram Rule
Tip to Tail Rule

(These are all the same rule, just different names.)

Tip to Tail Rule Recipe (Most Common Name)

STEP 1: Create a triangle using the given two vectors as components of $\vec{r}$. Move one vector's tail to the second vector's tip. It doesn't matter which vector you start with as they create the same triangle.

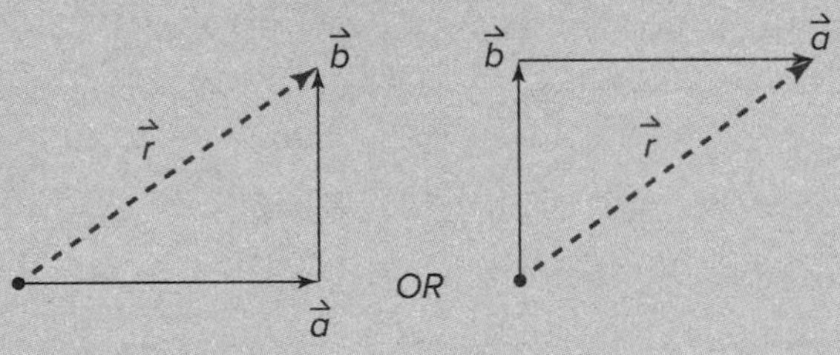

STEP 2: Identify as many angles in your triangle as possible.
Note: Sketch horizontal lines to make the angles more evident.

STEP 3: Calculate the magnitude of $\vec{r}$ using Pythagorean Theorem if triangle has a right angle or Law of Cosines if it does not.

STEP 4: Calculate the polar angle θ using the tangent function if the triangle has a right angle or Law of Sines if it does not.

STEP 5: Write your expression for $\vec{r}$ in polar form. Remember to reference your angle for $\vec{r}$ clearly.

TRIGONOMETRY REVIEW

Pythagorean Theorem

$$a^2 + b^2 = c^2$$

SOHCAHTOA

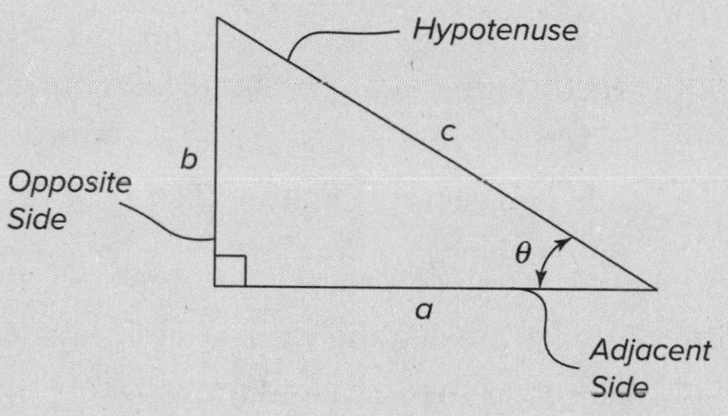

A mnemonic device (SOHCAHTOA) is used to remember the following trigonometry functions:

$$sin(\theta) = \frac{opp}{hyp} = \frac{b}{c}$$

$$cos(\theta) = \frac{adj}{hyp} = \frac{a}{c}$$

$$tan(\theta) = \frac{opp}{adj} = \frac{b}{a}$$

Note: For labeling triangles, the angle opposite a given side will always be the same name but capitalized.

The Law of Cosines

It's not just a good idea, it's the *law*!

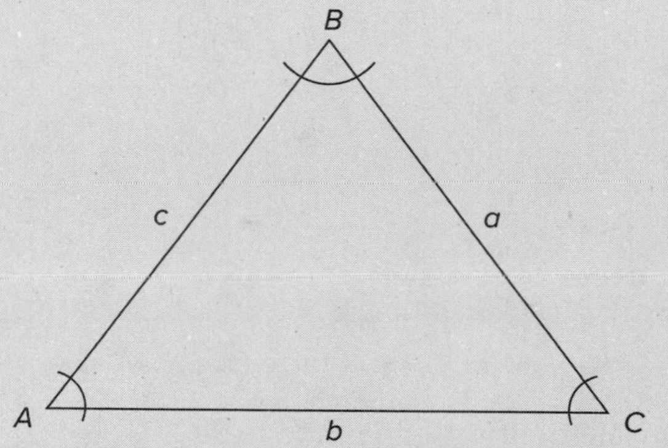

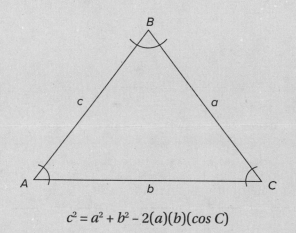

$$c^2 = a^2 + b^2 - 2(a)(b)(cos\ C)$$

Law of Sines (referred to as the side-angle-side)

$$\frac{a}{sin(A)} = \frac{b}{sin(B)} = \frac{c}{sin(C)}$$

Note: For the Law of Sines, you need to know one pair (i.e., side a and angle A) and one other thing to solve for all missing information.

CHALLENGE QUESTION

What if I accidentally write these equations upside down?

PRO TIP

We all remember the Pythagorean Theorem, which works for only one triangle, right triangles. The Law of Cosines works for *all* triangles, so let's remember this one instead! (*Note:* If angle C is 90°, then the Law of Cosines becomes the Pythagorean Theorem.)

ANSWER

It doesn't matter. They will still be equal because they are simply ratios.

GEOMETRY REVIEW

The interior angles in a triangle add to 180°.

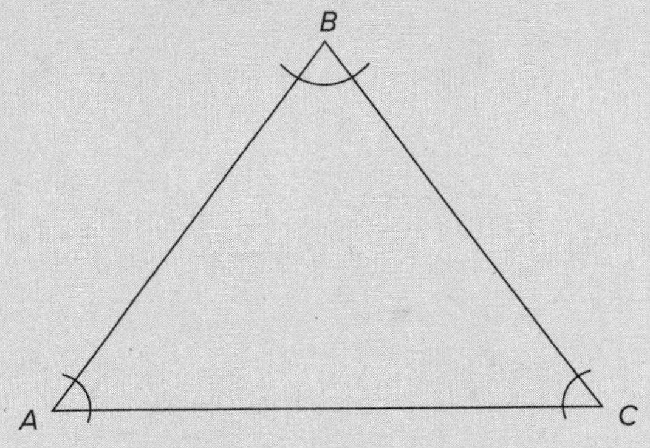

Complementary angles add to 90°.

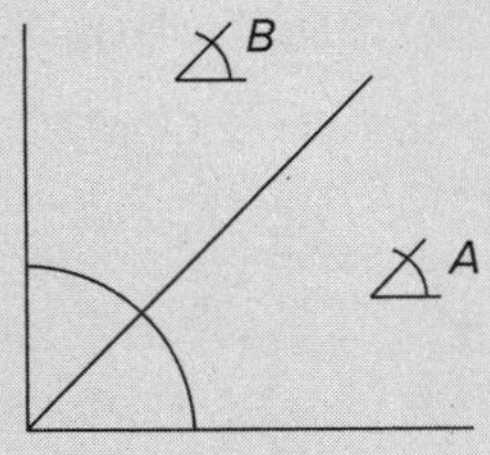

Supplementary angles add to 180°.

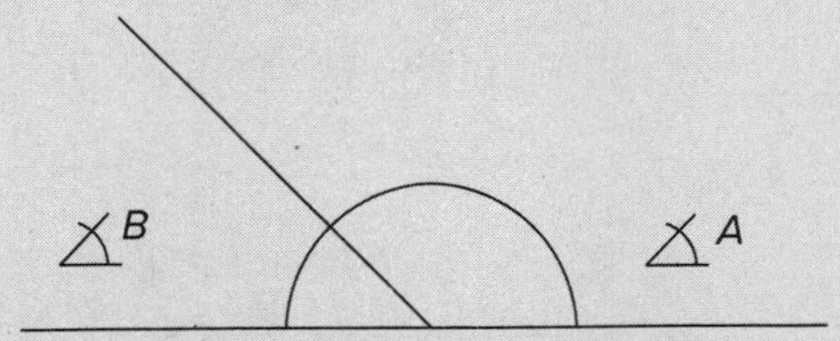

When two parallel lines are intersected by another transversal line, alternate interior angles and alternate exterior angles are congruent.

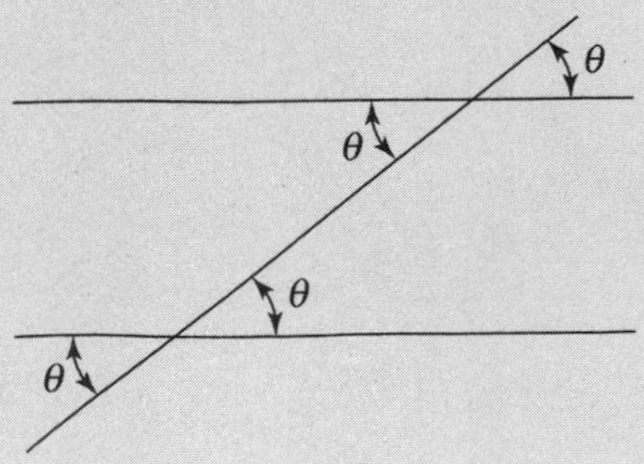

CASE 3

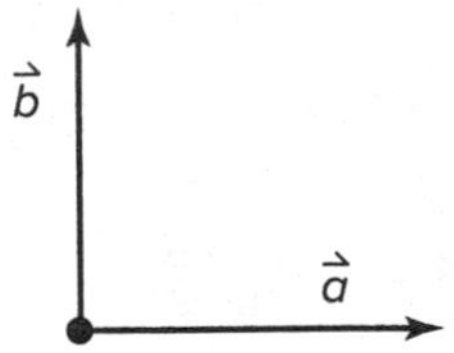

- Vectors are not on the same LOA but are at 90°
- Add two vectors together (see Tip to Tail Rule Recipe)

Note: Since we are dealing with a right triangle, use Pythagorean Theorem and the tangent function.

$$|\vec{r}| = \sqrt{(5\text{ lbs})^2 + (7\text{ lbs})^2} = 8.6\text{ lbs}$$
$$tan(\theta) = \tfrac{7}{5}\text{, making } \theta = 54.46°$$
$$\vec{r} = 8.6\text{ lbs} \measuredangle 54.5°$$

CASE 4

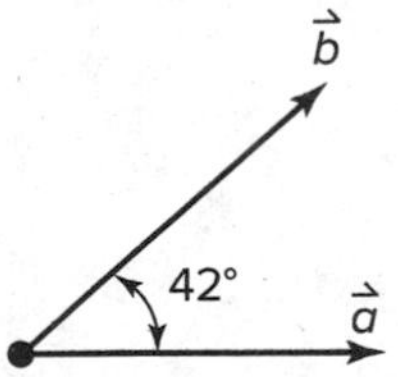

- Vectors are not on the same LOA and are at some oblique angle
- Add two vectors together (see Tip to Tail Rule Recipe)

Note: Since we are *not* dealing with a right triangle, use Law of Cosines and Law of Sines.

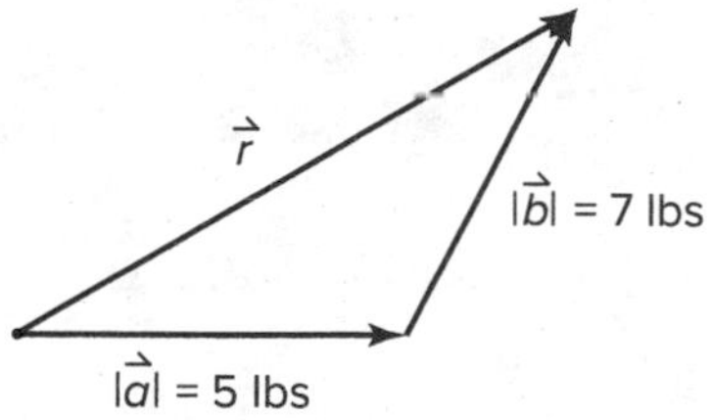

It seems like we may not have enough information to solve this . . . or do we?

PRO TIP

One helpful hint on finding angles and other missing information about a triangle is to sketch horizontal lines extended from the angles of the triangle. This will make angles more apparent.

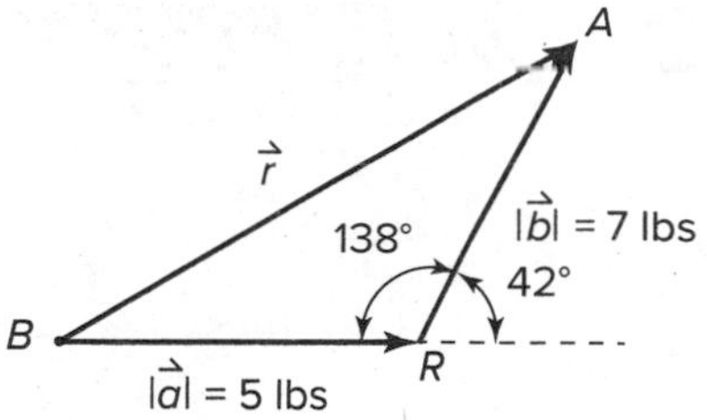

Using supplementary angles, the interior angle is 138°.

Using Law of Cosines

$$r^2 = 5^2 + 7^2 - 2(5)(7)cos(138°)$$

$$|\vec{r}| = 11.23 \text{ lbs}$$

Using Law of Sines

$$\frac{11.23 \text{ lbs}}{sin(138°)} = \frac{7 \text{ lbs}}{sin(B)}$$

$$\measuredangle B = 24.65°$$

$$\vec{r} = 11.23 \text{ lbs} \measuredangle 24.65°$$

EXAMPLE: TIP TO TAIL RULE

Find the resultant of the two vectors $\vec{F}_A$ and $\vec{F}_B$.

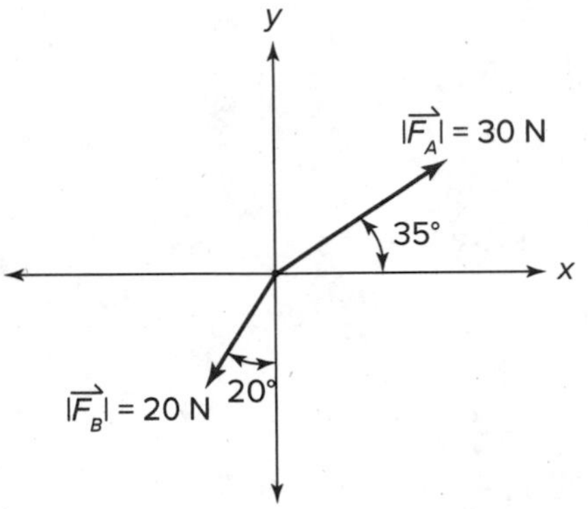

STEP 1: Create a triangle using the given two vectors.

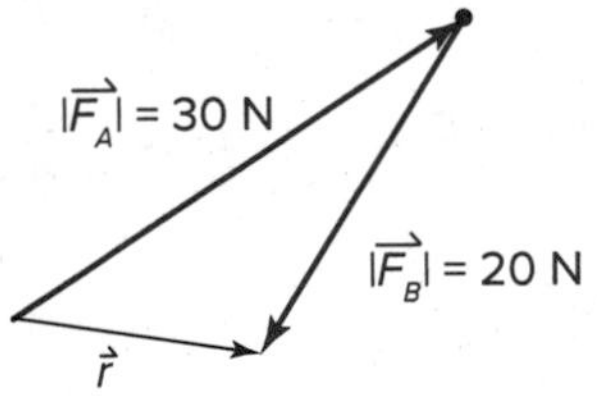

STEP 2: Identify as many angles in your triangle as possible.

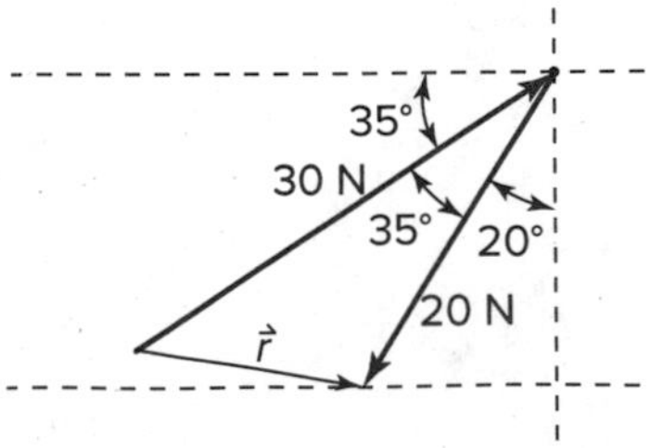

STEP 3: Solve for the remaining side. Notice here, we were only able to identify one angle. We know two sides and the angle between them, which sounds like a perfect job for the Law of Cosines.

$r^2 = 30^2 + 20^2 - 2(20)(30)\,cos(35°)$

$|\vec{r}| = 17.8$ N

STEP 4: Now that we know the magnitude of the third side, let's find the remaining angles using the Law of Sines.

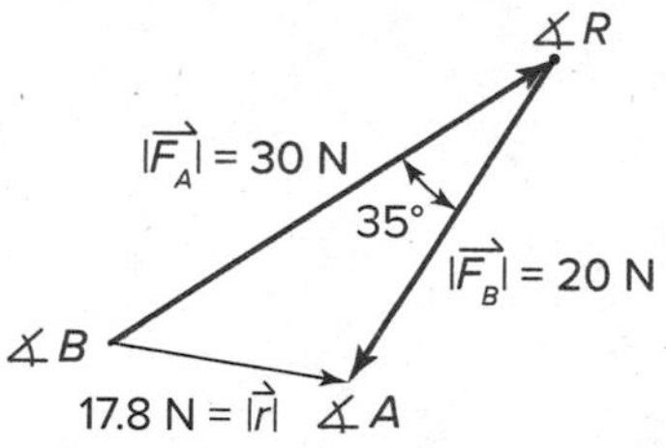

$$\frac{17.8N}{sin(35°)} = \frac{30}{sin(A)} = \frac{20}{sin(B)}$$

$\measuredangle A = 75.1° \; or \; 104.9°$

$\measuredangle B = 40.1°$

Danger: One of the shortcomings of the law of sines is that using it to find an angle is ambiguous since it doesn't differentiate between the angle and its supplementary angle. Since $\vec{F}_A$ is the longest of all the sides, we can deduce that angle A should be larger than 90°. Our two angle choices here are θ and $180° - \theta$. An easy way to check is to add the angles of your triangle to make sure they add to 180°.

STEP 5: Write your expression for $\vec{r}$ in polar form.

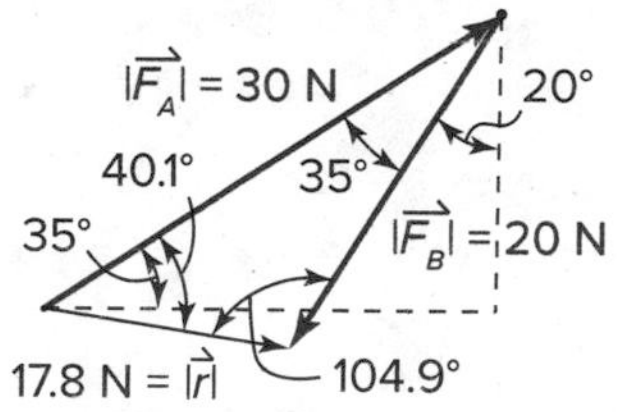

$|\vec{r}| = 17.8\,N \measuredangle 5.1°$

WATCH VIDEO

STATICS LESSON 4
Vector Addition, Triangle Rule, and Cartesian and Vector Notation

Add vectors $\vec{p}$ and $\vec{q}$ using Tip to Tail Rule to find the resultant force.

$|\vec{p}| = 40$ N (The magnitude of $\vec{p}$ is 40 N.)

$|\vec{q}| = 60$ N (The magnitude of $\vec{q}$ is 60 N.)

Press pause on the video once you get to the workout problem. Only press play if you get stuck.

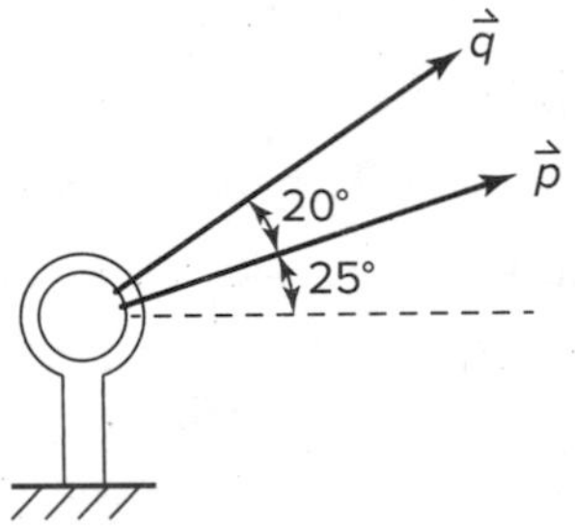

Sketch your triangle here.

PRO TIP

Make sure to draw pictures that are large enough that you can label. It shouldn't be able to be covered by a quarter!

CARTESIAN VECTORS

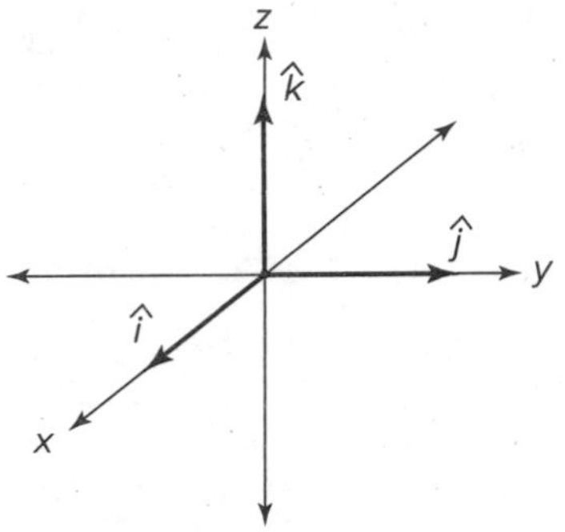

CHARACTERISTICS:

1. System is orthogonal (each axis is perpendicular to the next).
2. System is a right-handed coordinate system. (Fingers show the positive directions of each of the axes.)

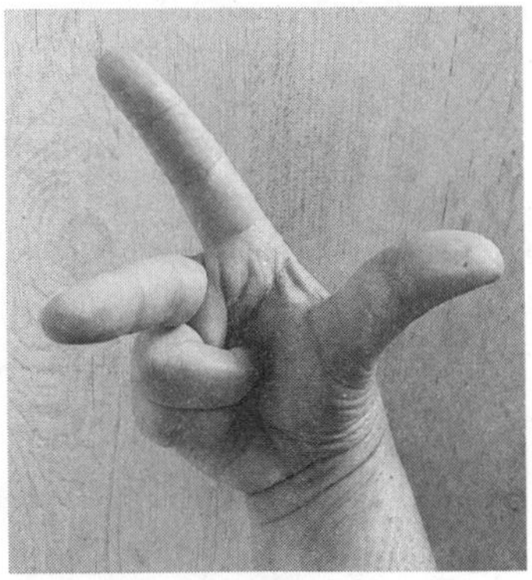

3. Contains $\hat{\imath}, \hat{\jmath}, \hat{k}$. Vectors written in Cartesian form look like the following:

   $\vec{F} = F_x\hat{\imath} + F_y\hat{\jmath} + F_z\hat{k}$

   Note anything with a "hat" on it is a *unit vector*.

What Is a Unit Vector?

- Vector with length of 1
- Simply assigns direction to a vector because anything can be multiplied by 1 and not change the quantity

PITFALL

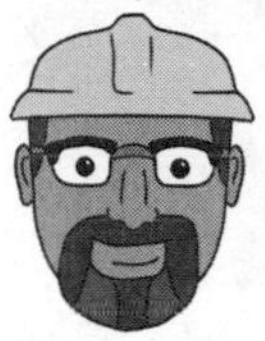

Make sure you can resolve a 2D force in Cartesian system into *x* and *y* components.

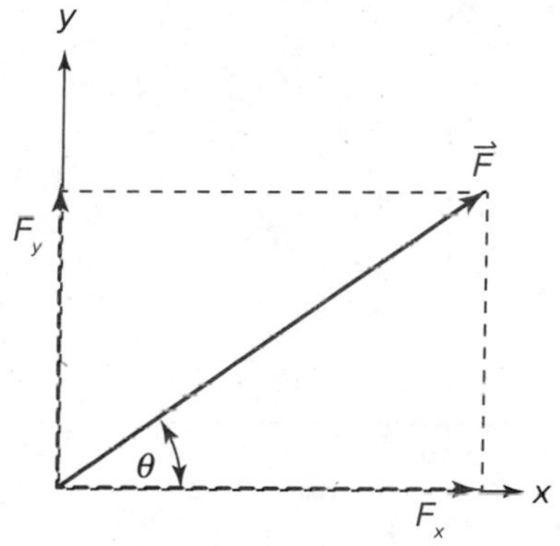

$\vec{F} = Fcos(\theta)\hat{\imath} + Fsin(\theta)\hat{\jmath}$
$F_x = Fcos(\theta)$
$F_y = Fsin(\theta)$
F: magnitude of the vector
θ: angle between *x*-axis and the vector
$\hat{\imath}$ and $\hat{\jmath}$: directions of the respective components

To find the magnitude of any 2D vector in Cartesian space use Pythagorean Theorem:

$$|\vec{F}| = \sqrt{F_x^2 + F_y^2}$$

Note: The components (also called the projections of the vectors) should be in the same quadrant as the resultant vector.

Solve this problem as earlier but solve using Cartesian coordinates instead of the Tip to Tail Rule.

$|\vec{p}| = 40$ N
$|\vec{q}| = 60$ N

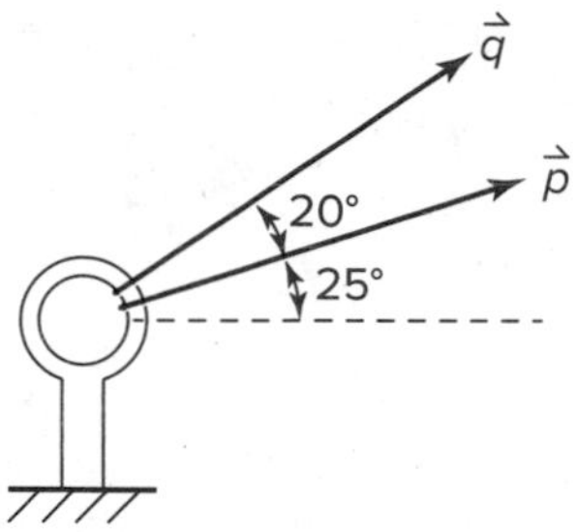

Sketch your work here.

✓ TEST YOURSELF 1.2

SOLUTION TO TEST YOURSELF: Introduction to Vector Mechanics

Resolve the following vectors into x and y components and express the vectors in Cartesian form.

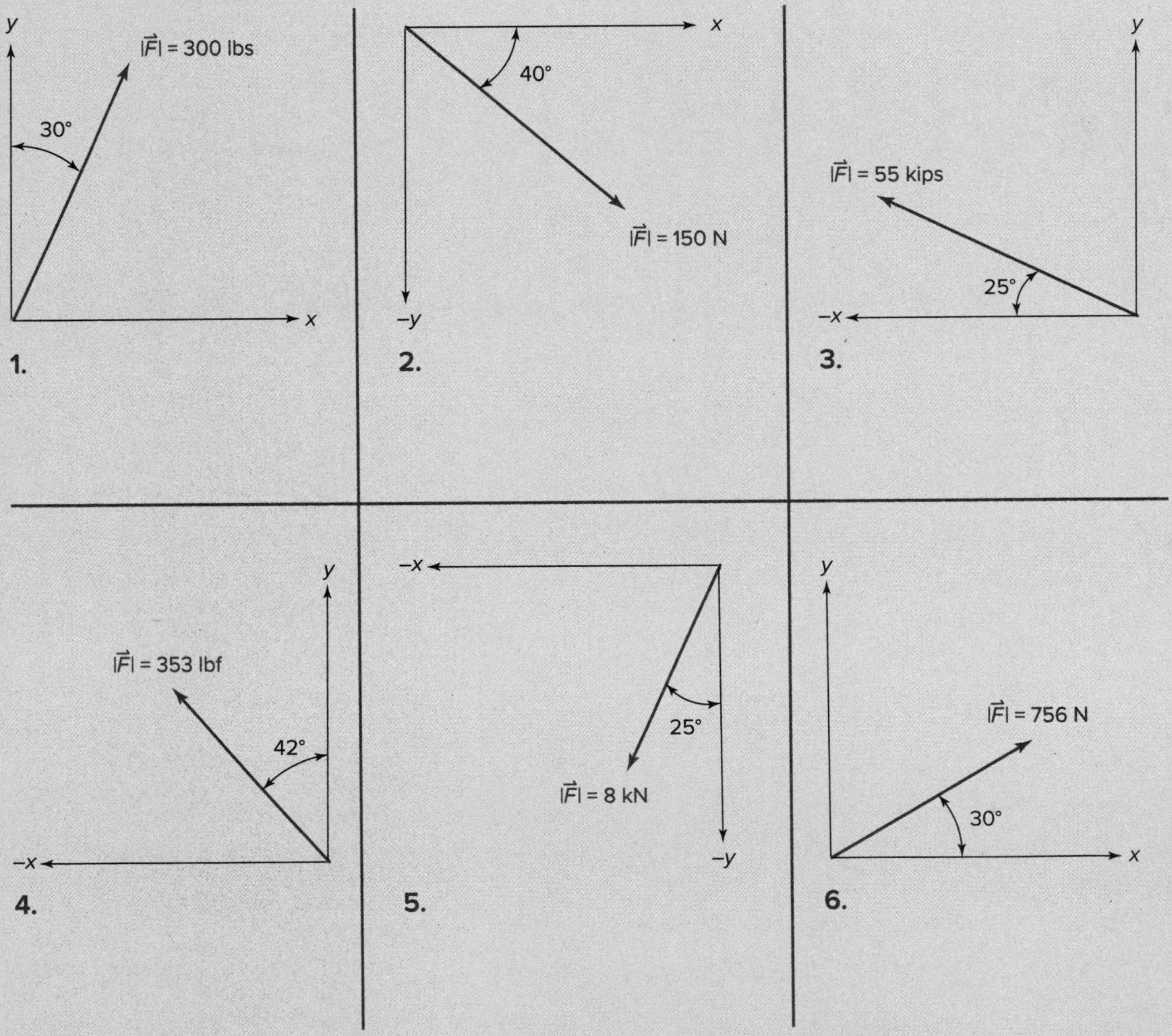

ANSWERS

1. $(150\hat{i} + 259.8\hat{j})$ lbs; **2.** $(114.9\hat{i} - 96.4\hat{j})$ N; **3.** $(-49.9\hat{i} + 23.2\hat{j})$ kips; **4.** $(-236.2\hat{i} + 262.3\hat{j})$ lbf; **5.** $(-3.38\hat{i} - 7.25\hat{j})$ kN; **6.** $(654.7\hat{i} + 378.0\hat{j})$ N

WATCH VIDEO

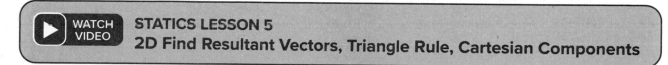

STATICS LESSON 5
2D Find Resultant Vectors, Triangle Rule, Cartesian Components

The crate has two forces being applied: (A) Find the resultant force on the crate in polar form using the triangle rule, and (B) Find the resultant force on the crate by resolving F_1 and F_2 into Cartesian components. Express the resultant vector in Cartesian form. (By the way, # is a shortcut notation for lbs.)

Press pause on the video once you get to the workout problem. Only press play if you get stuck.

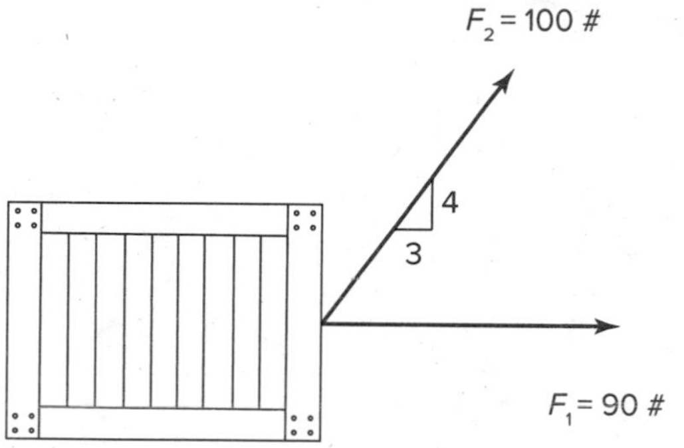

WATCH VIDEO

STATICS LESSON 6
Most Missed Topic in Statics, Cartesian Coordinates

Resolve each of the four forces into Cartesian coordinates.

Press pause on the video once you get to the workout problem. Only press play if you get stuck.

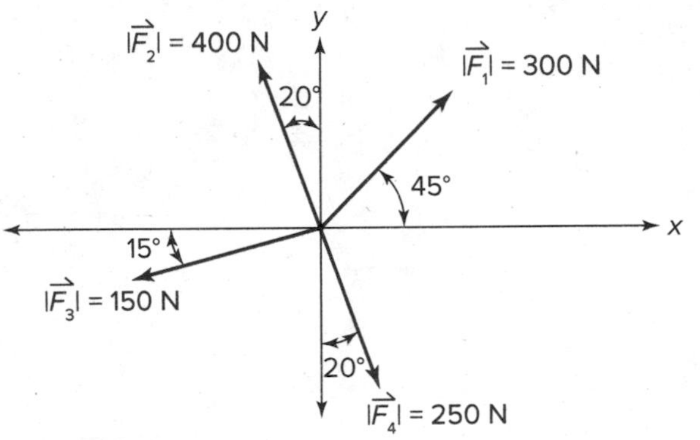

WATCH VIDEO

STATICS LESSON 7
Finding Vector Components in Non-Orthogonal Systems

Non-orthogonal coordinate systems contain axes that do not intersect at right angles. Typically textbooks will use u and v instead of x and y for these axes. For the purposes of discussion in this unbook, we will use u and v for non-orthogonal axes.

Finding Resultant Vectors in Non-Orthogonal Systems Recipe

STEP 1: Use the Tip to Tail Rule to construct a triangle. Remember the u and v components of the resultant vector must be parallel to their respective axes.

STEP 2: Use the Law of Sines to find u and v components.

PITFALL

When working problems with a non-orthogonal axis, the Cartesian component method *will not work* as it relies on the axes being 90° from each other!

EXAMPLE: NON-ORTHOGONAL

Find the magnitude of the components of the 30 lbs force in the given non-orthogonal coordinate system.

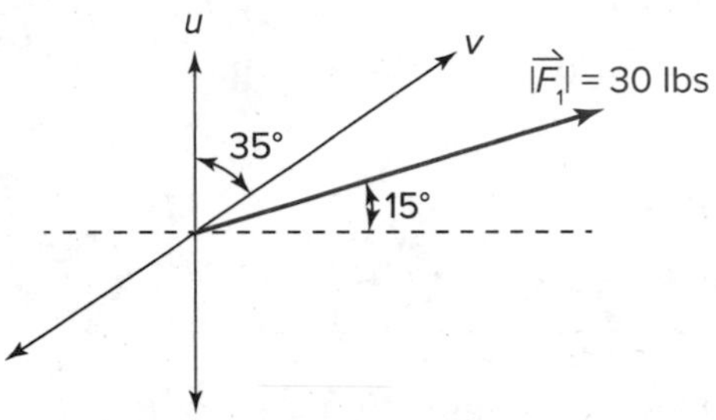

Since this problem is on a coordinate system that does not have the axes perpendicular to each other (non-orthogonal), you cannot use Cartesian coordinates to solve.

STEP 1: Use the Tip to Tail Rule to construct a triangle. Remember the *u* and *v* components must be parallel to their respective axes.

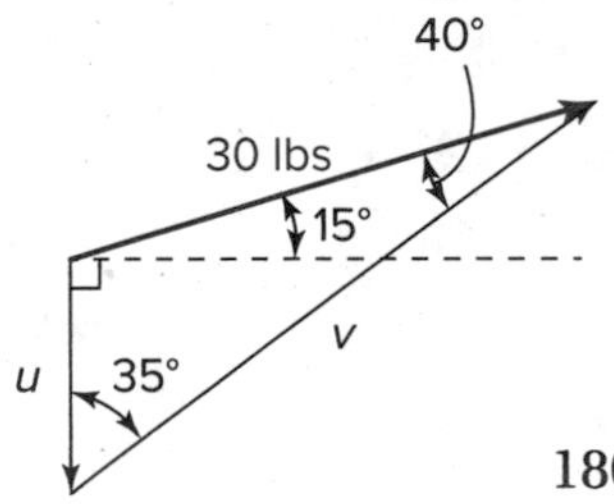

$$180° - 90° - 15° - 35° = 40°$$

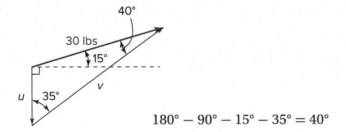

STEP 2: Use the Law of Sines to find *u* and *v* components:

$$\frac{30}{sin(35°)} = \frac{v}{sin(105°)}$$

$v = 50.52$ lbs

$$\frac{30}{sin(35°)} = \frac{u}{sin(40°)}$$

$u = 33.62$ lbs

Find the u and v components of the 30 # force using the triangle rule.

Press pause on video lesson 7 once you get to the workout problem. Only press play if you get stuck.

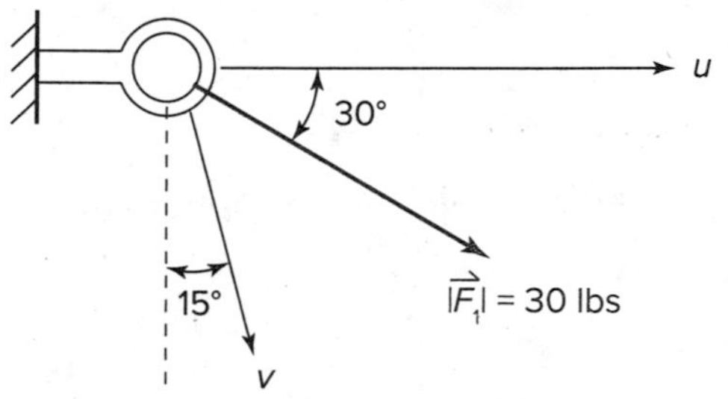

WATCH VIDEO

STATICS LESSON 8
Intro to 3D Vectors, Deriving Blue Triangle Eqns (Spherical Coordinates)

3D vectors can be represented via:

1. Two angles and a magnitude (referred to in the videos as the blue triangle vectors [spherical coordinates])
2. Three angles and a magnitude (directional cosine vectors)
3. Vectors expressed with dimensions ($\hat{\lambda}$ vectors)

BLUE TRIANGLE PROBLEMS (SPHERICAL COORDINATES)

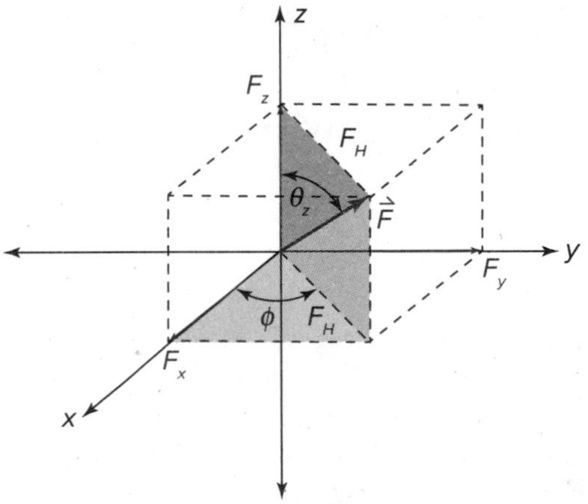

Writing a Cartesian Vector Using the Blue Triangle (Spherical Coordinates) Recipe

STEP 1: Identify the values for the following variables: $|\vec{F}|$ or F (magnitude of vector), ϕ (angle between the positive x-axis and the bottom edge of the plane containing vector $\vec{F}$) and θ_z (angle between the positive z-axis and the vector $\vec{F}$).

STEP 2: Plug the values for each of the parameters above into the following equations:

$$F_x = F\cos(\phi)\sin(\theta_z)$$
$$F_y = F\sin(\phi)\sin(\theta_z)$$
$$F_z = F\cos(\theta_z)$$

STEP 3: Use the equations above to write a Cartesian vector in the form: $\vec{F} = F_x\hat{\imath} + F_y\hat{\jmath} + F_z\hat{k}$.

EXAMPLE: BLUE TRIANGLE METHOD

Find the Cartesian vector of the force.

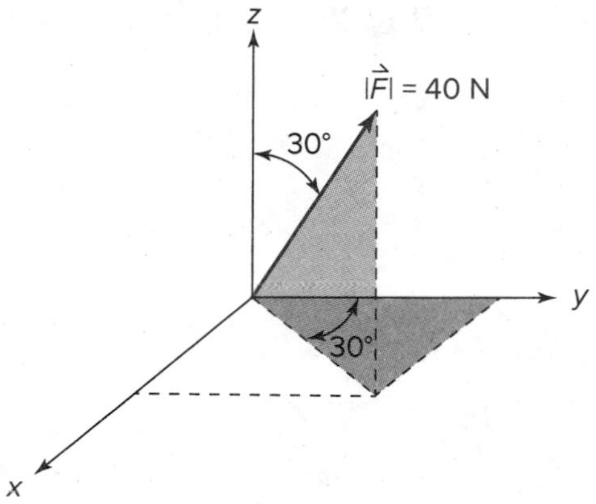

STEP 1: Identify the values for the following variables:

$|\vec{F}| = F = 40$ N

$\phi = 60°$ (***Note:*** Don't forget this is from the positive x-axis to the vector so $90° - 30° = 60°$.)

$\theta_z = 30°$

STEP 2: Substituting into the blue triangle equations:

$F_x = Fcos(\phi)sin(\theta_z) = 40cos(60°)sin(30°) = 10$ N

$F_y = Fsin(\phi)sin(\theta_z) = 40sin(60°)sin(30°) = 17.32$ N

$F_z = Fcos(\theta_z) = 40cos(30°) = 34.64$ N

STEP 3: Finally, write as a Cartesian vector:

$\vec{F} = F_x\hat{i} + F_y\hat{j} + F_z\hat{k} = (10\hat{i} + 17.32\hat{j} + 34.64\hat{k})$ N

PITFALL

When looking at the *x-y* plane for the blue triangle method, phi (ϕ) is positive when measured from the *x*-axis to the +*y*-axis and ϕ is negative when measured from the *x*-axis to the –*y*-axis.

WATCH VIDEO

STATICS LESSON 9
Drill Problems Practicing Blue Triangle (Spherical Coordinates)

Determine the angles ϕ and θ_z for each of the examples below and then write the Cartesian vector. Press pause on the video once you get to the workout problem. Only press play if you get stuck.

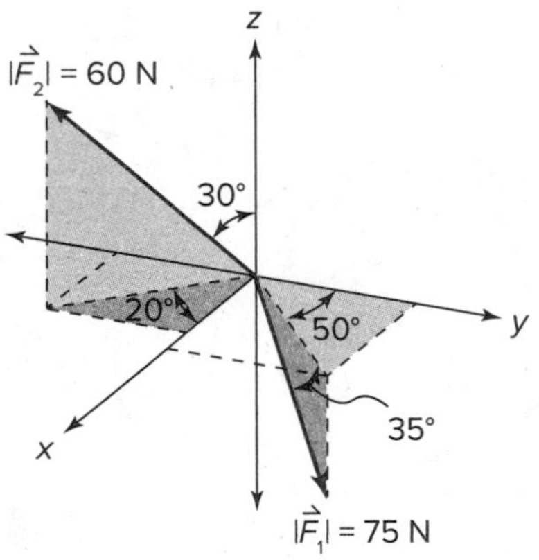

$|\vec{F}_1| =$

$\phi =$

$\theta_z =$

$\vec{F}_1 =$

$|\vec{F}_2| =$

$\phi =$

$\theta_z =$

$\vec{F}_2 =$

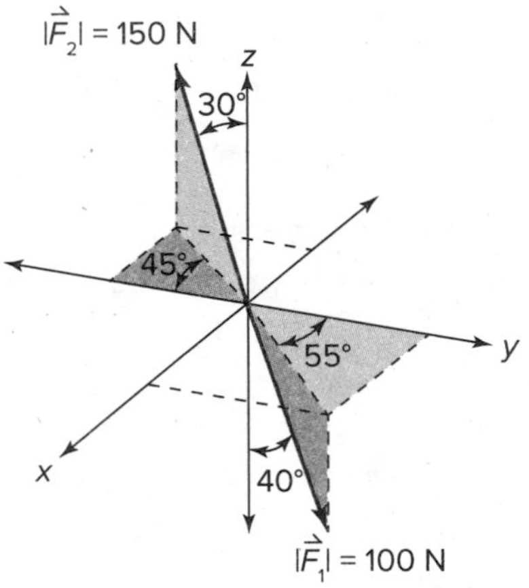

$|\vec{F}_1| =$

$\phi =$

$\theta_z =$

$\vec{F}_1 =$

$|\vec{F}_2| =$

$\phi =$

$\theta_z =$

$\vec{F}_2 =$

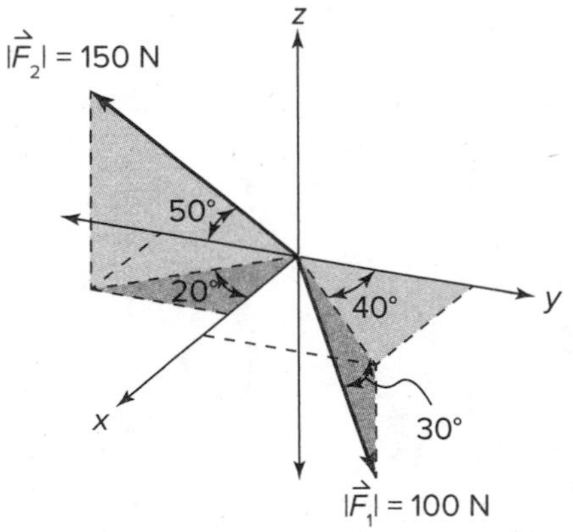

$|\vec{F}_1| =$

$\phi =$

$\theta_z =$

$\vec{F}_1 =$

$|\vec{F}_2| =$

$\phi =$

$\theta_z =$

$\vec{F}_2 =$

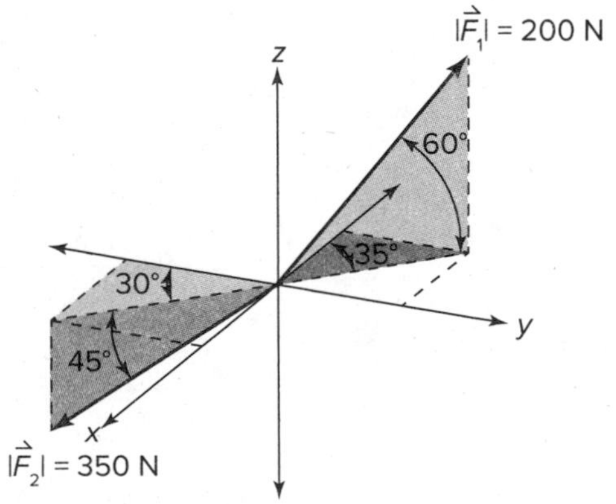

$|\vec{F}_1| =$

$\phi =$

$\theta_z =$

$\vec{F}_1 =$

$|\vec{F}_2| =$

$\phi =$

$\theta_z =$

$\vec{F}_2 =$

WATCH VIDEO

STATICS LESSON 10
Directional Cosines for 3D Vectors and Components

DIRECTIONAL COSINE VECTORS

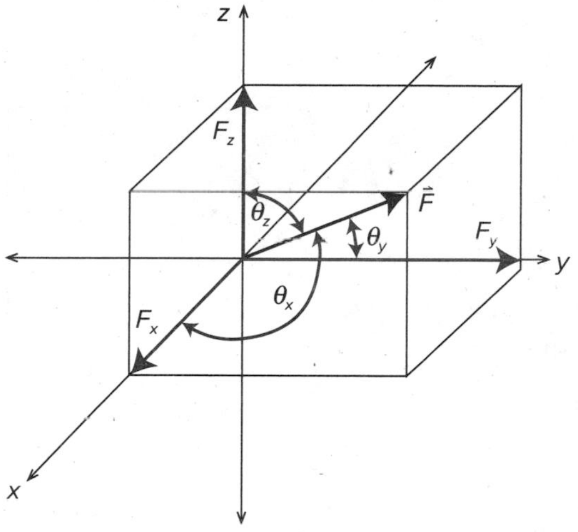

Writing a Cartesian Vector Using the Directional Cosine Recipe

STEP 1: Identify the values for the following variables: $|\vec{F}|$ or F (magnitude of vector), θ_x (angle between the positive *x*-axis and the vector $\vec{F}$), θ_y (angle between the positive *y*-axis and the vector $\vec{F}$), and θ_z (angle between the positive *z*-axis and the vector $\vec{F}$).

Note: Some textbooks present these coordinates angles with the following:

$\theta_x = \alpha$

$\theta_y = \beta$

$\theta_z = \gamma$

Note: For directional cosine vectors, the dimension lines describing the angles of the vector will always be from one of the coordinate axes and the vector $\vec{F}$.

(continued on next page)

Writing a Cartesian Vector Using the Directional Cosine Recipe (continued)

Also note: You may only be given two of the three coordinate angles! If so, find the missing angle via:

$$cos^2(\theta_x) + cos^2(\theta_y) + cos^2(\theta_z) = 1$$

STEP 2: Use directional cosine equations to find the x, y, and z components of the Cartesian vector.

$$F_x = Fcos(\theta_x) = Fcos(\alpha)$$
$$F_y = Fcos(\theta_y) = Fcos(\beta)$$
$$F_z = Fcos(\theta_z) = Fcos(\gamma)$$

STEP 3: Finally, write as a Cartesian vector.

$$\vec{F} = F_x\hat{\imath} + F_y\hat{\jmath} + F_z\hat{k}.$$

PRO TIP

For directional cosine vectors, the dimension lines describing the angles of the vector will always be from one of the coordinate axes and the vector $\vec{F}$.

EXAMPLE: DIRECTIONAL COSINE METHOD

Write the given vector as a Cartesian vector.

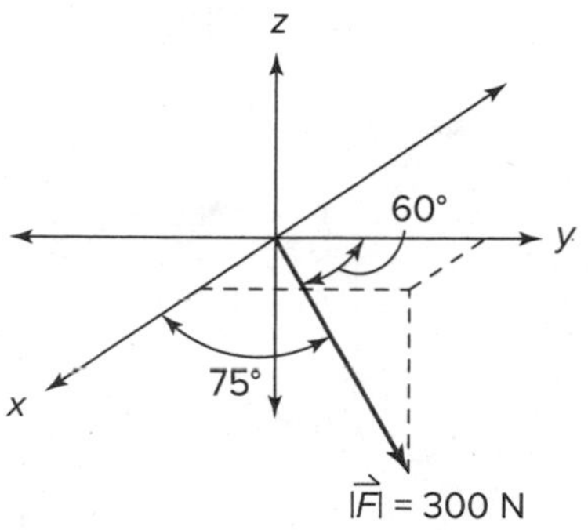

Note: Here only two of the three needed directional cosine angles are given. Use the following identity to find the third angle: $cos^2(\theta_x) + cos^2(\theta_y) + cos^2(\theta_z) = 1$.

STEP 1: Identify θ_x, θ_y, and θ_z.

From the problem statement, $\theta_x = 75°$ and $\theta_y = 60°$.

Let's find θ_z:

$cos^2(75°) + cos^2(60°) + cos^2(\theta_z) = 1$

$0.067 + 0.25 + cos^2(\theta_z) = 1$

$cos^2(\theta_z) = 0.683$

$cos(\theta_z) = \pm 0.826$

$\theta_z = 34.27°, 145.69°$

At this point, there is a choice regarding which angle to choose for θ_z. From the problem statement, θ_z has to be bigger than 90°. θ_z is always measured from the positive z-axis. Thus, $\theta_z = 145.69°$.

STEP 2: Use directional cosine equations to find the x, y, and z components of the Cartesian vector.

$F_x = Fcos(\theta_x)$; $F_x = 300cos(75°)$; $F_x = 77.65$ N

$F_y = Fcos(\theta_y)$; $F_y = 300cos(60°)$; $F_y = 150$ N

$F_z = Fcos(\theta_z)$; $F_z = 300cos(145.69°)$; $F_z = -247.8$ N

STEP 3: Write Cartesian vector:

$\vec{F} = (77.65\hat{\imath} + 150\hat{\jmath} - 247.8\hat{k})$ N

Find the resultant of vectors $\vec{F}_1$ and $\vec{F}_2$.

Press pause on video lesson 10 once you get to the workout problem. Only press play if you get stuck.

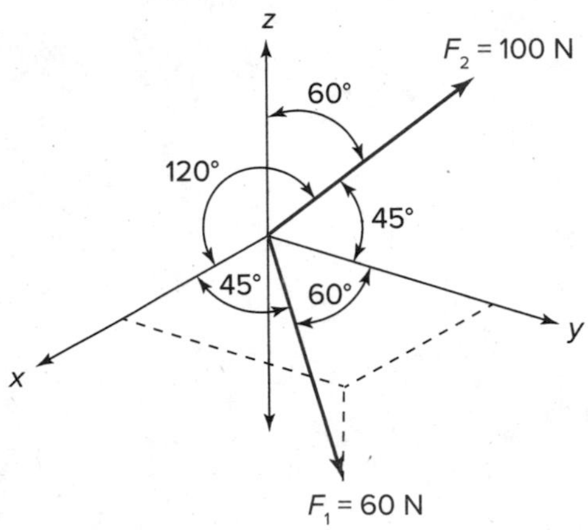

WATCH VIDEO **STATICS LESSON 11**
Finding 3D Vectors When Given Coordinates

$\hat{\lambda}$ Method (Vectors Given with Dimensions)

Some textbooks use the notation ***U*** instead of $\hat{\lambda}$.

$\vec{F} = F\hat{\lambda}$ or $\vec{F} = F\boldsymbol{U}$

F is the magnitude of the force

$\hat{\lambda}$ is a unit vector in direction of the force vector: $\hat{\lambda} = \dfrac{\overrightarrow{AB}}{|\overrightarrow{AB}|}$

Writing a Cartesian Vector Using the $\hat{\lambda}$ Recipe

STEP 1: Identify the coordinates of the points A and B.

STEP 2: Find $\overrightarrow{AB}$ by taking point B in ($\hat{\imath}$, $\hat{\jmath}$, and $\hat{k}$) and subtracting point A ($\hat{\imath}$, $\hat{\jmath}$, and $\hat{k}$) (end point minus start point). $\overrightarrow{AB}$ is any 3D Cartesian vector in the following form: $\overrightarrow{AB} = AB_x\hat{\imath} + AB_y\hat{\jmath} + AB_z\hat{k}$ (this is the position vector).

STEP 3: Find the magnitude of $\overrightarrow{AB}$

$$|\overrightarrow{AB}| = \sqrt{(AB_x)^2 + (AB_y)^2 + (AB_z)^2}$$

STEP 4: $\hat{\lambda}$ can be found via taking the vector from step 2 and dividing by the vector magnitude from step 3:

$$\hat{\lambda} = \frac{\overrightarrow{AB}}{|\overrightarrow{AB}|}$$

STEP 5: Write the force vector in Cartesian form using the unit vector via: $\vec{F} = F\hat{\lambda}$

PRO TIP

Remember, you can always check the accuracy of your unit vector $\hat{\lambda}$ by computing its magnitude. This should always result in a value of 1!

EXAMPLE: $\hat{\lambda}$ METHOD

Write $\overrightarrow{F_{AB}}$ in Cartesian form.

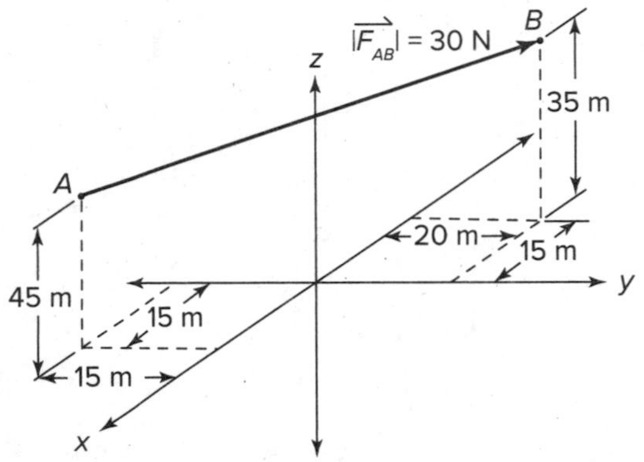

STEP 1: Identify the coordinate points for both ends of the vector:

$A = (15, -15, 45)$

$B = (-15, 20, 35)$

STEP 2: Subtract the start point of the vector from the end point. (***Remember:*** It is this way if the vector goes from A to B, then it is $B - A$.)

$(-15, 20, 35) - (15, -15, 45) = (-30\hat{\imath} + 35\hat{\jmath} - 10\hat{k})$

STEP 3: Find the magnitude of the vector found in the previous step:

$\sqrt{(-30)^2 + 35^2 + (-10)^2} = \sqrt{2{,}225} = 47.17$

STEP 4: Divide the vector from step 2 (called the *position vector*) by the magnitude

$$\frac{(-30\hat{\imath} + 35\hat{\jmath} - 10\hat{k})}{47.17} = (-0.636\hat{\imath} + 0.742\hat{\jmath} - 0.212\hat{k})$$

This vector is called $\hat{\lambda}_{AB}$. It is a unit vector, and this can be verified by taking the square root of the sum of its squares. This should yield a value of 1.

$$\sqrt{(-0.636)^2 + 0.742^2 + (-0.212)^2} = 1$$

STEP 5: Write the unit vector using the following formula $\overrightarrow{F_{AB}} = F\hat{\lambda}_{AB}$.

$$\overrightarrow{F_{AB}} = 30\text{ N}(-0.636\hat{\imath} + 0.742\hat{\jmath} - 0.212\hat{k})$$

$$\overrightarrow{F_{AB}} = (-19.08\hat{\imath} + 22.26\hat{\jmath} - 6.36\hat{k})\text{ N}$$

Another sanity check can be completed by taking the square root of the sum of the squares. This should result in the magnitude of the force vector.

$$\sqrt{(-19.08)^2 + 22.26^2 + (-6.36)^2} = 30\text{ N}$$

Find the force $\vec{F}_{AB}$ acting on point A as a Cartesian vector using the $\hat{\lambda}$ method.
Press pause on video lesson 11 once you get to the workout problem. Only press play if you get stuck.

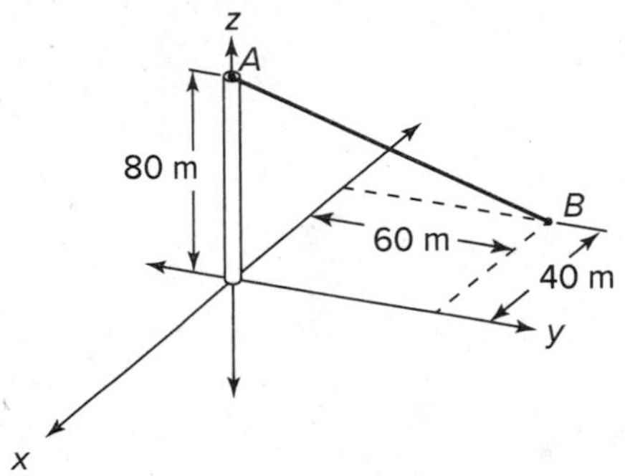

✓ TEST YOURSELF 1.3

SOLUTION TO TEST YOURSELF: Introduction to Vector Mechanics

For the given vectors, find the unit vector $\hat{\lambda}_{AB}$, $\hat{\lambda}_{AC}$, $\hat{\lambda}_{AD}$, and $\hat{\lambda}_{AE}$.

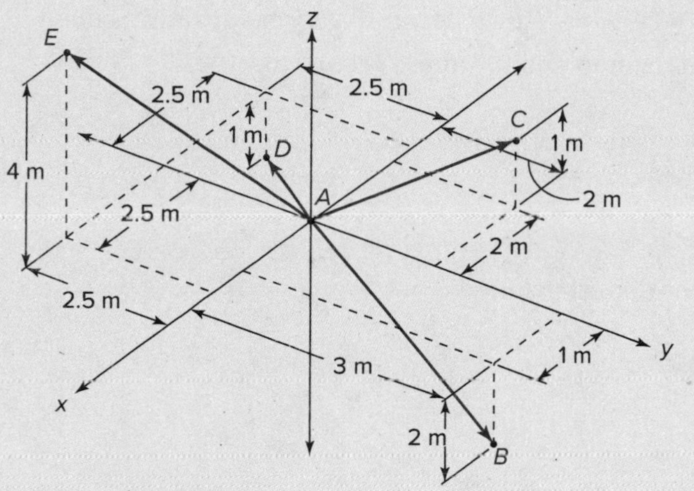

ANSWERS

$\hat{\lambda}_{AB} = 0.267\hat{i} + 0.802\hat{j} - 0.534\hat{k}$

$\hat{\lambda}_{AC} = -0.667\hat{i} + 0.667\hat{j} + 0.333\hat{k}$

$\hat{\lambda}_{AD} = -0.680\hat{i} - 0.680\hat{j} - 0.272\hat{k}$

$\hat{\lambda}_{AE} = 0.468\hat{i} - 0.468\hat{j} + 0.749\hat{k}$

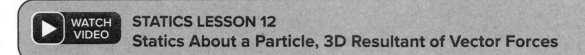

WATCH VIDEO **STATICS LESSON 12**
Statics About a Particle, 3D Resultant of Vector Forces

Why Do I Need to Know This?

Once we have all vectors in Cartesian form ($\vec{F} = F_x\hat{\imath} + F_y\hat{\jmath} + F_z\hat{k}$), it becomes very easy to add as many vectors as we want with ease. We will also use this later to multiply these vectors together.

Breakthrough!

At this point, we should be able to:

- Find the resultant of 2D vectors in polar or Cartesian form
- Find the resultant of 3D vectors via:
  - Blue triangles (spherical coordinates)
  - Directional cosines
  - $\hat{\lambda}$ method

Now that we have learned all three of these methods individually: blue triangle, directional cosines, and $\hat{\lambda}$, let's try a single problem that utilizes all of these methods.

Find the resultant of vectors $\vec{F}_1$, $\vec{F}_2$, and $\vec{F}_3$ and find the directional cosine angles of the resultant.

Press pause on video lesson 12 once you get to the workout problem. Only press play if you get stuck.

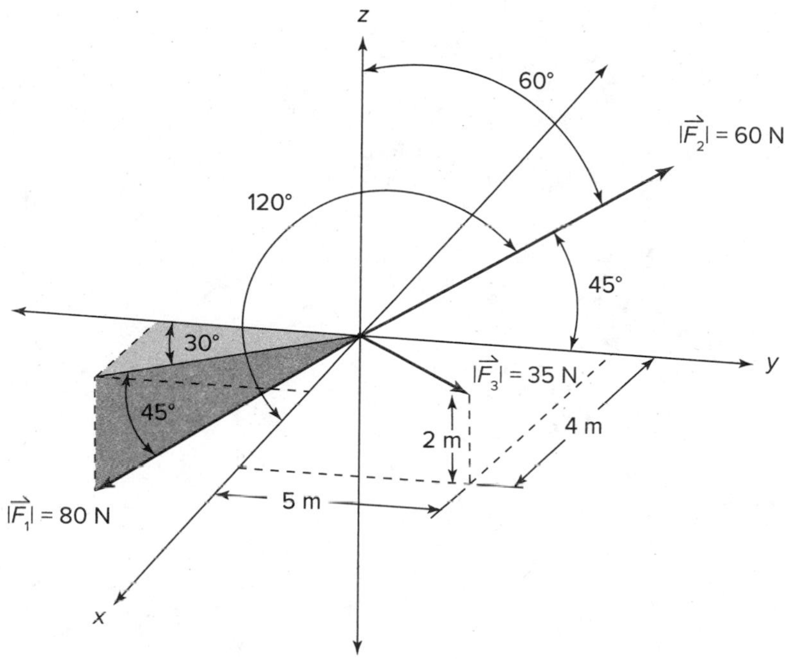

✓ TEST YOURSELF 1.4

SOLUTION TO TEST YOURSELF: Introduction to Vector Mechanics

Can you find the resultant of the given vectors? Assume the following magnitudes of the vectors:
$|\vec{F}_1| = 200$ N, $|\vec{F}_2| = 350$ N, $|\vec{F}_3| = 120$ N, and $|\vec{F}_4| = 225$ N

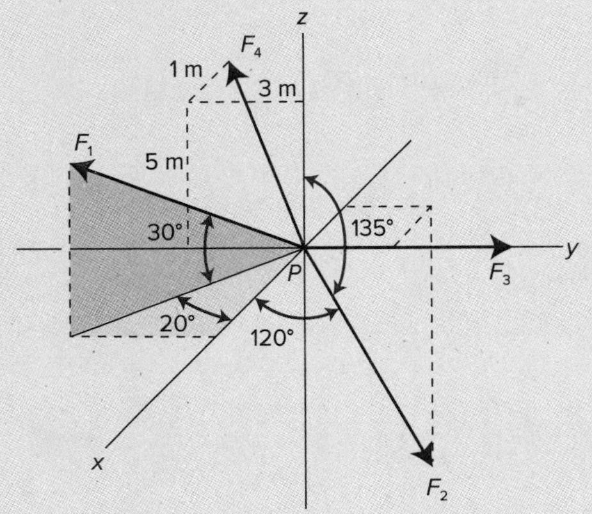

ANSWER
$\vec{F}_R = (-50.3\hat{i} + 121.7\hat{j} + 42.6\hat{k})$ N

WATCH VIDEO

STATICS LESSON 13
Dot Product for Angles Between Vectors and Projections

Dot product is used when dealing with vectors in two main ways:

1. When calculating the angle θ between any two concurrent, 3D vectors:

   $$cos(\theta) = \frac{\vec{A} \cdot \vec{B}}{|\vec{A}||\vec{B}|}$$

2. For projecting a component from one vector onto a second vector

   $$F_{\text{component}} = \vec{F} \cdot \hat{\lambda}$$

   $F_{\text{component}}$ is the projection of $\vec{F}$ onto some line.
   $\vec{F}$ is the vector doing the projecting or casting a "shadow."
   $\hat{\lambda}$ is the unit vector of the line being projected onto.

Dot Product for Angles Between Vectors and Projections Recipe

STEP 1: Write both vectors in Cartesian form using blue triangle, directional cosines, or $\hat{\lambda}$ method.

STEP 2: Use the vectors found in the previous step and plug them into the dot product equation to calculate the angle.

$$cos(\theta) = \frac{\vec{A} \cdot \vec{B}}{|\vec{A}||\vec{B}|}$$

- First compute the numerator by using the dot product. Simply multiply the like terms together (x times x, y times y, etc.). Then add together to get a scalar.
- Calculate the denominator by multiplying the magnitude of the first vector by the magnitude of the second vector.
- Calculate the angle θ.

STEP 3: Calculate the projection of one vector onto a second vector:

$$F_{\text{component}} = \vec{F} \cdot \hat{\lambda}$$

EXAMPLE: DOT PRODUCT

Find the angle between $\vec{F}_1$ and $\vec{F}_2$. Also find the projection of $\vec{F}_1$ onto $\vec{F}_2$.

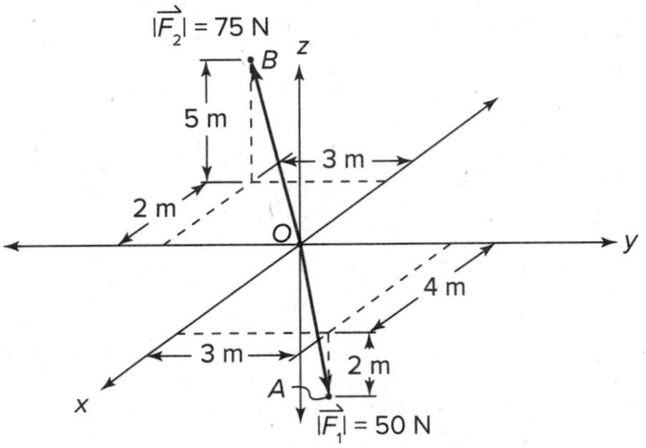

STEP 1: Write both vectors in Cartesian form. Here, both vectors are given with dimensions, so let's use the $\hat{\lambda}$ method.

$$F_1 = A - O = (4, 3, -2) - (0, 0, 0) = (4\hat{\imath} + 3\hat{\jmath} - 2\hat{k})$$
$$F_2 = B - O = (-2, -3, 5) - (0, 0, 0) = (-2\hat{\imath} - 3\hat{\jmath} + 5\hat{k})$$

$$\hat{\lambda}_1 = \frac{(4\hat{\imath} + 3\hat{\jmath} - 2\hat{k})}{\sqrt{4^2 + 3^2 + (-2)^2}} = (0.743\hat{\imath} + 0.557\hat{\jmath} - 0.371\hat{k})$$

$$\hat{\lambda}_2 = \frac{(-2\hat{\imath} - 3\hat{\jmath} + 5\hat{k})}{\sqrt{(-2)^2 + (-3)^2 + 5^2}} = (-0.324\hat{\imath} - 0.487\hat{\jmath} + 0.811\hat{k})$$

Finally using $\vec{F} = F\hat{\lambda}$

$$\vec{F}_1 = F_1\hat{\lambda}_1 = 50(0.743\hat{\imath} + 0.557\hat{\jmath} - 0.371\hat{k}) = (37.15\hat{\imath} + 27.85\hat{\jmath} - 18.55\hat{k})\text{ N}$$
$$\vec{F}_2 = F_2\hat{\lambda}_2 = 75(-0.324\hat{\imath} - 0.487\hat{\jmath} + 0.811\hat{k}) = (-24.3\hat{\imath} - 36.53\hat{\jmath} + 60.83\hat{k})\text{ N}$$

STEP 2: Use the vectors found in the previous step and plug them into the dot product equation to find θ.

$$cos(\theta) = \frac{\vec{F}_1 \cdot \vec{F}_2}{|\vec{F}_1||\vec{F}_2|}$$

- First compute the numerator ($\vec{F}_1 \cdot \vec{F}_2$) by using the dot product. Simply multiply the like terms together (x times x, y times y, etc.). Then add together to obtain a scalar.

  $\vec{F}_1 \cdot \vec{F}_2 = (37.15\hat{\imath} + 27.85\hat{\jmath} - 18.55\hat{k}) \cdot (-24.3\hat{\imath} - 36.53\hat{\jmath} + 60.83\hat{k}) = (37.15 \times -24.3) + (27.85 \times -36.53) + (-18.55 \times 60.83) = (-902.75) + (-1{,}017.36) + (-1{,}128.40) = -3{,}048.5$

- Calculate the denominator ($|\vec{F}_1|\,|\vec{F}_2|$) by multiplying the magnitude of the first vector by the magnitude of the second vector.

  $|\vec{F}_1| = 50\text{ N}, |\vec{F}_2| = 75\text{ N}$

  $|\vec{F}_1|\,|\vec{F}_2| = (50)(75) = 3{,}750$

- Calculate the angle θ.

$$\theta = cos^{-1}\left(\frac{\vec{F}_1 \cdot \vec{F}_2}{|\vec{F}_1||\vec{F}_2|}\right) = cos^{-1}\left(\frac{-3{,}048.5}{3{,}750}\right) = 144.38°$$

Note: This is the angle between vector $\vec{F}_1$ and $\vec{F}_2$.

STEP 3: Calculate the projection of $\vec{F}_1$ onto $\vec{F}_2$.

$$\vec{F}_{\text{Projection}} = \vec{F}_1 \cdot \hat{\lambda}_2$$

$$\vec{F}_{\text{Projection}} = \vec{F}_1 \cdot \hat{\lambda}_2 = (37.15\hat{\imath} + 27.85\hat{\jmath} - 18.55\hat{k}) \cdot (-0.324\hat{\imath} - 0.487\hat{\jmath} + 0.811\hat{k}) =$$
$$(37.15 \times -0.324) + (27.85 \times -0.487) + (-18.55 \times 0.811)$$
$$= (-12.04) + (-13.56) + (-15.04) = -40.64\text{ N}$$

Remember this projection, or component, is a scalar.

Find the angle between vectors $\vec{F}_1$ and $\vec{F}_2$ as well as the projection of vector $\vec{F}_2$ onto $\vec{F}_1$. Press pause on video lesson 13 once you get to the workout problem. Only press play if you get stuck.

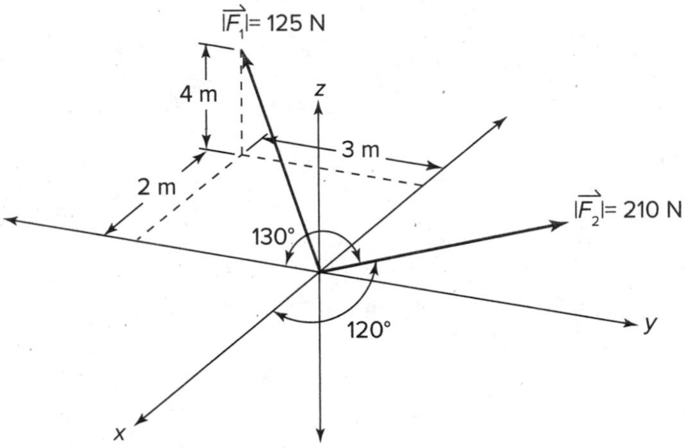

KEY TAKEAWAYS

Vectors can be written with polar coordinates (magnitude and angle) or with Cartesian coordinates ($\hat{\imath}, \hat{\jmath}, \hat{k}$).

Main takeaways from Level 1 includes how to add:

- 2D vectors
- 3D vectors

2D Vector Summary

- To determine the resultant vector in polar form when adding 2D vectors:
  - Use the Tip to Tail Rule to create a vector triangle.
  - Calculate the magnitude of $\vec{r}$ using Pythagorean Theorem if triangle has a right angle or Law of Cosines if it does not:
    $c^2 = a^2 + b^2 - 2(a)(b)cos(C)$
  - Calculate the angle θ using the tangent function if triangle has a right angle or Law of Sines if it does not:
    $$\frac{a}{sin(A)} = \frac{b}{sin(B)} = \frac{c}{sin(C)}$$
- To determine the resultant vector in a Cartesian form when adding 2D vectors:
  - Resolve each individual vector into x and y components (Cartesian vector) using trigonometric identities (SOHCAHTOA).
  - Add the vectors together by simply adding together the "like components" (i.e., add all of the $\hat{\imath}$ terms together with each other and add all of the $\hat{\jmath}$ terms together with each other).
  - ***Note:*** You can always find the magnitude of this resultant vector by squaring each of its terms, summing the squares together, and taking the square-root (i.e.: $|\vec{F}| = \sqrt{F_x^2 + F_y^2}$).

KEY TAKEAWAYS

3D Vector Summary

Note: For our purposes here for 3D vectors, we typically only use Cartesian form and rarely polar form as Cartesian form is way easier!

- Add 3D vectors in a similar fashion to the 2D Cartesian form method, except instead of only having an x and y component, now there is a z component too!
- Add the vectors together by simply adding together the "like components" (i.e., add all of the $\hat{\imath}$ terms together with each other, add all of the $\hat{\jmath}$ terms together with each other, add all of the $\hat{k}$ terms together with each other).
- Note you can always find the magnitude of this resultant vector by squaring each of its terms, summing the squares together, and taking the square-root (i.e.: $|\vec{F}| = \sqrt{Fx^2 + Fy^2 + Fz^2}$).

Recall there are three ways 3D vectors can be presented to you:

1. Blue triangles (spherical coordinates)—Two angles (ϕ and θ_z) and a magnitude (F):
   $F_x = Fcos(\phi)sin(\theta_z)$
   $F_y = Fsin(\phi)sin(\theta_z)$
   $F_z = Fcos(\theta_z)$
2. Directional cosines—Three angles (θ_x, θ_y, θ_z) and a magnitude (F):
   $F_x = Fcos(\theta_x)$
   $F_y = Fcos(\theta_y)$
   $F_z = Fcos(\theta_z)$
3. $\hat{\lambda}$ unit vector method (vector expressed with a dimension):
   $\vec{F} = F\hat{\lambda}$
   Note: You can find $\hat{\lambda}$ via $\hat{\lambda} = \frac{\overrightarrow{AB}}{|\overrightarrow{AB}|}$.

KEY TAKEAWAYS

RECIPES

- **Tip to Tail Rule Recipe**
- **Finding Resultant Vectors in Non-Orthogonal Systems Recipe**
- **Writing a Cartesian Vector Using the Blue Triangle (Spherical Coordinates) Recipe**
- **Writing a Cartesian Vector Using the Directional Cosine Recipe**
- **Writing a Cartesian Vector Using the $\hat{\lambda}$ Recipe**
- **Dot Product for Angles Between Vectors and Projections Recipe**

PRO TIPS

2D and 3D Vector Tips

- Remember if you have a vector on one side of the equal sign, you must have a vector on the other.
- Make sure to draw pictures that are large enough that you can label. It shouldn't be able to be covered by a quarter!

2D Vector Tips

- One helpful hint on finding angles and other missing information about a triangle is to sketch horizontal lines extended from the angles of the triangle. This will make angles more apparent.
- We all remember the Pythagorean Theorem, which works for only one triangle, right triangles. The Law of Cosines works for *all* triangles, so let's remember this one instead! (*Note:* If angle C is 90°, then the Law of Cosines becomes the Pythagorean Theorem.)

3D Vector Tips

- For directional cosine vectors, the dimension lines describing the angles of the vector will always be from one of the coordinate axes and the vector $\bar{F}$.
- You can always check the accuracy of your unit vector $\hat{\lambda}$ by computing its magnitude. This should always result in a value of 1!

PITFALLS

2D Vector Tips

- Don't forget for position vector $\overrightarrow{AB}$ you take $B - A$ and for position vector $\overrightarrow{BA}$ you take $A - B$.

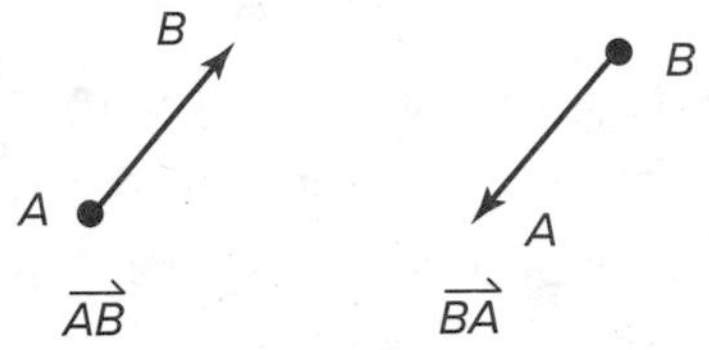

- Make sure you can resolve a 2D force in Cartesian system into x and y components.
- When working problems with a non-orthogonal axis, the Cartesian component method *will not work* as it relies on the axes being 90° from each other!

3D Vector Tips

- When looking at the x-y plane for the blue triangle method, phi (ϕ) is positive when measured from the x-axis to the $+y$-axis and ϕ is negative when measured from the x-axis to the $-y$-axis.

Level 2
Particle Equilibrium

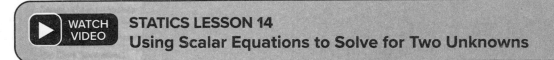

WATCH VIDEO **STATICS LESSON 14**
Using Scalar Equations to Solve for Two Unknowns

Introduction to Particle Equilibrium

This section discusses statics on a particle (an infinitesimally small point) with multiple forces acting on a point.

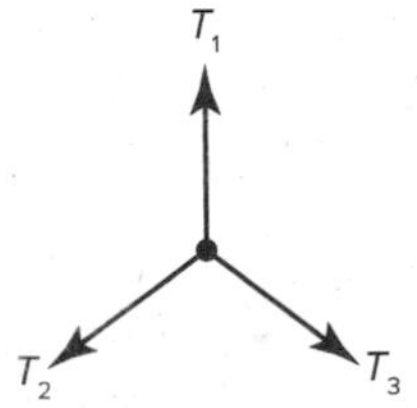

(Typical FBD of particle)

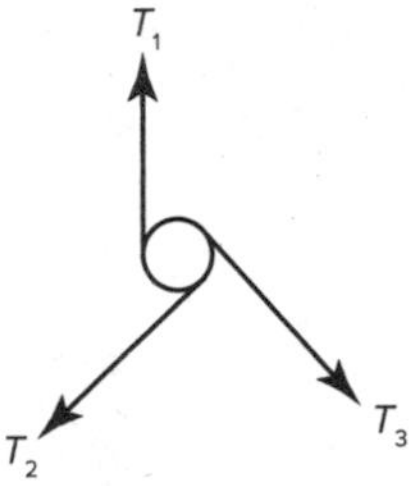

(If a particle wasn't very small)

If the point wasn't small, you might have rotation of the body—in this section, there will be no rotation!

Equilibrium—this means a body (or a single particle) is not moving. All forces are balanced.

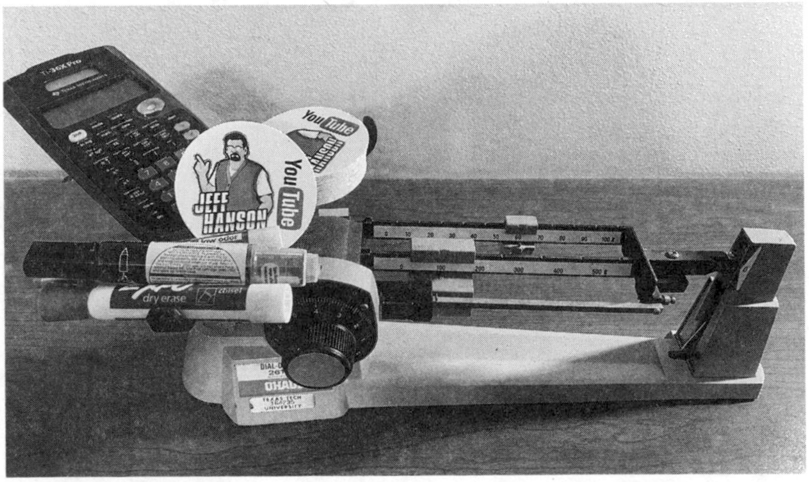

Totally balanced system!

If all forces are balanced, then:

- $\Sigma F_x = 0$; Add up all $\hat{\imath}$ components from all Cartesian vectors in the system.
- $\Sigma F_y = 0$; Add up all $\hat{\jmath}$ components from all Cartesian vectors in the system.
- $\Sigma F_z = 0$; Add up all $\hat{k}$ components from all Cartesian vectors in the system.

Since we have multiple equations, we can solve for multiple unknowns!

- In 2D can solve for two unknowns as there are two equations:

$$\Sigma F_x = 0, \Sigma F_y = 0$$

- In 3D can solve for three unknowns as there are three equations:

$$\Sigma F_x = 0, \Sigma F_y = 0, \Sigma F_z = 0$$

STAR CONCEPT

THE FREE BODY DIAGRAM (FBD)

- Diagram of the body of interest
- "Freeing" up a "body" from a system to examine it and all the forces acting upon it

PRO TIP

In this topic, the "body" for the FBD is always a single point or particle! Then, I like to think of myself as the "free" body. If I was the particle, what would I be feeling?

Solving 2D Statics About a Particle Recipe

STEP 1: Determine what is the point of interest.

STEP 2: "Cut" that body away from the system and "free" it up.

STEP 3: Draw the body (in this level, a single point).

STEP 4: Add all forces to the FBD acting on that body.

Remember: Ropes, cables, chains, or cords are always in tension (i.e., going away from or pulling on the body). You can't push a rope but can pull on it!

STEP 5: Label all components of the FBD.

STEP 6: Resolve all forces that are at an angle into Cartesian components.

Note: When drawing an FBD and resolving forces into components, if the vector goes away from a point, the components should go away from the point; if the vector goes to the point, the components should point toward the point.

STEP 7: Write your equations of equilibrium.

$$\Sigma F_x = 0$$
$$\Sigma F_y = 0$$

Note: When it comes to a body in equilibrium (not moving), remember that the "up stuff" has to equal the "down stuff," and the "left stuff" has to equal the "right stuff."

STEP 8: Solve for your unknowns!

Note: This will require algebra or the use of your system solver on your calculator.

EXAMPLE: 2D PARTICLE STATICS

Find the tension in cables *AC* and *AB*.

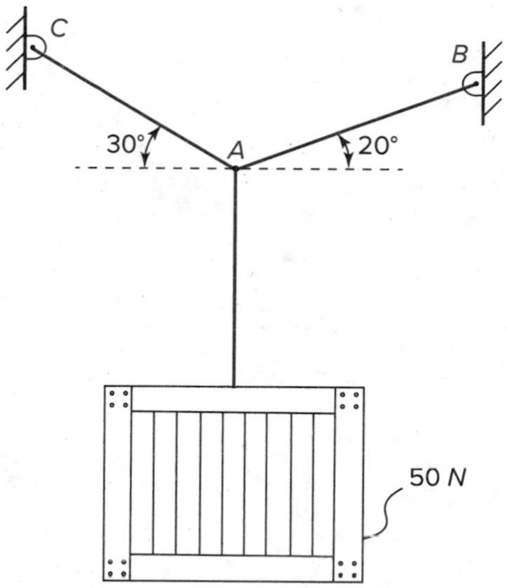

STEPS 1–4: Determine point of interest, "cut body away," draw body, and then add all forces.

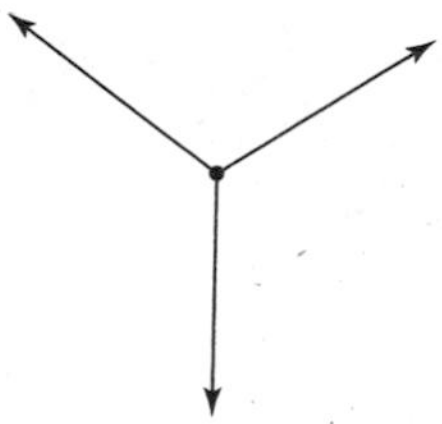

STEP 5: Label all components.

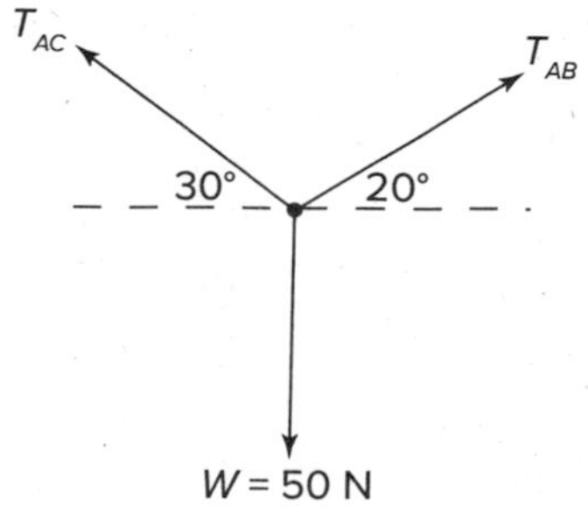

STEP 6: Resolve forces into Cartesian components.

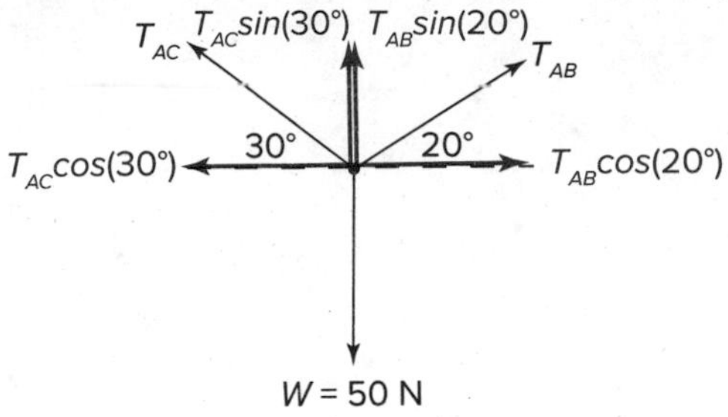

STEP 7: Write equations of equilibrium.

$$\Sigma F_x = 0 = T_{AB}cos(20°) - T_{AC}cos(30°)$$

$$\Sigma F_y = 0 = T_{AB}sin(20°) + T_{AC}sin(30°) - 50$$

STEP 8: Solve for your unknowns.

$$T_{AB} = 56.5 \text{ N}$$

$$T_{AC} = 61.3 \text{ N}$$

Find the angle θ and the magnitude of vector F_3 such that the system is in equilibrium.
Press pause on video lesson 14 once you get to the workout problem. Only press play if you get stuck.

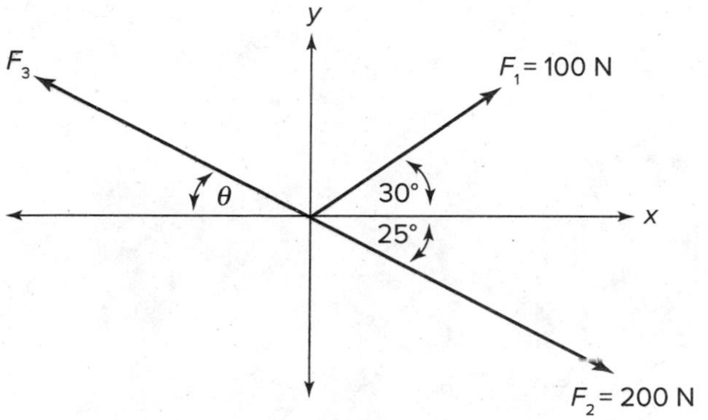

✓ TEST YOURSELF 2.1

SOLUTION TO TEST YOURSELF: Particle Equilibrium

Draw the FBD for these real-world examples:

1.

2.

3.

4.

ANSWERS

See Test Yourself 2.1 solutions at the back of the book.

WATCH VIDEO

STATICS LESSON 15
Proving There's No Such Thing as Straight Strings

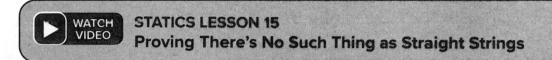

SIDE TRIP! Follow along with this video as Dr. Hanson proves there is no such thing as a straight string! Oh, you have to see this!

Find the tension in the cables.

Press pause on the video once you get to the workout problem. Only press play if you get stuck.

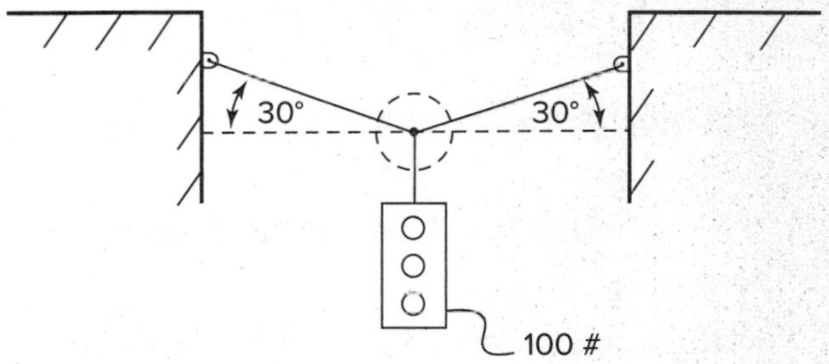

WATCH VIDEO

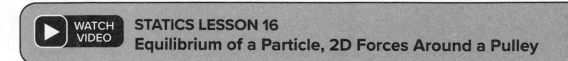

STATICS LESSON 16
Equilibrium of a Particle, 2D Forces Around a Pulley

All pulleys primarily consist of a wheel with a rope around it. Depending on the pulley arrangement, a pulley may either allow:

- A heavy object to be lifted with less force, called mechanical advantage
- Simply redirect the force applied in a different direction
- A combination of both of the above

Assuming pulleys are frictionless, the rope or cable wrapped around the pulley has the exact same force on each side of the pulley (i.e., the rope is in tension).

What force do you as an individual need to supply to the pulley via *DE* to keep the block-pulley system in its current position?

Note: If no force is applied to the pulley then the box would move on its own and come to rest halfway between points *A* and *C* since the tension in *AB* has to be the same as the tension in *BC*.

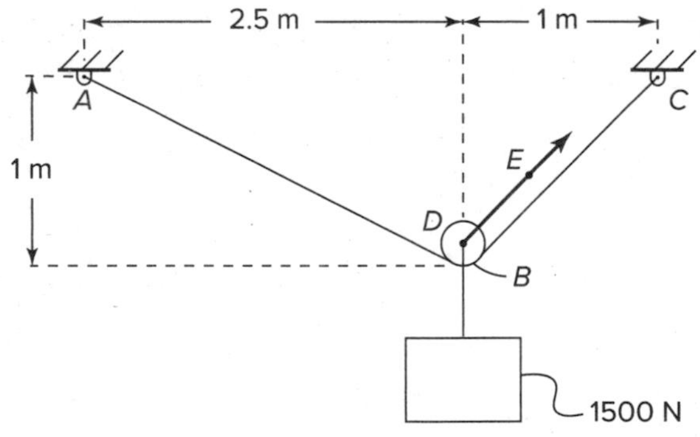

WATCH VIDEO **STATICS LESSON 17**

2D Equilibrium of a Particle with Springs

Forces expressed as springs:

- Common mistake is to use $F = kx^2$ (potential energy in a spring from physics). This equation does not exist as it is $E = kx^2$!
- Use equation $F = kx$ where k is the spring constant and x is the stretch in the spring!

PITFALL

Commonly, in spring problems, the problem will have units given in mm, but the spring constant *k* will be given in N/m. *Watch* for mixed units! Don't fall for that trick!

PRO TIP

***Do not* be scared of springs! They are simply another way to express a force. Springs are just a force!**

If the angle θ was originally 45 degrees, find the weight of each block.

Press pause on video lesson 17 once you get to the workout problem. Only press play if you get stuck.

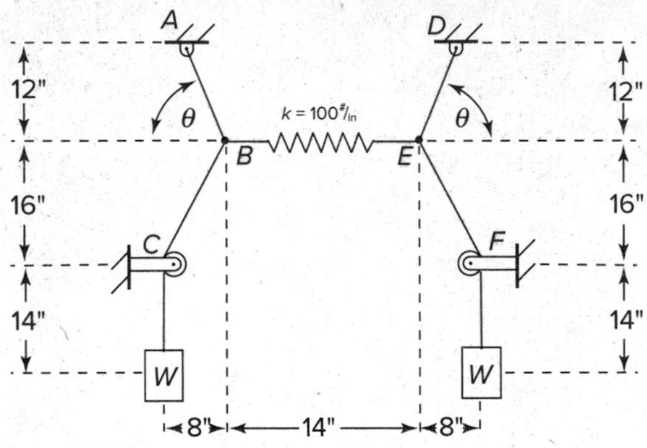

WATCH VIDEO

STATICS LESSON 18
2D Statics on a Particle, Multiple Free Body Diagrams

If a problem gives a maximum force value for any member in the system:

- Pick one member to set to this maximum.
- Solve for the remaining forces.
- If another member turns out to be larger than the maximum, then you picked the wrong member to set to max.

Maximum Condition with Multiple FBDs Recipe

STEP 1: Select a point to start and draw FBD of that point. (Generally, it is best to start far away and work toward the thing you are trying to find.) Resolve all forces that are at an angle into Cartesian components. ***Note:*** This is identical to steps 1–6 of Solving 2D Statics About a Particle Recipe.

STEP 2: Set only one member to the maximum.

STEP 3: Write equations of equilibrium and solve.

STEP 4: If multiple points, draw FBD of the next point.

Note: When you have multiple FBDs, force vectors that appear on both FBDs will always have opposite directions on the two FBDs.

STEP 5: Write equations of equilibrium for this FBD and solve.

STEP 6: Repeat as many times as necessary to solve for all unknowns.

PITFALL

For those problems that provide a minimum or maximum condition, never ever set more than one vector at a time to the maximum value when solving.

EXAMPLE: MULTIPLE FBDs

Find the max weight of the lamp if no single rope can hold more than 100 lbs before breaking.

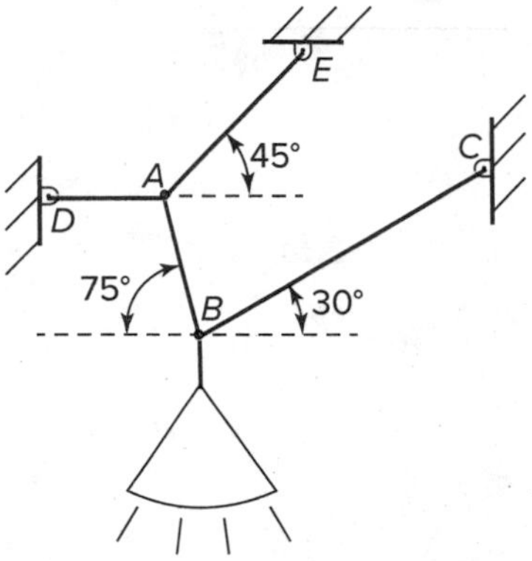

STEP 1: Select a point to start and draw FBD of that point. (Generally, it is best to start far away and work toward the thing you are trying to find.) Let's start at point A.

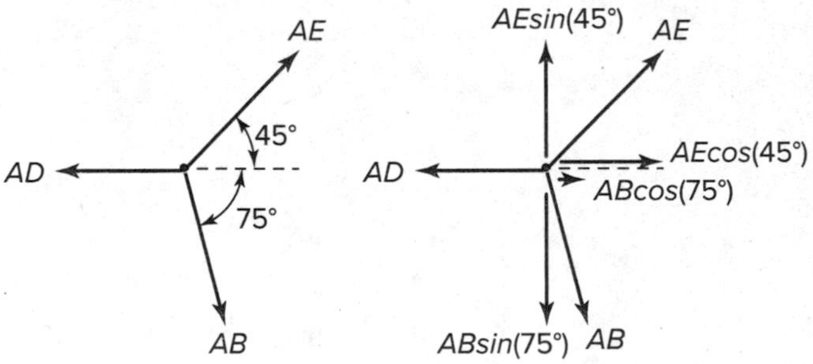

STEP 2: Set ***only one*** rope to max (in this case, 100 lbs) and solve for the other two tensions. If we picked the correct rope, the others will be less than 100 lbs. If not, we will have to resolve, setting the rope with the most tension to max.

Let's choose AE to set to maximum (100 lbs).

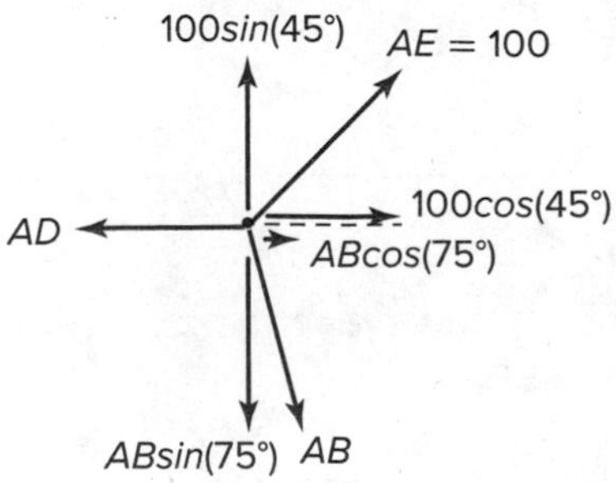

STEP 3: Write equations of equilibrium and solve:

$\Sigma F_x = 0 = 100cos(45°) + ABcos(75°) - AD$

$\Sigma F_y = 0 = 100sin(45°) - ABsin(75°)$

$AD = 89.66$ lbs

$AB = 73.19$ lbs

Note since *AB*, *AD* are less than 100 lbs we guessed correctly so far. As long as *BC* is less than 100 lbs, we are golden.

STEP 4: Draw FBD of the next point.

Don't forget that since *AB* appears on the previous diagram, its direction needs to be opposite on this FBD.

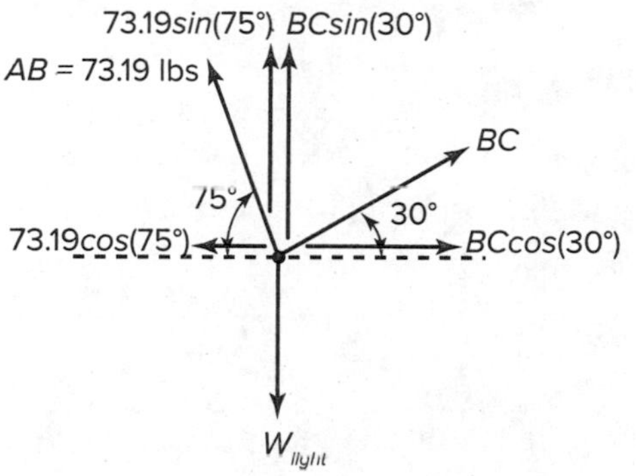

STEP 5: Write equations of equilibrium for this FBD and solve.

$\Sigma F_x = 0 = BCcos(30°) - 73.19cos(75°)$

$\Sigma F_y = 0 = BCsin(30°) + 73.19sin(75°) - W_{light}$

Using simple substitution,

$BC = 21.87$ lbs

$W_{light} = 81.63$ lbs max

Note: *BC* is also less than 100 lbs so we are all good!

PARTICLE EQUILIBRIUM

Think of these problems as a playground tug-of-war game. If the three forces on a point were kids pulling, who would lose? One of the kids will be the weakest link, and that vector would be set to the max value. Then solve for the others and make sure they don't exceed the maximum. Thus, these are "guess and check" type problems.

Solve for all the force values of each member in the given system. Press pause on video lesson 18 once you get to the workout problem. Only press play if you get stuck.

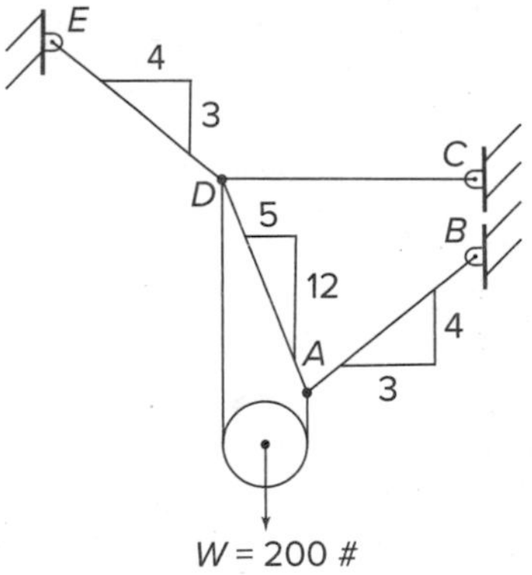

✓ TEST YOURSELF 2.2

SOLUTION TO TEST YOURSELF: Particle Equilibrium

For the problem shown below, the two weights are supported by cables; one of which runs over an ideal pulley.

1. Cable *AC* is horizontal and connects to a ring at *A*. Another ring at *B* connects cables *DB*, *EB*, and *AB*. What are the forces in cables *DB* and *EB*?

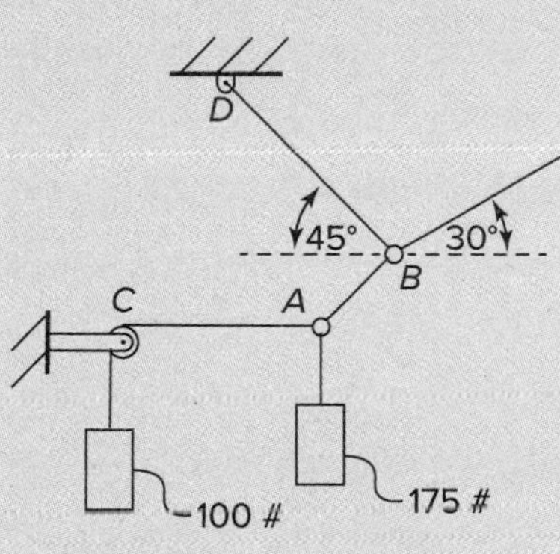

2. If the 175 # weight is now unknown, and the maximum allowable tension in cable *AB* is 500 #, what is the maximum weight at *A*?

ANSWERS

DB = 105.2 lbs

EB = 201.3 lbs

Weight = 490 lbs

WATCH VIDEO

STATICS LESSON 19
3D Statics About a Particle, Calculating Unit Vectors

Solving 3D Statics About a Particle Recipe

STEP 1: Draw FBD of the system. (See 2D Statics About a Particle Recipe steps 1-5 if you don't remember how to do this.)

STEP 2: Write each of the vectors in Cartesian form. (Review the 3D methods from Level 1 (blue triangle, directional cosine, or $\hat{\lambda}$.)

STEP 3: Write the equations of equilibrium ($\Sigma F_x = 0$, $\Sigma F_y = 0$, $\Sigma F_z = 0$).

STEP 4: Solve using either of these methods:

- Method 1: Brute force algebra using the substitution method
- Method 2: Use your system solver function on your calculator (recommended)

EXAMPLE: 3D PARTICLE STATICS

Find the tension in cables *AD*, *AC*, and *AB* knowing that the balloon has a buoyant force of 8 kN.

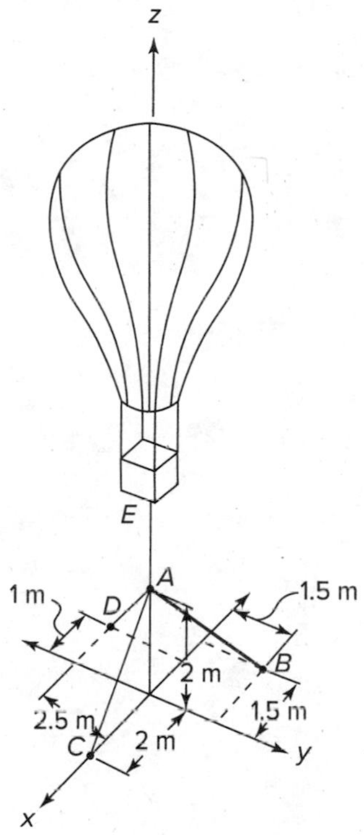

STEP 1: Draw an FBD of the system. (See 2D Statics About a Particle Recipe steps 1–5 if you don't remember how to do this.)

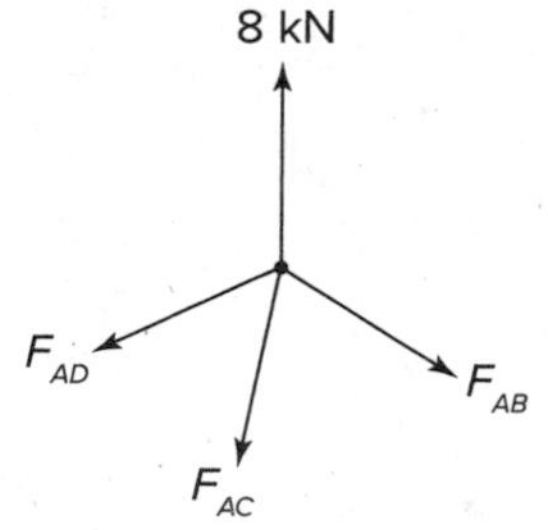

STEP 2: Write each of the four vectors in Cartesian form (review the $\hat{\lambda}$ Method from Level 1):

$$\vec{F}_{AB} = (-1.5, 1.5, 0) - (0, 0, 2) = (-1.5\hat{\imath} + 1.5\hat{\jmath} - 2\hat{k})$$

$$|\vec{F}_{AB}| = \sqrt{(-1.5)^2 + 1.5^2 + (-2)^2} = 2.915$$

$$\vec{F}_{AC} = (2, 0, 0) - (0, 0, 2) = (2\hat{\imath} + 0\hat{\jmath} - 2\hat{k})$$

$$|\vec{F}_{AC}| = \sqrt{2^2 + 0^2 + (-2)^2} = 2.828$$

$$\vec{F}_{AD} = (-1, -2.5, 0) - (0, 0, 2) = (-1\hat{\imath} - 2.5\hat{\jmath} - 2\hat{k})$$

$$|\vec{F}_{AD}| = \sqrt{(-1)^2 + (-2.5)^2 + (-2)^2} = 3.354$$

$$F_{AE} = 8\hat{k}$$

$$F_{AB} = -0.515F_{AB}\hat{\imath} + 0.515F_{AB}\hat{\jmath} - 0.686F_{AB}\hat{k}$$

$$F_{AC} = 0.707F_{AC}\hat{\imath} - 0.707F_{AC}\hat{k}$$

$$F_{AD} = -0.299F_{AD}\hat{\imath} - 0.746F_{AD}\hat{\jmath} - 0.597F_{AD}\hat{k}$$

STEP 3: Write the equations of equilibrium:

$$\Sigma F_x = -0.515F_{AB} + 0.707F_{AC} - 0.299F_{AD} = 0$$

$$\Sigma F_y = 0.515F_{AB} - 0.746F_{AD} = 0$$

$$\Sigma F_z = -0.686F_{AB} - 0.707F_{AC} - 0.597F_{AD} + 8 = 0$$

STEP 4: Solve:

$$F_{AB} = 4.40 \text{ kN}$$

$$F_{AC} = 4.49 \text{ kN}$$

$$F_{AD} = 3.04 \text{ kN}$$

Solve for all the tensions in the cables in the given system if the slab weighs 2,500 N and point A is 2 m above the center of the plate.

Press pause on video lesson 19 once you get to the workout problem. Only press play if you get stuck.

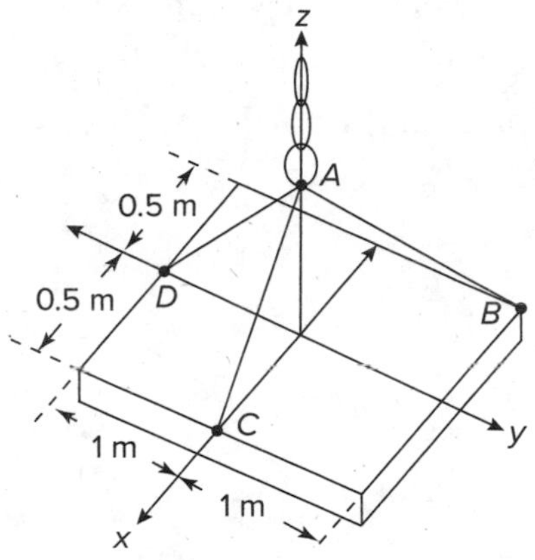

WATCH VIDEO

STATICS LESSON 20
Three Equations and Unknowns, 3D Vectors

Solve for all the tensions in the three cables in the given system if the box weighs 600 N. Points *A*, *B*, and *C* are all on the same plane, and point *D* is located 2 m below the origin.

Press pause on the video once you get to the workout problem. Only press play if you get stuck.

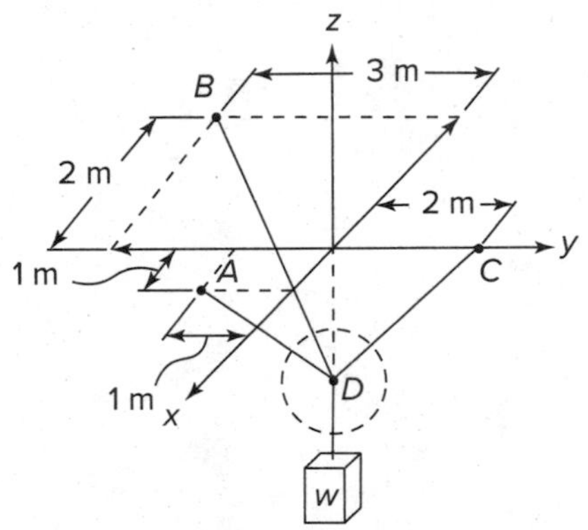

✓ TEST YOURSELF 2.3

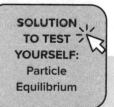

SOLUTION TO TEST YOURSELF: Particle Equilibrium

A thin, uniform square plate, 3 m on each side, has a weight 10 kN. It is supported as shown below by three cables. Point A is 4 m above the origin.

1. What are the forces in the three supporting cables?
2. If the maximum force in AB is 5 kN, what is the maximum weight of the plate that can be lifted?

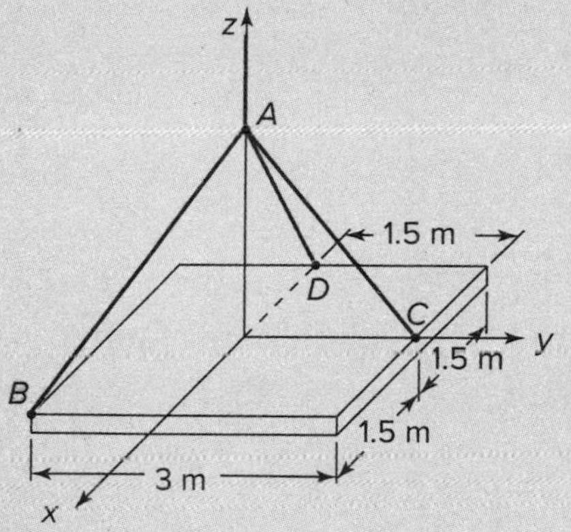

PARTICLE EQUILIBRIUM

ANSWERS

$AB = 3.77$ kN

$AC = 3.56$ kN

$AD = 3.56$ kN

$W = 13.25$ kN

KEY TAKEAWAYS

Constructing Free Body Diagrams (FBDs)

- It is important to master free body diagrams as they are crucial to solving statics problems.
- Free body diagrams will generally come in two forms: in a form of a particle (single point) or a rigid body (beam, car, etc.).
- In both cases, the easiest way to think of a free body diagram is to imagine yourself as the free body and then ask yourself: What is the world doing to me? (i.e., pulling, pushing, rotating).

2D Statics on a Particle

- All of these problems can be solved using $\Sigma F_x = 0$, $\Sigma F_y = 0$.
- On problems with more than one concurrent point, multiple free body diagrams are required.
- On any problem requiring a maximum value, pick the weakest link, set that member to max, and solve for the rest of the members. (Make sure none of the rest of the members exceed that maximum.)

3D Statics on a Particle

- All of these problems can be solved using $\Sigma F_x = 0$, $\Sigma F_y = 0$, $\Sigma F_z = 0$.
- Write all vectors in Cartesian form.
- Learn how to use your system solver to solve these systems of equations. It will save your grade from algebra mistakes!

RECIPES

- **Solving 2D Statics About a Particle Recipe**
- **Maximum Condition with Multiple FBDs Recipe**
- **Solving 3D Statics About a Particle Recipe**

KEY TAKEAWAYS

PRO TIPS

2D and 3D Problem Tips

- **In this topic, the "body" for the FBD will always be a single point or particle! Then, I like to think of myself as the "free" body. If I was the particle, what would I be feeling?**
- ***Do not* be scared of springs! They are simply another way to express a force. Springs are just a force!**

PITFALLS

2D and 3D Problem Tips

- **Commonly in spring problems, the problem will have units given in mm, but the spring constant *k* will be given in N/m. *Watch* for mixed units! Don't fall for that trick!**
- **For those problems that provide a minimum or maximum condition, never ever set more than one vector at a time to the maximum value when solving.**

Level 3

Statics on a Rigid Body

WATCH VIDEO

STATICS LESSON 21
Introduction to Moments and Torque

Up to this point, we have been discussing statics on a particle or a single point. So, what happens if the forces don't all go through a single point?

"Rigid Body" Problems

- A body that doesn't bend, stretch, or deform when loads are applied to it.
- It doesn't happen in real life, but for the sake of finding the forces acting on a body, ignore these deformations. (***Note:*** Future Strength of Materials course delves into solving for these deformations.)
- It can be a person, a book, a car, or anything that isn't a single point.
- For a body to be in static equilibrium, $\Sigma F = 0$, and $\Sigma M = 0$, in all directions (i.e., the body doesn't move or rotate).

CHALLENGE QUESTION

Is this system in static equilibrium?

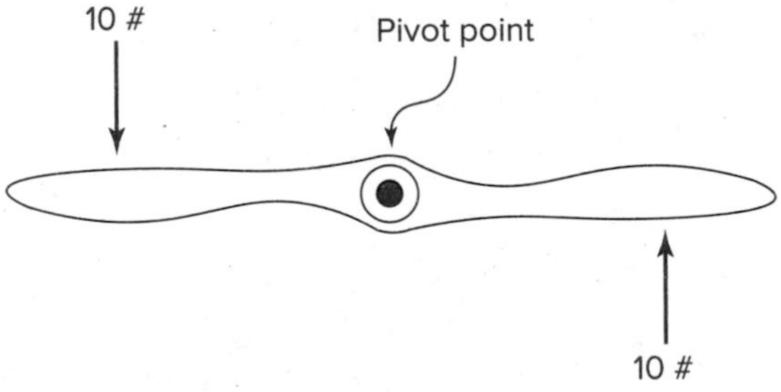

ANSWER

No, it is not in static equilibrium. The two 10 # forces will act to spin the object around the pivot point.

THREE NEW TERMS TO INTRODUCE: TORQUE, MOMENT, COUPLE

- All three of these terms can simply be defined as a tendency to cause rotation about an axis.
- Moment is the general term, but torque is a special type of a moment that causes twisting.
- External moments are represented by curved arrows or a double arrow according to the "right hand rule." In this text, we will only be using the swirl arrow.
- A couple is two forces of equal magnitude in opposite dirctions, on parallel lines of action.

A blacksmith applies a force to twist the bar. The forces of his hands with the distance causes a moment in the heated region. For this case the moment applied causes a torque (i.e., twisting) in the heated shaft.

"Hey man, you're an engineer, right?"
"Yep."
"Then, what is torque?"
" . . . Just a moment . . . "
Buddy waiting for response . . . *ha.*

CHALLENGE QUESTION

Which of these are a coupled moment?

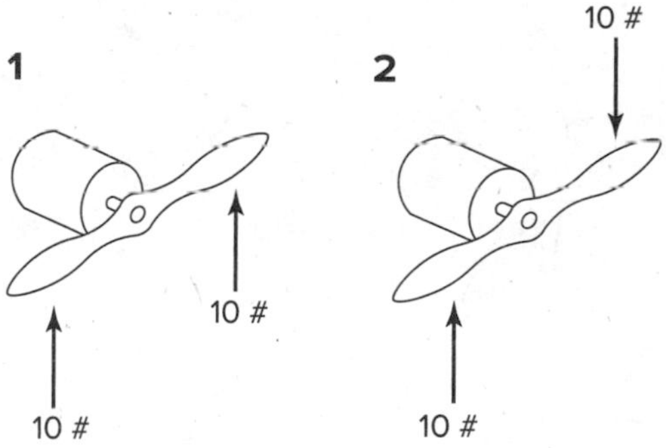

ANSWER

Only the second figure (on the right) represents a couple moment. Recall, a couple moment is two equal and parallel forces that act in opposite directions. Note that in the first figure, on the left, the two forces are parallel but act in the same direction, therefore that is not a couple moment.

Equations for Calculating Torques and Moments

$M = Fd$ $\quad F$: force applied

d: perpendicular distance

PITFALL

To use *M* = *Fd* equation, the distance *must* be perpendicular to the force vector!

$\vec{M} = \vec{r} \times \vec{F}$ $\quad \vec{r}$ (position vector): A vector whose tail is at the point where we wish to take a moment and whose tip is on a point *anywhere* along the line of action (LOA) of the force. Also top half of the $\hat{\lambda}$ formula.

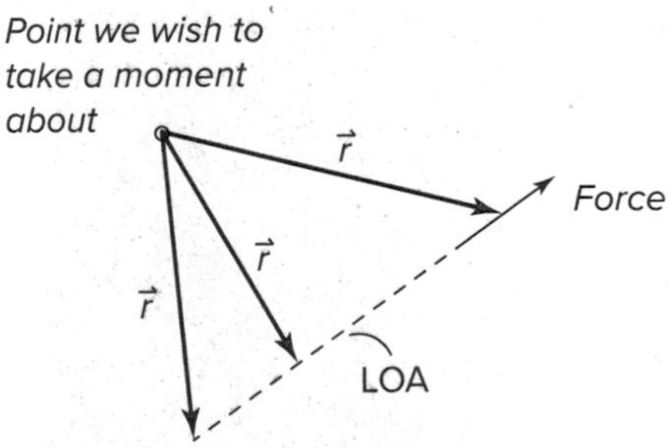

PRO TIP

Think of vector $\vec{r}$ as how to get to Grandma's house. You live at the point we wish to take the moment, and Grandma lives at the point on the LOA. It's just the direction how to get from here to there.

PITFALL

***Do not* divide the position vector by its magnitude like we did when calculating $\hat{\lambda}$! That would always make a vector of length 1!**

The easiest way to increase torque (moment) is to increase the perpendicular distance to the applied force. If you ever have difficulty removing a lug nut, use a longer wrench to increase the torque!

2D moment problems can be solved one of two ways:

1. Using $M = \vec{r} \times \vec{F}$: Requires writing position and force vectors in Cartesian form and using Kramer's Rule to solve for the moment.
2. Using $M = Fd$: Resolve the force vector into Cartesian components, then simply multiply each component by its respective perpendicular distance.

How Do You Calculate ($\vec{r} \times \vec{F}$)? Recall How to Cross Multiply Vectors (Kramer's Rule)

Determine the Cross Product (Kramer's Rule) Recipe

STEP 1: Create a matrix with ($\hat{\imath}, \hat{\jmath}, \hat{k}$) in the first row. In the second row, fill in the columns with the numerical values from your first vector. In the third row, fill in the columns with the values from your second vector.

STEP 2: Cover $\hat{\imath}$ column and cross multiply the $\hat{\jmath}$ of the first vector with the $\hat{k}$ of the second vector and vice versa, subtracting the latter from the former.

Note: When writing the cross product, use brackets and parentheses as it is super easy to make sign errors when calculating these.

STEP 3: Move to $\hat{\jmath}$. Cover the $\hat{\jmath}$ column and cross multiply $\hat{\imath}$ of the first vector and $\hat{k}$ of the second vector and vice versa, subtracting the latter from the former. The $\hat{\jmath}$ term always has a negative out in front when using this method.

STEP 4: Move to $\hat{k}$. Cover the $\hat{k}$ column and cross multiply $\hat{\imath}$ of the first vector and $\hat{\jmath}$ of the second vector, subtracting the latter from the former.

STEP 5: Compute the final vector.

Note: Most calculators will do this for you!

EXAMPLE: KRAMER'S RULE

Find $\vec{a} \times \vec{b}$:

$$\vec{a} = 5\hat{\imath} - 7\hat{\jmath} + 2\hat{k}$$
$$\vec{b} = 6\hat{\imath} + 8\hat{\jmath} - 7\hat{k}$$

STEP 1: Create a matrix with $(\hat{\imath}, \hat{\jmath}, \hat{k})$ in the first row, $\vec{a}$ in the second row, and $\vec{b}$ in the third row.

$$\begin{array}{c|ccc|} & \hat{\imath} & \hat{\jmath} & \hat{k} \\ \vec{a} & 5 & -7 & 2 \\ \vec{b} & 6 & 8 & -7 \end{array}$$

STEP 2: Cover $\hat{\imath}$ column and cross multiply the $\hat{\jmath}$ with the $\hat{k}$ and vice versa, subtracting the latter from the former.

$$\begin{array}{c|ccc|} & \hat{\imath} & \hat{\jmath} & \hat{k} \\ \vec{a} & 5 & -7 & 2 \\ \vec{b} & 6 & 8 & -7 \end{array}$$

$[(-7 \times -7) - (2 \times 8)]\hat{\imath}$

STEP 3: Move to $\hat{\jmath}$. Cover the $\hat{\jmath}$ column and cross multiply $\hat{\imath}$ and $\hat{k}$ and vice versa, subtracting the latter from the former. The $\hat{\jmath}$ term is always negative when using this method.

$$\begin{array}{c|ccc|} & \hat{\imath} & \hat{\jmath} & \hat{k} \\ \vec{a} & 5 & -7 & 2 \\ \vec{b} & 6 & 8 & -7 \end{array}$$

$-[(5 \times -7) - (2 \times 6)]\hat{\jmath}$

STEP 4: Move to $\hat{k}$. Cover the $\hat{k}$ column and cross multiply $\hat{\imath}$ and $\hat{\jmath}$, subtracting the latter from the former. This term is always positive.

$$\begin{array}{c|ccc|} & \hat{\imath} & \hat{\jmath} & \hat{k} \\ \vec{a} & 5 & -7 & 2 \\ \vec{b} & 6 & 8 & -7 \end{array}$$

$+[(5 \times 8) - (-7 \times 6)]\hat{k}$

STEP 5: Compute the final vector.

$[(49) - (16)]\,\hat{\imath} - [(-35) - (12)]\,\hat{\jmath} + [(40) - (-42)]\,\hat{k} = 33\hat{\imath} + 47\hat{\jmath} + 82\hat{k}$

✓ TEST YOURSELF 3.1

SOLUTION TO TEST YOURSELF: Statics on a Rigid Body

Find $\vec{a} \times \vec{b}$.

1. $\vec{a} = 3\hat{\imath} - 2\hat{\jmath} + 6\hat{k}$

   $\vec{b} = 1\hat{\imath} + 8\hat{\jmath} - 7\hat{k}$

2. $\vec{a} = 8\hat{\imath} - 6\hat{\jmath} + 7\hat{k}$

   $\vec{b} = -10\hat{\imath} + 5\hat{\jmath} - 3\hat{k}$

3. $\vec{a} = 12\hat{\imath} + 6\hat{\jmath} - 8\hat{k}$

   $\vec{b} = -8\hat{\imath} - 7\hat{\jmath} - 6\hat{k}$

4. $\vec{a} = -9\hat{\imath} + 4\hat{\jmath} - 1\hat{k}$

   $\vec{b} = 8\hat{\imath} - 3\hat{\jmath} - 2\hat{k}$

ANSWERS

$-34\hat{\imath} + 27\hat{\jmath} + 26\hat{k}$

$-17\hat{\imath} - 46\hat{\jmath} - 20\hat{k}$

$-92\hat{\imath} + 136\hat{\jmath} - 36\hat{k}$

$-11\hat{\imath} - 26\hat{\jmath} - 5\hat{k}$

Find the moment at point *A*, using both the $M = Fd$ and $M = \vec{r} \times \vec{F}$ methods.

Press pause on video lesson 21 once you get to the workout problem. Only press play if you get stuck.

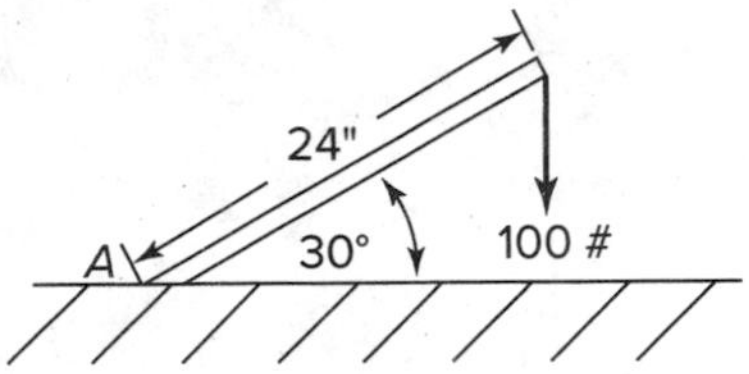

WATCH VIDEO

STATICS LESSON 22
2D Moments About a Point, Two Methods

2D moment problems can be solved one of two ways using $M = \vec{r} \times \vec{F}$ or $M = Fd$. Use $M = \vec{r} \times \vec{F}$ when the problem statement provides forces in Cartesian form. $M = Fd$ can only be used with the perpendicular distance from the point you are taking the moment to the line of the action of the force. In the $\vec{r} \times \vec{F}$ method, you don't need to know the perpendicular distance as the math already does that for you.

Remember my song to remind you of which distance to use for the $M = Fd$ method (distance is from the point we are taking the moment to the line of action of the force):

iiiif the force is in the x . . . the distance is in the y.
iiiif the force is in the y . . . the distance is in the x.

PITFALL

Remember, it is $\vec{r} \times \vec{F}$ not $\vec{F} \times \vec{r}$.
If you cross multiply backwards you will get the exact same moment vector, but all your signs will be backwards.

PRO TIP

For 2D problems, the moment is always about a point. For 3D problems, the moment is always about an axis. It is recommended to use $M = Fd$ for 2D problems and to use $\vec{M} = \vec{r} \times \vec{F}$ for 3D problems.

If you use $M = Fd$ for 3D problems, these are some helpful tips:

$M_x = F_y \cdot d_z$ and $F_z \cdot d_y$ (a force in the x direction cannot cause a moment around the x-axis)

$M_y = F_x \cdot d_z$ and $F_z \cdot d_x$ (a force in the y direction cannot cause a moment around the y-axis)

$M_z = F_x \cdot d_y$ and $F_y \cdot d_x$ (a force in the z direction cannot cause a moment around the z-axis)

(Notice how the subscripts on the force and distance are not the same as the moment.)

RIGHT HAND RULE FOR MOMENTS

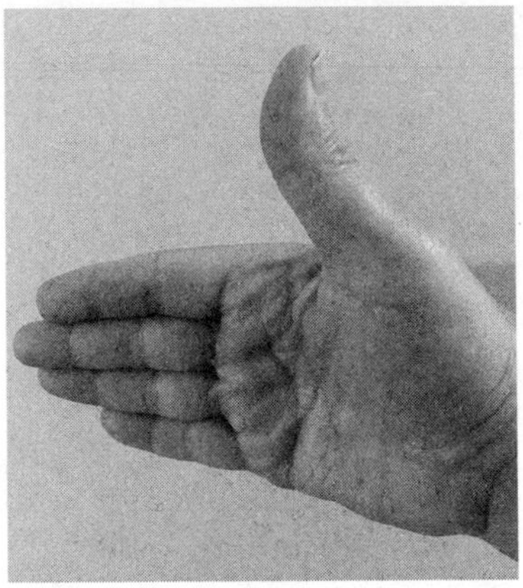

1. Using your right hand, point your fingers in the direction of the positive vector.
2. Curl your fingers in the direction of the force vector.

3. Your thumb will indicate the axis and direction (positive or negative) of the moment vector.

PITFALL

If you write with your right hand, don't forget to drop your pencil when determining the moment's direction via the *right* hand rule.

PRO TIP

In two dimensions (*xy*), the moment is *always* around the *z*-axis or the $\hat{k}$ direction.

Find the moment at point A, using both the $M = Fd$ and $M = \vec{r} \times F$ methods.

Press pause on video lesson 22 once you get to the workout problem. Only press play if you get stuck.

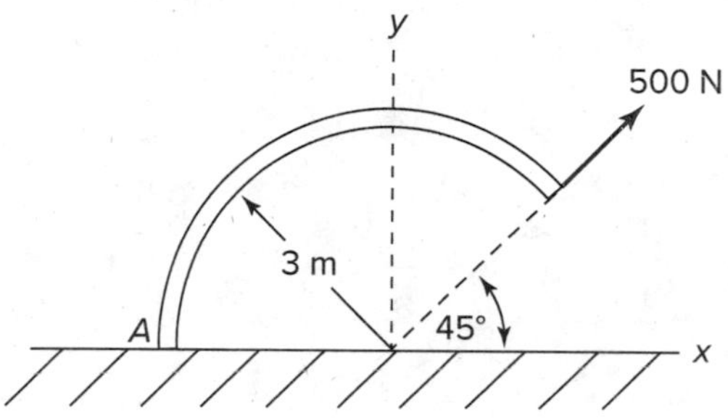

TEST YOURSELF 3.2

SOLUTION TO TEST YOURSELF: Statics on a Rigid Body

Compute the moment of the 200 lbs force about point A using both the $M = Fd$ and $M = \vec{r} \times \vec{F}$ methods.

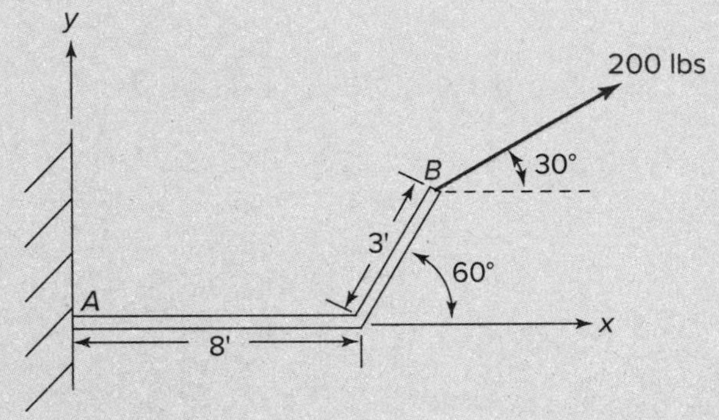

ANSWER

$500\hat{k}$ lb·ft

WATCH VIDEO

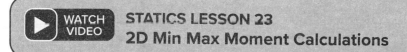

STATICS LESSON 23
2D Min Max Moment Calculations

Max moment always occurs when the position vector is perpendicular to the force vector.

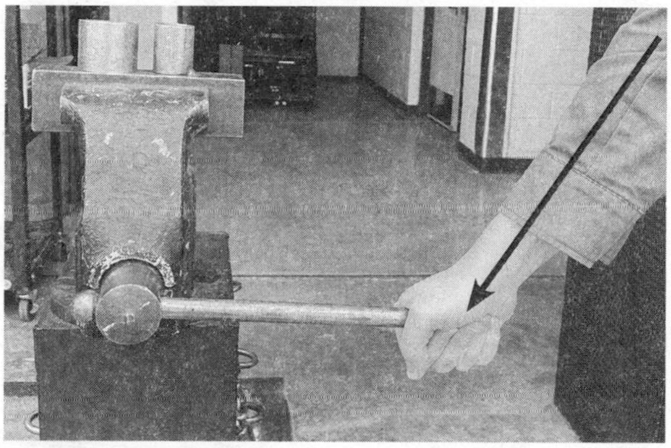

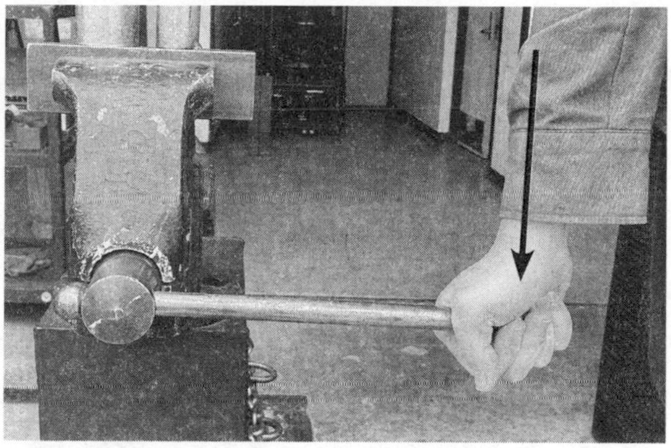

For providing the maximum moment (clamping force on vice), the force vector you apply with your arm should always be perpendicular to the bar. The converse is also true.

For creating the minimum moment, instead of pushing at a right angle on the bar, push where the force vector passes through the point of rotation for the system. This will result in a moment of zero.

The tension in the cable is 2,500 #. What is the moment produced at point *A*? If the cable could pull from any direction, what is the maximum moment possible at point *A*?

Press pause on video lesson 23 once you get to the workout problem. Only press play if you get stuck.

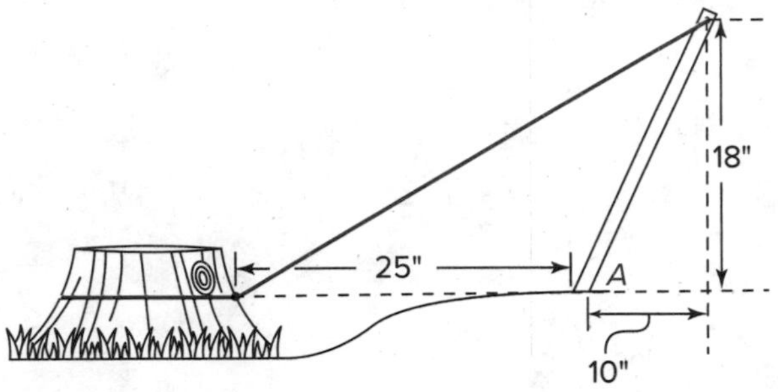

WATCH VIDEO

STATICS LESSON 24
3D Moment About a Point and $\vec{r} \times \vec{F}$ Example

In 3D, $M = \vec{r} \times \vec{F}$ is better suited (until we get more experience and level up) to handle these problems.

Find the moment generated by the 39 # force about point *A*.

Press pause on the video once you get to the workout problem. Only press play if you get stuck.

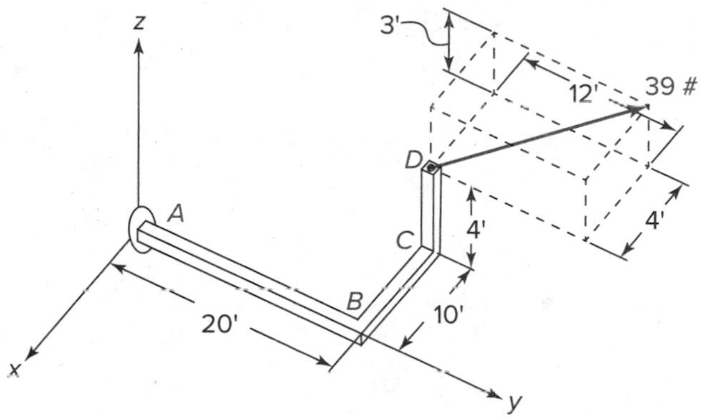

✓ TEST YOURSELF 3.3

SOLUTION TO TEST YOURSELF: Statics on a Rigid Body

Find the moment of the 500 lbs force shown in the diagram about the base (point A).

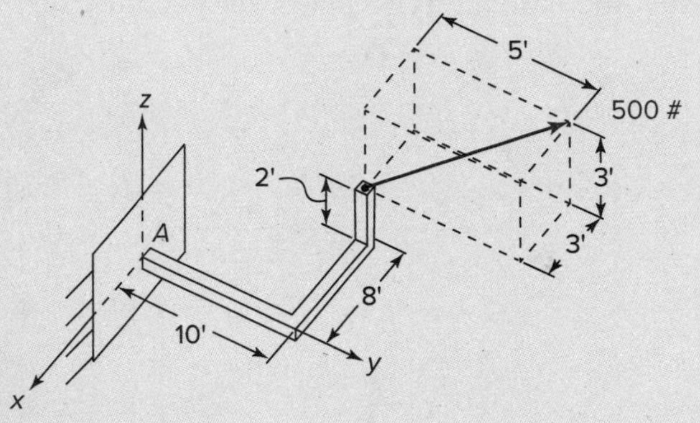

ANSWER

$M_A = 1{,}525\hat{i} + 1{,}373\hat{j} - 762.5\hat{k}$ lb•ft

WATCH VIDEO

STATICS LESSON 25
Moment About a Specified Axis (Wacky Axis)

Hanson refers to any axis other than the x, y, or z as a "wacky axis (W.A.)" (any random axis of rotation).

The equation for solving the moment around the W.A. is $M_{W.A.} = (\vec{r} \times \vec{F}) \cdot \hat{\lambda}_{W.A.}$ You can always use $M_{W.A.} = (\vec{r} \times \vec{F}) \cdot \hat{\lambda}_{W.A.}$, even if not a wacky axis, and simply, for instance, around the x-axis. The $\hat{\lambda}$ for the x-axis would be $(1, 0, 0)$ so you would get just the $\hat{\imath}$ component of $\vec{r} \times \vec{F}$.

Moment Calculations About a Wacky Axis Recipe

STEP 1: Write the position vector in Cartesian form.

- Identify a position vector by selecting any point on the axis of rotation.
- Place the starting point for the tail of the position vector on the axis of rotation.
- Place the tip of the positive vector on any point on the line of action of the force.

STEP 2: Write the force vector in Cartesian form.

STEP 3: Complete a matrix calculation of $\vec{r} \times \vec{F}$.

Note: This will yield a Cartesian moment vector.

STEP 4: Write a $\hat{\lambda}$ unit vector of the wacky axis in Cartesian form.

STEP 5: Dot the Cartesian product (dot product) of the $(\vec{r} \times \vec{F})$ matrix with the $\hat{\lambda}_{W.A.}$ vector.

Note: As learned earlier, the dot product will yield a scalar quantity! This scalar will be the entire moment quantity rotating around the wacky axis.

PRO TIP

Remember that a negative scalar moment means you are rotating clockwise around the wacky axis, and a positive moment means you are rotating counterclockwise around the wacky axis.

EXAMPLE: WACKY AXIS MOMENT

Find the moment produced around the pipe segment *AB* by the given force.

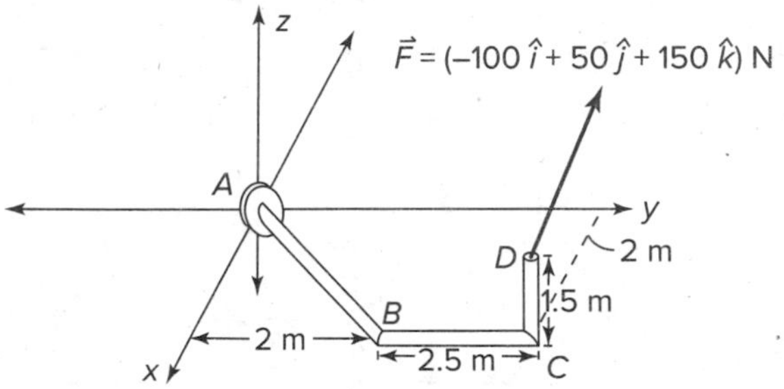

STEP 1: Let's find $\vec{r}$ (position vector). We can calculate $\vec{r}_{AD}$ or $\vec{r}_{BD}$ (we will get the same answer using either position vector).

$\vec{r}_{AD} = D - A = (2, 4.5, 1.5) - (0, 0, 0) = (2\hat{\imath} + 4.5\hat{\jmath} + 1.5\hat{k})$

STEP 2: The force vector is already given in Cartesian form $\vec{F} = (-100\hat{\imath} + 50\hat{\jmath} + 150\hat{k})$ N:

STEP 3: Calculate $(\vec{r} \times \vec{F})$.

	$\hat{\imath}$	$\hat{\jmath}$	$\hat{k}$
$\vec{r}$	2	4.5	1.5
$\vec{F}$	-100	50	150

$= [(4.5 \times 150) - (1.5 \times 50)]\hat{\imath}$
$- [(2 \times 150) - (1.5 \times -100)]\hat{\jmath}$
$+ [(2 \times 50) - (4.5 \times -100)]\hat{k}$

$= (675 - 75)\hat{\imath} - (300 + 150)\hat{\jmath} + (100 + 450)\hat{k}$
$= (600\hat{\imath} - 450\hat{\jmath} + 550\hat{k})$ N•m

STEP 4: Calculate $\hat{\lambda}_{AB}$.
$\vec{r}_{AB} = B - A = (2, 2, 0) - (0, 0, 0) = (2\hat{\imath} + 2\hat{\jmath} + 0\hat{k})$

$$\frac{2\hat{\imath} + 2\hat{\jmath} + 0\hat{k}}{\sqrt{2^2 + 2^2 + 0^2}} = 0.707\hat{\imath} + 0.707\hat{\jmath} + 0\hat{k} = \hat{\lambda}_{AB}$$

STEP 5: Calculate the dot product of the moment vector from Step 2 with $\hat{\lambda}_{AB}$ from Step 3.

$M_{AB} = (600\hat{\imath} - 450\hat{\jmath} + 550\hat{k}) \cdot (0.707\hat{\imath} + 0.707\hat{\jmath} + 0\hat{k})$
$M_{AB} = [(600 \times 0.707) + (-450 \times 0.707) + (0 \times 550)]$ N•m
$M_{AB} = (424.2 - 318.15)$ N•m $= 106.05$ N•m

If you were to look at segment *AB* (looking toward the origin), you would see a counter-clockwise moment of 106.05 N•m around line *AB*.

Find the moment about the OB axis due to force F.

Press pause on video lesson 25 once you get to the workout problem. Only press play if you get stuck.

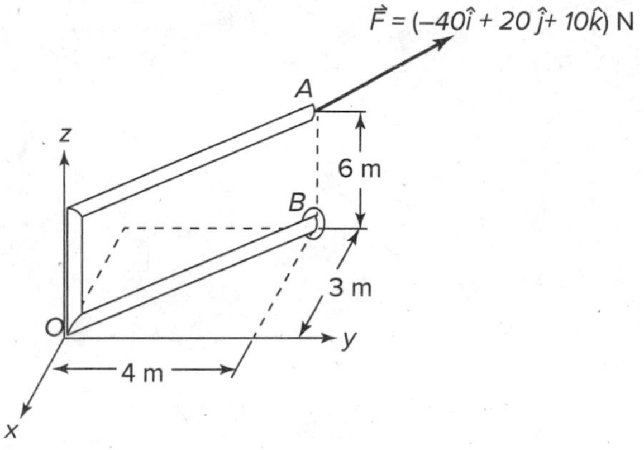

✓ TEST YOURSELF 3.4

SOLUTION TO TEST YOURSELF: Statics on a Rigid Body

1. Find the moment of the 1,000 N force F about the line given about vector C which is at $\theta = 60°$ in the xy plane. Point A is at (9, 3, 6) and B is at (7, 5, 11).

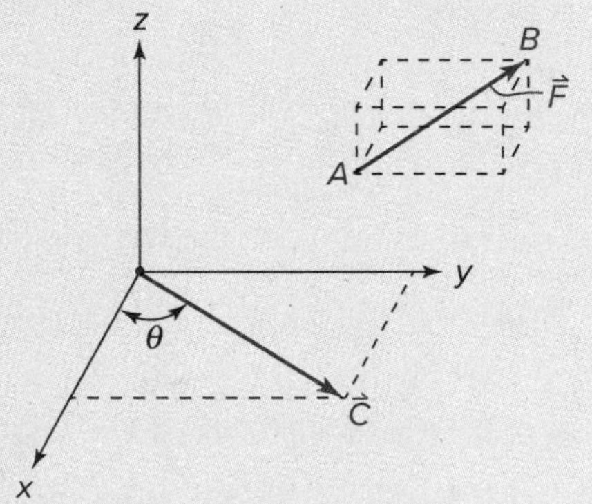

ANSWER
$M_C = -8,328$ N•m

2. The magnitude of the force is 100 lbs; what is the moment of this force about the line going from B to A?

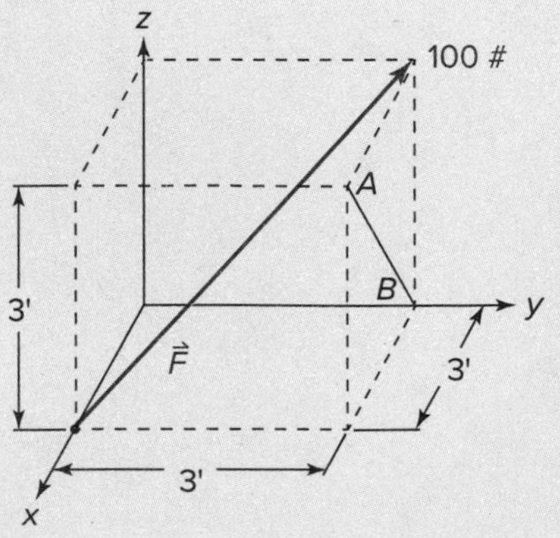

ANSWER
$M_{BA} = -122.5$ lb•ft

WATCH VIDEO

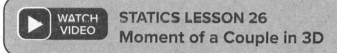

STATICS LESSON 26
Moment of a Couple in 3D

Recall: A couple is two forces, of equal magnitude, in opposite directions, on parallel lines of action. A couple is simply a different way that moments are sometimes given.

PRO TIPS

- **A simple trick for turning couple forces into an applied moment is to simply take one of the couple forces (doesn't matter which one as they are both the same) and multiply that force by the perpendicular distance between the two forces' lines of action.**
- **If all distances are known, the forces in a "moment of a couple" problem can simply be treated as ordinary forces, and the problem can then be solved like any other moment problem.**

Determine the distance (d) between A and B so that the resultant moment has a ***magnitude*** of 20 N•m.

Press pause on video lesson 26 once you get to the workout problem. Only press play if you get stuck.

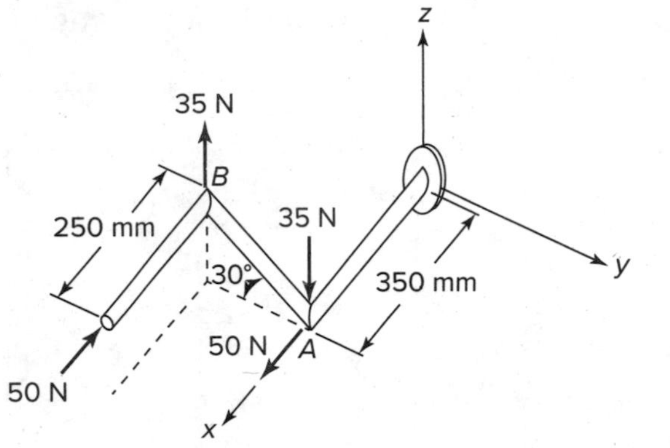

✓ TEST YOURSELF 3.5

SOLUTION TO TEST YOURSELF: Statics on a Rigid Body

Compute the total moment of the two couples. The 400 lbs forces point toward the opposite corner on the top and bottom face, respectively. The 300 lbs forces point toward the opposite corner on the right and left face, respectively.

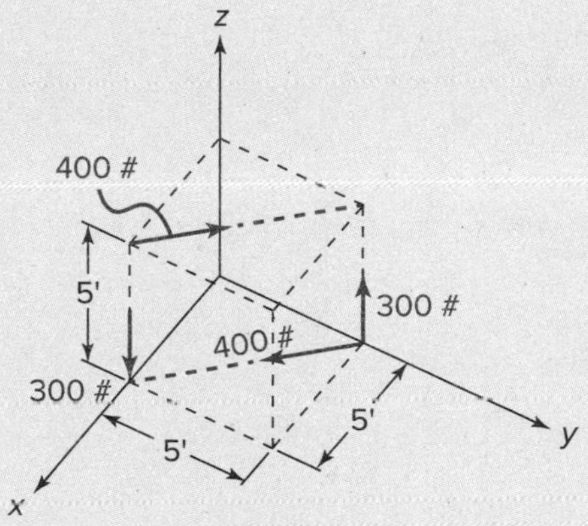

ANSWER

$\vec{M} = 85.8\hat{i} + 85.8\hat{j}$ lb·ft

WATCH VIDEO

STATICS LESSON 27
Equivalent Systems Simplification, Burrito Force!

Equivalent Systems—could also be called a *force simplification*.

- Involve taking a very complicated system of multiple forces and/or coupled moments and rewriting them in a simpler form as either just a single force whose distance is at a specific location to supply the correct moment or a force-couple at some point.
- Easily identified by the wording in the problem statement, such as:
  - REPLACE the following . . .
  - REDUCE the following . . .
  - SIMPLIFY the following . . .
  - REWRITE the following . . .
  - Etc.
- Can refer to these as a System I–System II problem, in that System I will be equivalent to System II.

This means that

$\Sigma F_{x_I} = \Sigma F_{x_{II}}$ The sum of forces in the x-direction is the same in both systems.

$\Sigma F_{y_I} = \Sigma F_{y_{II}}$ The sum of forces in the y-direction is the same in both systems.

$\Sigma M_I = \Sigma M_{II}$ The sum of the moments about point of interest is the same in both systems.

PRO TIP

The Burrito Force!

A burrito force is a concentrated moment on a body, drawn with a swirl arrow:

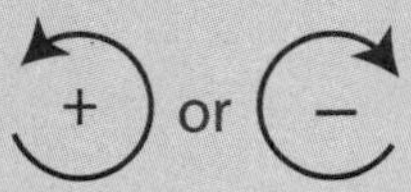

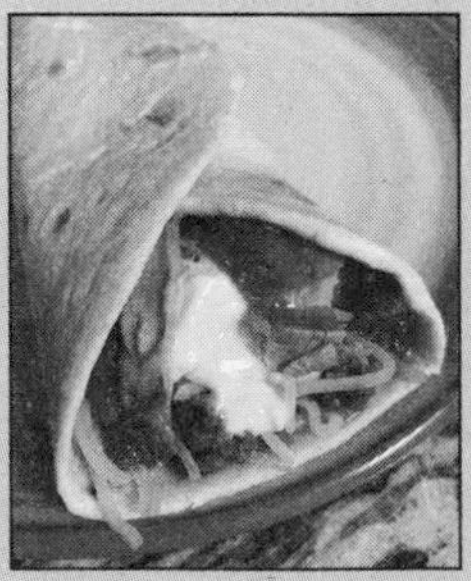

The "swirl arrow" is rolled up just like the ends on a burrito.

When including a concentrated moment in a sum of the moments equation, a very common mistake is to multiply that moment by a distance. . . . Wrong! Because the distance is already in the burrito! So, what's in a burrito? *All* the good stuff. It includes the force *and* distance (N·m or #·ft).

Solving Equivalent System Problems Recipe

STEP 1: ΣF_{x_I} Sum the forces in the x-direction in System I.

STEP 2: ΣF_{y_I} Sum the forces in the y-direction in System I.

STEP 3: ΣM_I Sum the moments about the point of interest in System I.

STEP 4: Draw System II (same body as in System I) and draw the forces and moment, found in Steps 1 through 3, at the point of interest.

STEP 5: Draw System III and further simply the system by sliding the force along the beam so that the force would create a moment in the correct direction about the point of interest. Use the moment equation $M = Fd$ to solve for the distance d. From the previous steps, M is known and F is known.

Note: Only use this step if the question says to "replace with a force only system." If it states force-couple system, stop after Step 4.

PITFALL

On equivalent systems problems, do not solve for the reaction forces (you're finding the equivalent load, not anything else!).

EXAMPLE: EQUIVALENT SYSTEMS

Replace the given loaded beam with (a) a force-couple at point *A* and (b) a force only system and specify the distance from point *A* that force's line of action intersects *AB*.

In this case, point *A* would be our point of interest.

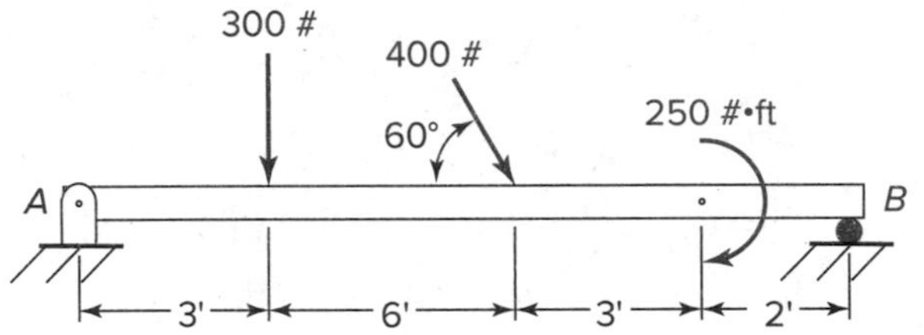

System I (the given system)

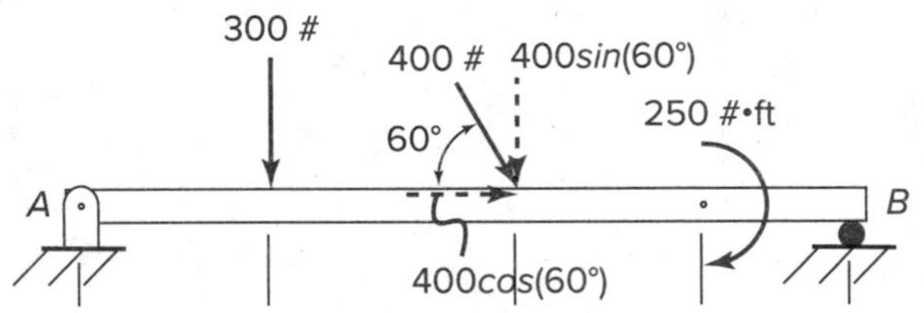

STEP 1: $\Sigma F_{x_I} = 400cos(60°) = 200$ #

STEP 2: $\Sigma F_{y_I} = -300 - 400sin(60°) = -646.4$ #

STEP 3: $\Sigma M_{A_I} = -300(3) - 400sin(60°)(9) - 250 = -4{,}267.7$ #ft

STEP 4: Draw System II (same body as in System I) and locate the forces and moment to the point of interest.

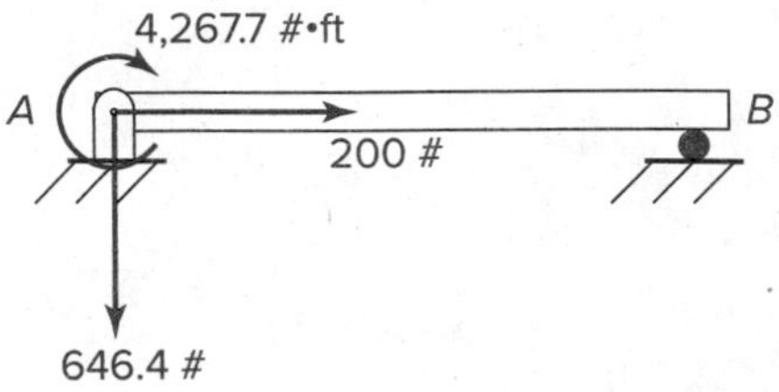

At this point, part (a) is finished. (Remember the *x* and *y* components at point *A* can easily be converted into a single force using the Pythagorean Theorem.)

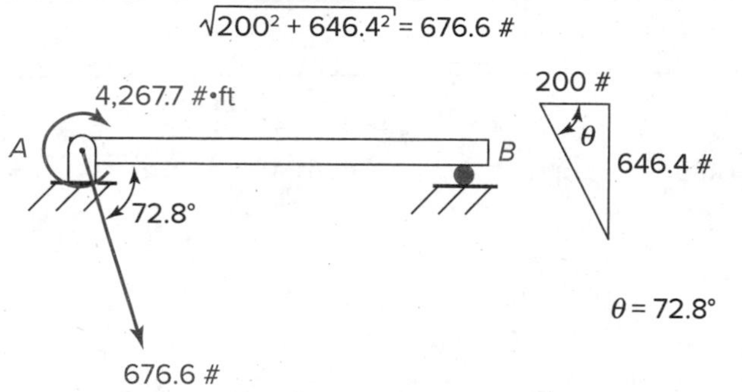

For part (b), pick up where we left off at part (a).

STEP 5: Draw a System III and slide the forces only along the beam until the desired moment about the point of interest is produced.

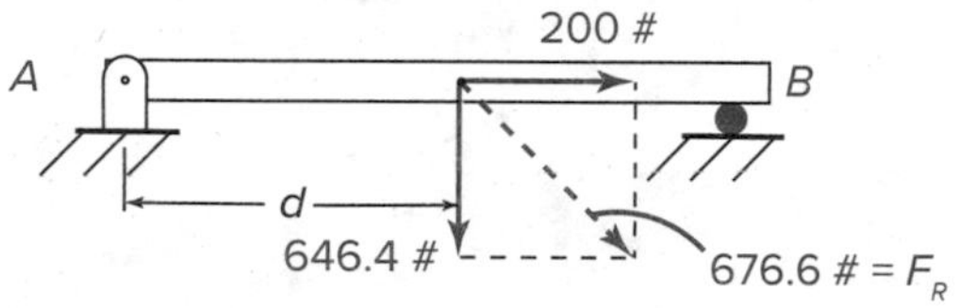

Calculate the distance (d) by using the formula: $M = Fd$.
$d = M/f = 4{,}267.7/646.4 = 6.6$ ft to the right of point A.

Note: The most common mistake made here is to use the resultant force. *Always* use the perpendicular component to the distance: $M = Fd$.

Moving the force to the left of point A would have produced a counterclockwise moment. We moved to the right of point A, which produced a clockwise moment, which is what we needed as 4,267.7 #•ft was clockwise!

PRO TIP

There will be times where the point at which the force intersects the line of interest, to create an equivalent moment, that is not on the part. Don't panic if you come across that. It may look weird, but it may also be correct.

STATICS ON A RIGID BODY

Find an equivalent force only system and find where that force's LOA intersects line *BE* measured from point *B*.

Press pause on video lesson 27 once you get to the workout problem. Only press play if you get stuck.

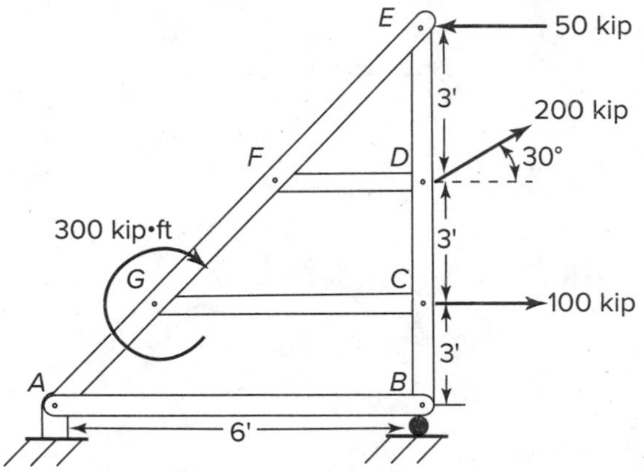

WATCH VIDEO

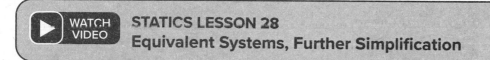

STATICS LESSON 28
Equivalent Systems, Further Simplification

Replace the giving loading with a single equivalent force. What is the magnitude and direction of the force? Where does it intersect line *AB* measured from *A*?

Press pause on the video once you get to the workout problem. Only press play if you get stuck.

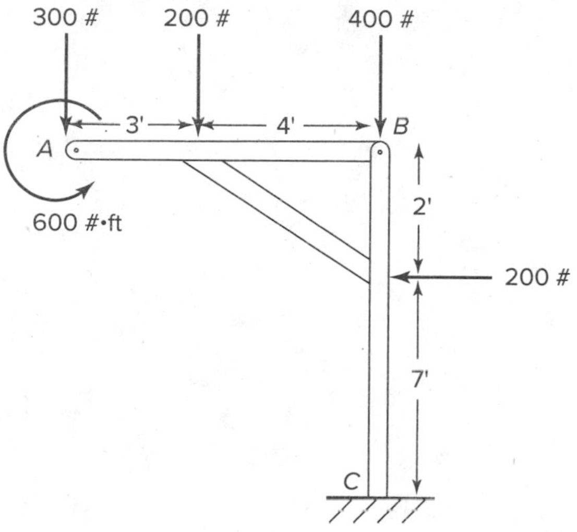

✓ TEST YOURSELF 3.6

SOLUTION TO TEST YOURSELF: Statics on a Rigid Body

For the loads on the beam below, $F_1 = 10{,}000$ lbs, $F_2 = 8{,}000$ lbs, $\theta = 30°$, and $M_1 = 15{,}000$ ft•lb. Find the equivalent load on the beam and the distance from the left end (point A).

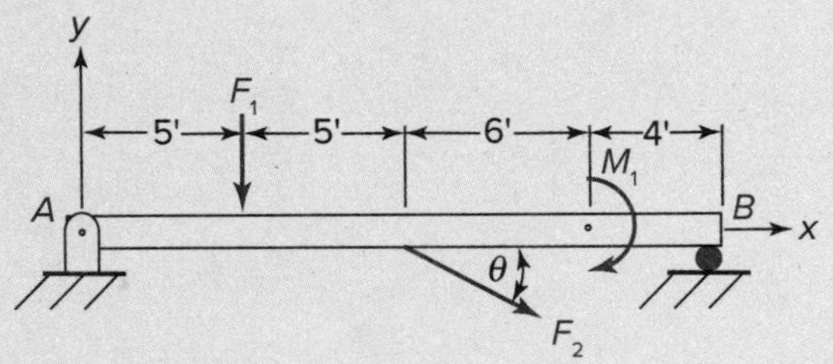

ANSWERS

$F = (6{,}928\hat{i} - 14{,}000\hat{j})$ lbs

Distance = 7.5 ft

KEY TAKEAWAYS

Rotation of a rigid body includes torques, moments, and couples. Rotation has a magnitude and a direction and therefore is a vector with Cartesian components $(\hat{\imath}, \hat{\jmath}, \hat{k})$.

Main takeaways from Level 3 include how to calculate:

- 2D moments
- 3D moments
- Equivalent systems

2D Moment Summary

- Remember that in 2D, moments are taken about a point.
- 2D moments are always about the z-axis when the force and distance are in xy plane.
- To determine moment in 2D there are two methods:
  - $M = \vec{r} \times \vec{F}$
    - $\vec{r}$ (position vector): A vector whose tail is at the point where we wish to take a moment and whose tip is on a point anywhere along the line of action (LOA) of the force. Also top half of the $\hat{\lambda}$ formula.
    - $\vec{F}$ is the force vector with Cartesian components $(\hat{\imath}, \hat{\jmath}, \hat{k})$.
    - Recall Kramer's Rule to complete this calculation.
  - $M = Fd$
    - Don't forget "*d*" *must* be the perpendicular distance between the point you wish to take moments about and the force vector.
    - Once you practice enough, this should become your preferred method for 2D moments.
    - When using this equation, put your finger on the point of interest and then determine which direction the force makes you spin about that point.
      - A force causing a clockwise rotation (CW) is a negative moment.
      - A force causing a counterclockwise rotation (CCW) is a positive moment.

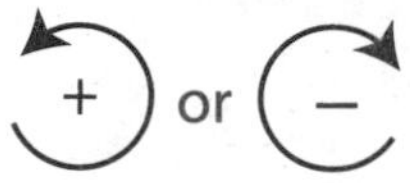

KEY TAKEAWAYS

3D Moment Summary

- Remember that in 3D moments are taken about an axis
- To determine moment in 3D use Kramer's rule in this level (once you level-up you can use $M = Fd$ but have to find all the parallel distances!). Using Kramer's rule, you don't have to find all the parallel distances, and it gives you the correct sign.
    - $\hat{\imath}$ component are things that make me spin about the x-axis
    - $\hat{\jmath}$ component are things that make me spin about the y-axis
    - $\hat{k}$ component are things that make me spin about the z-axis
- Wacky axis problems involve taking a moment about a specified axis other than the x, y, or z axis.
    - $\vec{M}_{W.A.} = (\vec{r} \times \vec{F}) \bullet \hat{\lambda}_{W.A.}$ (See Moment Calculations About a Wacky Axis Recipe.)
    - Remember this utilizes a dot product which yields a scalar rather than a vector.

Equivalent Systems

- Identify these problems by recognition of the key words: replace, reduce, simplify, rewrite, etc.
- Determine which of the following systems it is asking you to identify:
    - Force and couple system
        - Requires you only to use System I–System II.
    - Force only system
        - Requires you to use System I–System II–System III following the recipe (to rid yourself of the moment by moving the force to a location that produces an equivalent moment).

RECIPES

- **Determine the Cross Product (Kramer's Rule) Recipe**
- **Moment Calculations About a Wacky Axis Recipe**
- **Solving Equivalent System Problems Recipe**

KEY TAKEAWAYS

PRO TIPS

Moment Tips

- Think of vector $\vec{r}$ as how to get to Grandma's house. You live at the point we wish to take the moment, and Grandma lives at the point on the LOA. It's just the direction how to get from here to there.
- For 2D problems, the moment is always about a point. For 3D problems, the moment is always about an axis. It is recommended to use $M = Fd$ for 2D problems and to use $\vec{M} = \vec{r} \times \vec{F}$ for 3D problems.

2D Moment Tips

- In two dimensions (*xy*), the moment is *always* around the *z*-axis or the $\hat{k}$ direction.

3D Moment Tips

- Remember that a negative scalar moment means you are rotating clockwise around the wacky axis, and a positive moment means you are rotating counterclockwise around the wacky axis.

2D and 3D Moment Tips

- A simple trick for turning couple forces into an applied moment is to simply take one of the couple forces (doesn't matter which one as they are both the same) and multiply that force by the perpendicular distance between the two forces' lines of action.
- If all distances are known, the forces in a "moment of a couple" problem can simply be treated as ordinary forces, and the problem can then be solved like any other moment problem.

Equivalent System

- There will be times where the point at which the force intersects the line of interest, to create an equivalent moment, that is not on the part. Don't panic if you come across that. It may look weird, but it may also be correct.

PRO TIP

The Burrito Force!

A burrito force is a concentrated moment on a body, drawn with a swirl arrow:

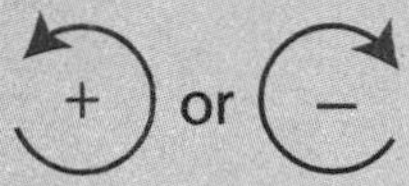

The "swirl arrow" is rolled up just like the ends on a burrito.

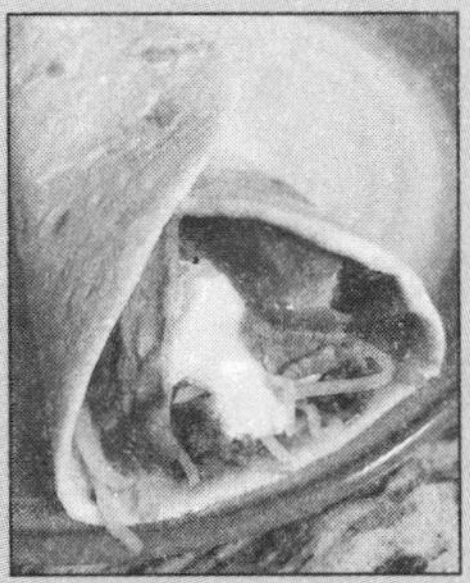

When including a concentrated moment in a sum of the moments equation, a very common mistake is to multiply that moment by a distance. . . . Wrong! Because the distance is already in the burrito! So, what's in a burrito? *All* the good stuff. It includes the force *and* distance (N·m or the #·ft).

PITFALLS

2D Moment Tips

- To use the $M = Fd$ equation, the distance *must* be perpendicular to the force vector.

3D Moment Tips

- *Do not* divide the position vector by its magnitude like we did when calculating $\hat{\lambda}$! That would always make a vector of length 1!
- Remember, it is $\vec{r} \times \vec{F}$ not $\vec{F} \times \vec{r}$. If you cross multiply backwards you will get the exact same moment vector, but all your signs will be backwards.
- If you write with your right hand, don't forget to drop your pencil when determining the moment's direction via the *right* hand rule.
- On equivalent systems problems, do not solve for the reaction forces (you're finding the equivalent load, not anything else!).

Level 4

Global Equilibrium Reactions

WATCH VIDEO **STATICS LESSON 29**
2D Reaction at Supports, Example Problem

Rigid Body

- A car, a person, a beam, a tractor, or any body that is not merely a single point or particle
- Does not deflect, bend, expand, or deform (note: you will learn to evaluate deformable bodies in Solid Mechanics)

Reaction Forces and Moments

- Exist because of Newton's Third Law (for every action there is an equal and opposite reaction)
- Forces or moments that prevent motion or rotation of a system
- Sometimes referred to as "Global Equilibrium" because these are points where the rigid body is connected to the "world" or the "globe"
- Generally forces are expressed as components, not as a single magnitude (i.e., force at an angle)
- To determine moment reactions for both 2D and 3D, from here on we will use $M = Fd$ and ***not*** $M = \vec{r} \times \vec{F}$

The most common types of 2D connections within a system that generate reaction forces:

Types of Connections	Reactions	Description of Reaction(s)
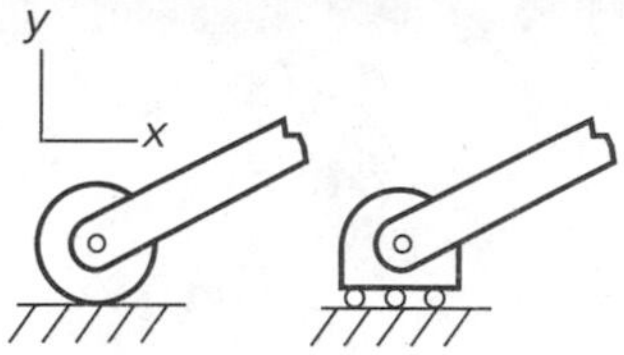	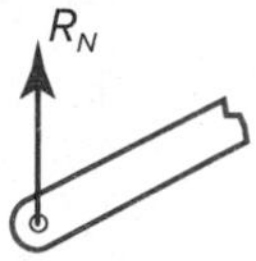	**Roller Connections** Only one reaction force. Always normal to the plane of contact.
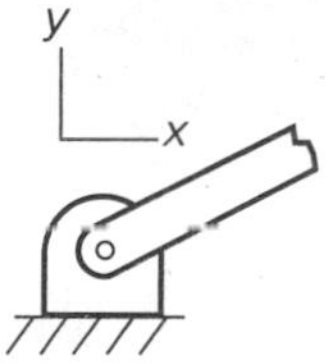	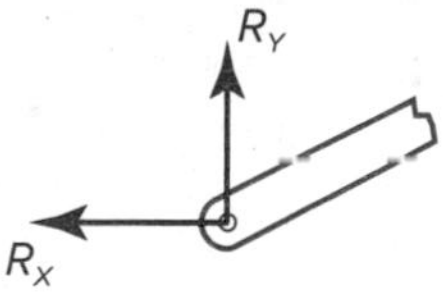	**Pin Connection** Has two reaction forces, an *x* component and a *y* component.
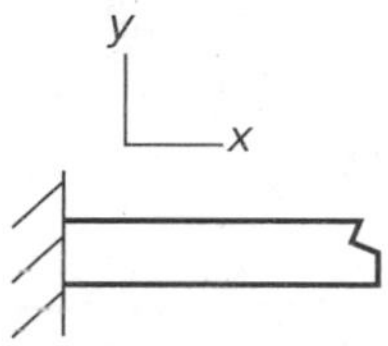	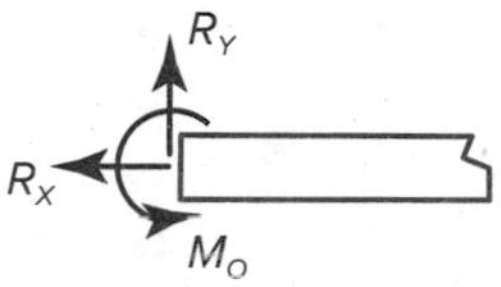	**Cantilever Connection or Fixed Connection** Has an *x* component and a *y* component, as well as a reaction moment in the *z* direction.
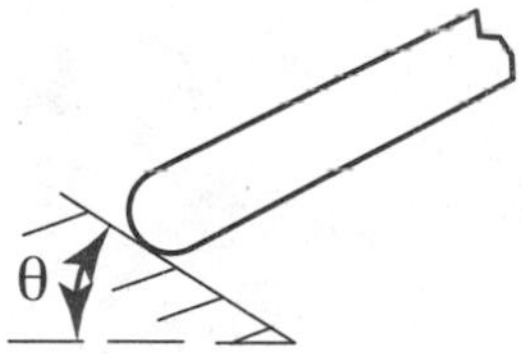	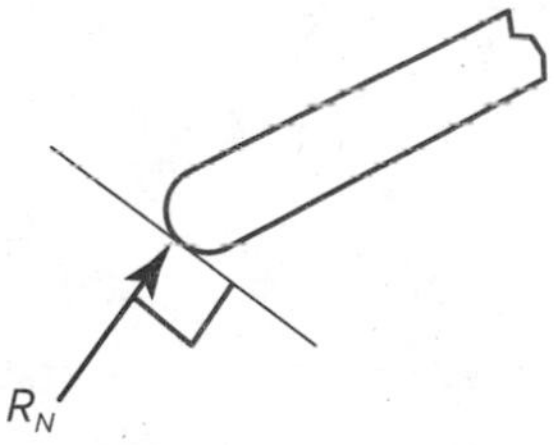	**Single Point of Contact** Has a normal reaction force perpendicular to the plane of contact.
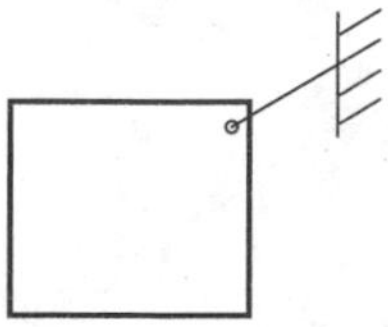	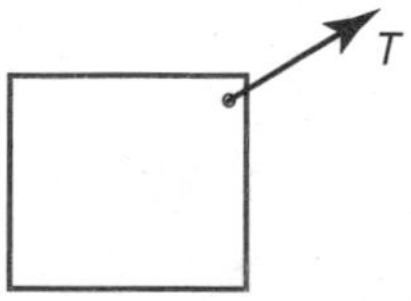	**Simple Cable Connection** Always drawn in tension along the line of action of the cable.
	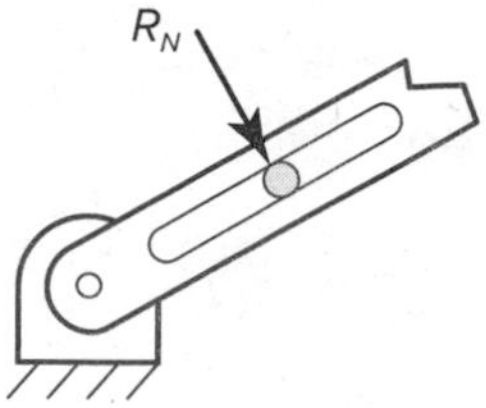	**Smooth Pin in a Slot** Has single normal force reaction always perpendicular to the slot.

(continued on next page)

Types of Connections	Reactions	Description of Reaction(s)
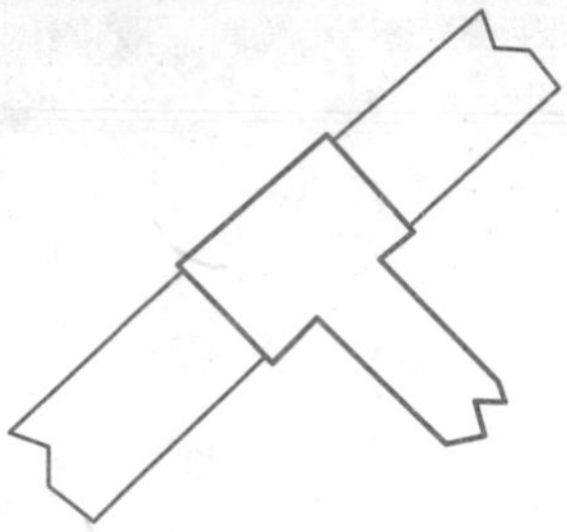	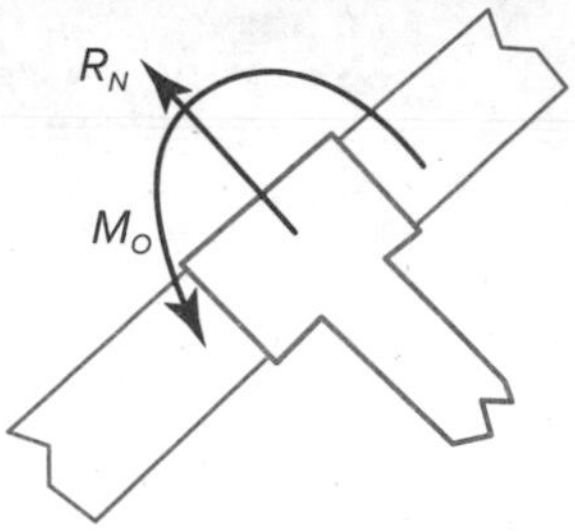	**Smooth Collar** Has a single normal reaction force perpendicular to the pipe, as well as a reaction moment.

Do I need to memorize this table?

No! Instead simply ask yourself a few questions . . .

- Can I move the system in the x-direction?
- Can I move the system in the y-direction?
- Can I rotate this system (i.e., spin around the z-axis)?

If the answer is ever *no* to the "move" question, then there must be a reaction force and if *no* to the "rotate" question, there must be a moment preventing that action.

- Why do I need to know these reaction forces? These forces allow us to have equilibrium.
- To have 2D equilibrium in a system:

$$\Sigma F_x = 0$$
$$\Sigma F_y = 0$$
$$\Sigma M_z = 0$$

Normal Forces

- Think of a these as a "bathroom scale" force.
- Always act perpendicular to the plane of contact.

TEST YOURSELF 4.1

SOLUTION TO TEST YOURSELF: Global Equilibrium Reactions

For these real world pictures, which type of 2D connections are these?

1

2

3

4

5

6

7

8

9

ANSWER

See Test Yourself 4.1 solutions at the back of the book.

Solve for 2D Reaction at the Supports Recipe

STEP 1: Draw the free body diagram by asking yourself the "can I move" and "can I rotate" questions. Recall, if you get the answer "no," then there must be a reaction.

Note: If you are not sure which direction the reaction arrows are pointed, just guess a direction. If you guessed incorrectly, you will simply get a negative answer. To get the direction of the reaction correct, recall "the up stuff has to equal the down stuff; the left stuff has to equal the right stuff." Always try to get the reaction direction correct. If you expect a positive answer but get a negative, double check your work.

STEP 2: Write equations of equilibrium ($\Sigma F_x = 0$, $\Sigma F_y = 0$, and $\Sigma M_z = 0$).

Note: Start your equations of equilibrium with the moment equation. Taking the moment at the point with the most unknowns will "knock out" these unknowns, generally leaving only one unknown, which makes solving much easier!

STEP 3: Solve for unknowns.

Note: If you get a negative sign for any of your unknowns, it is opposite to what you assumed.

EXAMPLE: 2D SUPPORT REACTIONS

Find the reactions at the supports for a simply supported beam.

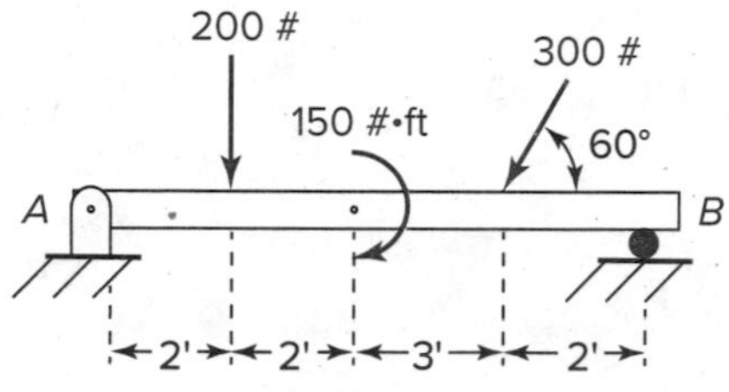

STEP 1: Draw the free body diagram.

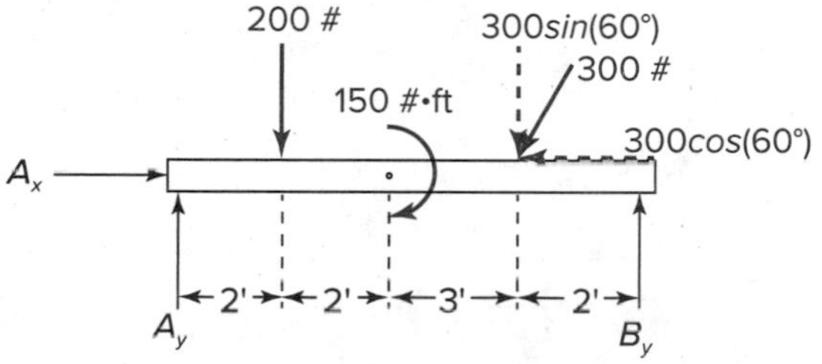

STEP 2: Write equations of equilibrium ($\Sigma F_x = 0$, $\Sigma F_y = 0$, and $\Sigma M_A = 0$) and solve.

Note: It is wise to take the moment either around point A or point B as this gets rid of an unknown. If you take the moment about any other point on the beam, then you will have to include both A_y and B_y in the equation.

Also note: A_x is not included in the moment equation since it doesn't have a perpendicular distance.

$$\circlearrowleft + \; \Sigma M_A = 0 = -200(2) - 150 - 300sin(60°)(7) + B_y(9)$$

Note: Counterclockwise moments are positive.

$$\Sigma F_x = 0 = A_x - 300cos(60°)$$
$$\Sigma F_y = 0 = -200 - 300sin(60°) + A_y + B_y$$

STEP 3: Solve for unknowns.

$B_y = 263.18$ #
$A_x = 150$ #
$A_y = 196.62$ #

Solve for the reaction forces at point *A* and point *B*.

Press pause on video lesson 29 once you get to the workout problem. Only press play if you get stuck.

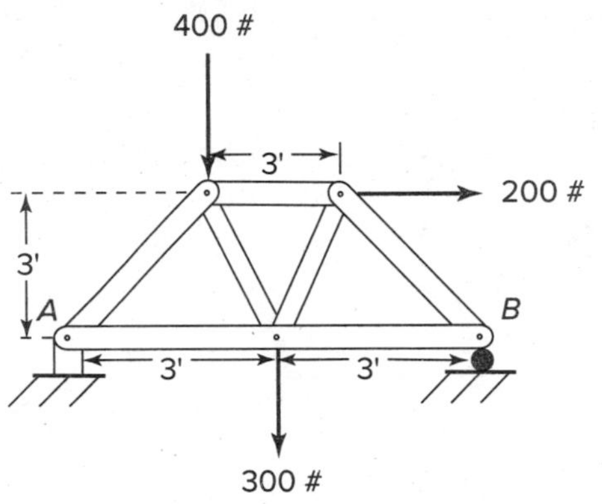

WATCH VIDEO

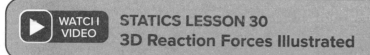

STATICS LESSON 30
3D Reaction Forces Illustrated

The most common types of 3D reactions:

Support	Reaction(s)	Description of Reaction(s)
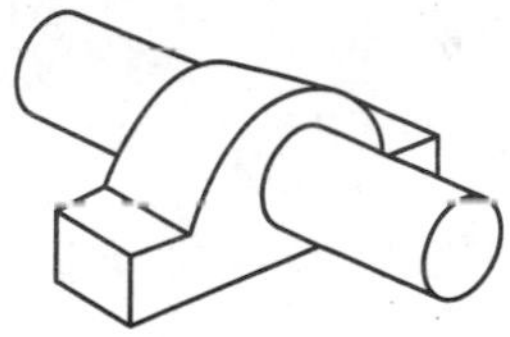	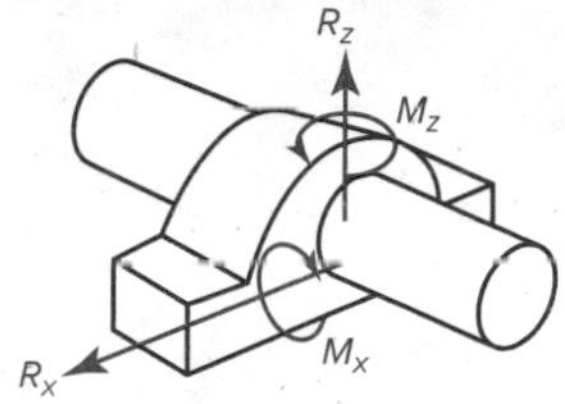	**Pillow Block or Smooth Journal Bearing** Cannot resist axial thrust so it has two force reactions and two moment reactions.
	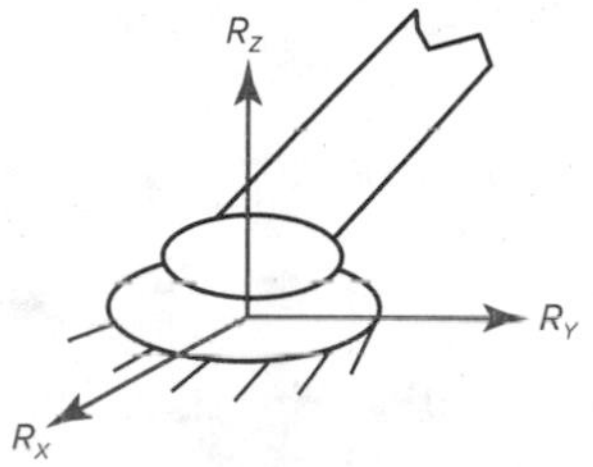	**Ball and Socket Joint** Since it can rotate in any direction, it has only three force reactions, no moment reactions.
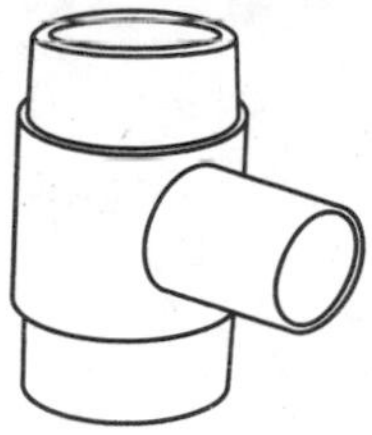	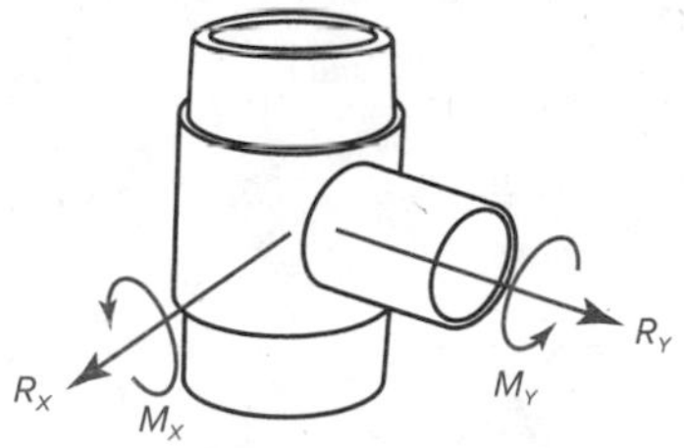	**3D Smooth Collar on Round Shaft** Has two force reactions and two moment reactions.
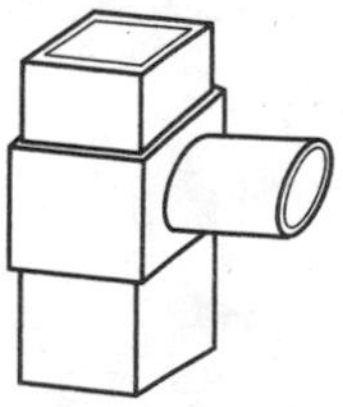	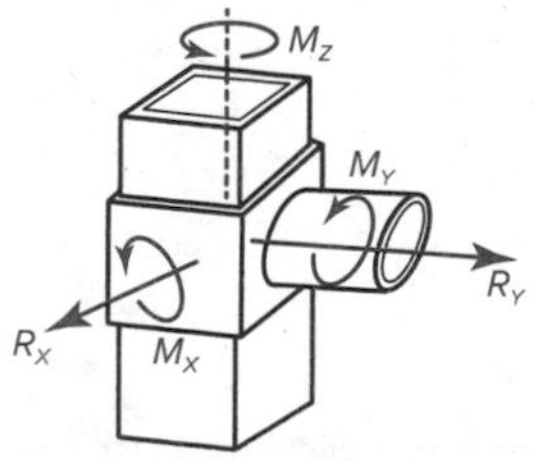	**3D Smooth Collar on a Square Shaft** Has two force reactions and three moment reactions.
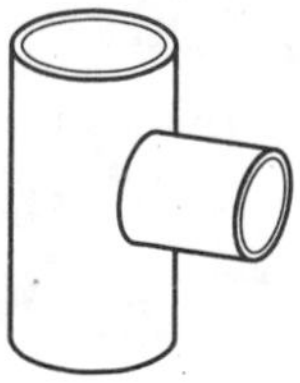	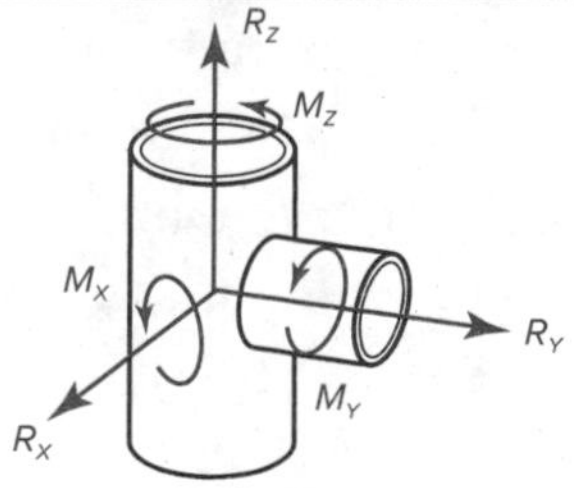	**3D Fixed Connection** Has three force reactions as well as three moment reactions. Any fixed connection to a wall, pipe, or beam will have the exact same reactions as shown in this diagram.

CHALLENGE QUESTION

What type of 3D connection is holding the sides of the bell?

Just like in 2D, determine the reaction of the 3D supports by asking the following questions:

- Can I move the system in the x-direction?
- Can I move the system in the y-direction?
- Can I move the system in the z-direction?
- Can I rotate the system about the x-axis?
- Can I rotate the system about the y-axis?
- Can I rotate the system about the z-axis?

As discussed earlier, if the answer to any of these questions is ever *no*, there is a reaction force or moment preventing that motion.

PRO TIP

In 3D, it is almost impossible to determine the direction of the reaction forces, so save some time and just assume the positive direction for all forces and moments, and let the math tell if you assumed correctly or incorrectly. An incorrect assumption will result in a negative answer.

ANSWER
Pillow block

WATCH VIDEO

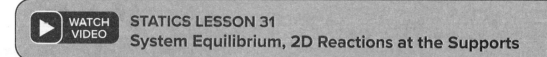

STATICS LESSON 31
System Equilibrium, 2D Reactions at the Supports

Two Force Members

- Members with pin connections on each end and have no other external forces or moments applied to them along the length.
- Cables, ropes, chains, cords, etc. are always two force members in tension.
- Springs and hydraulic/pneumatic cylinders are two force members that can be in tension or compression.
- Identifying is important as the direction (i.e., angle) is always known (the line of action always passes through the two pin connections) thus eliminating an unknown.

Examples of common two force members are shown below:

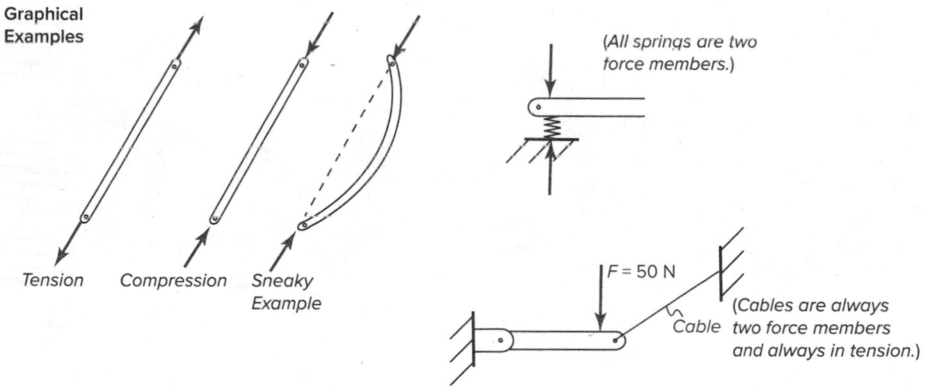

✓ TEST YOURSELF 4.2

SOLUTION TO TEST YOURSELF: Global Equilibrium Reactions

For the following problems, identify which member(s) are two force members.

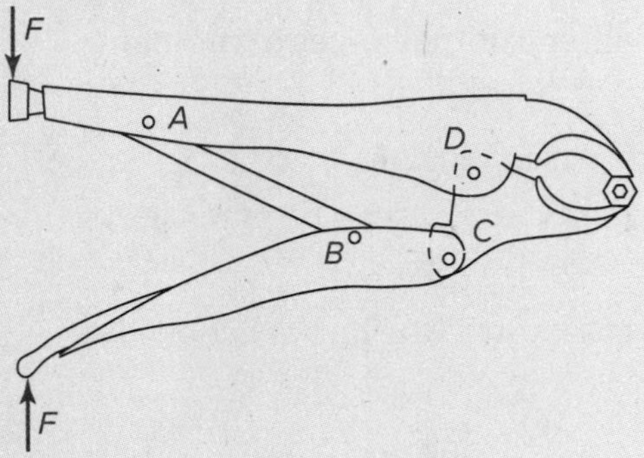

1.

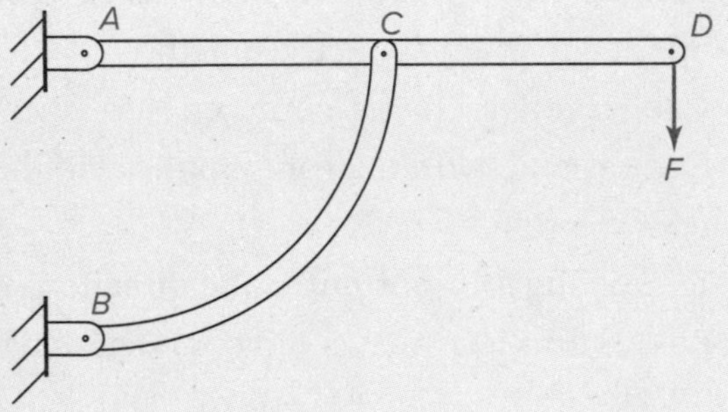

2.

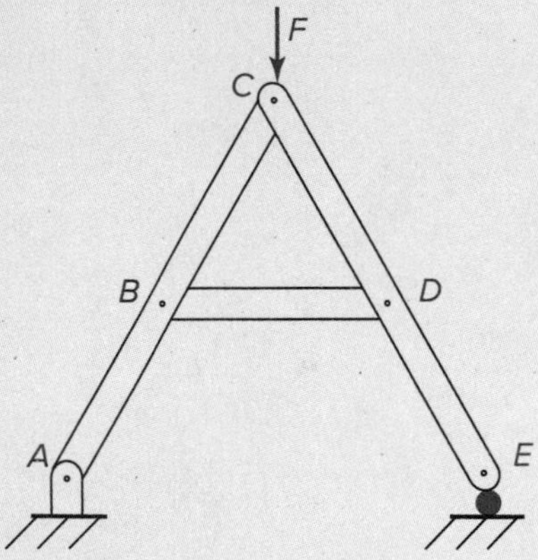

3.

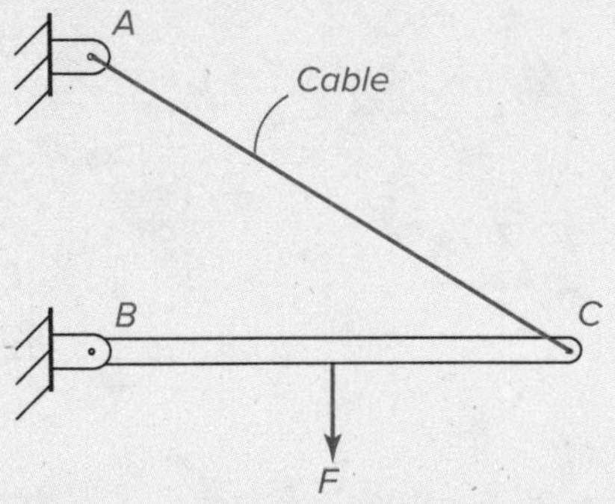

4.

ANSWERS

1. Member *AB* and the nut in the jaws of the pliers
2. Member *BC*
3. Member *BD*
4. Cable *AC*

Find the reactions at point A and point C.

Press pause on video lesson 31 once you get to the workout problem. Only press play if you get stuck.

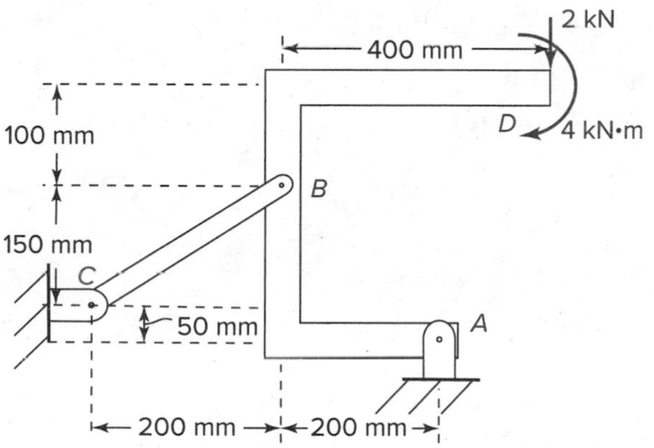

WATCH VIDEO

STATICS LESSON 32
Fixed Support 2D Reaction Force Problem

If the radius of each pulley is 0.4 inches, find the reactions at point *C*.

Press pause on the video once you get to the workout problem. Only press play if you get stuck.

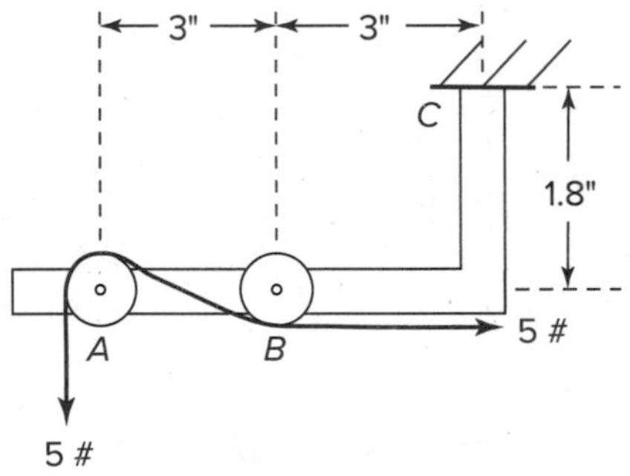

PITFALL

One of the most common mistakes made in statics is leaving the reaction moment off of FBDs of fixed or cantilevered supports.

WATCH VIDEO

STATICS LESSON 33
2D Reactions at the Supports with a Pulley

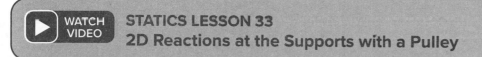

If the weight on the bar is 500 N, and the uniform bar itself weighs 300 N, find the tension in the rope and the reactions at point *A*.

Press pause on the video once you get to the workout problem. Only press play if you get stuck.

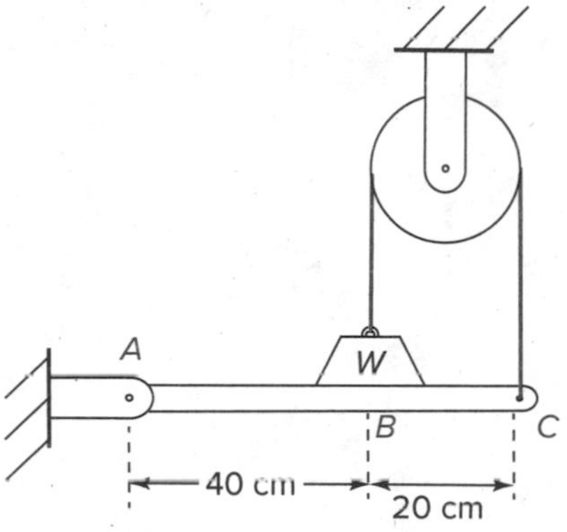

✓ TEST YOURSELF 4.3

SOLUTION TO TEST YOURSELF: Global Equilibrium Reactions

For the beam shown below, compute the reactions at the two supports.

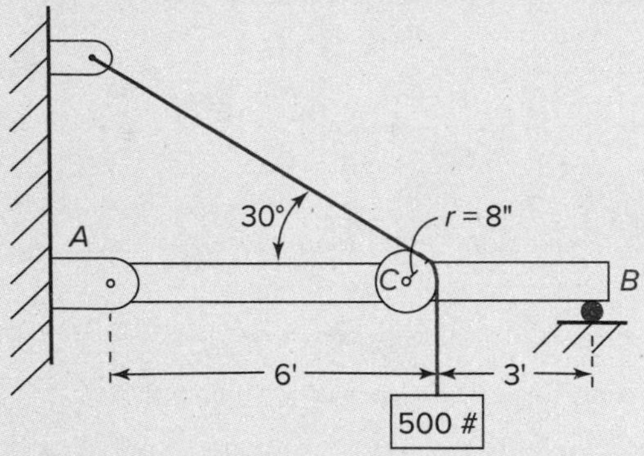

ANSWERS

$A_x = 433$ #

$A_y = 101.85$ #

$B_y = 148.15$ #

WATCH VIDEO

STATICS LESSON 34
Tipping Problems, Reactions in 2D

A common type of problem when solving reaction forces are referred to as tipping problems.

- Always involve normal forces connecting the system to the "world."
- At the point of tipping, one of these normal forces (bathroom scale forces) will go to zero.
  - Think what will happen to the system as loads increase or decrease.
  - Usually involve calculating the max load a system can handle before equilibrium will be lost.

PRO TIP

To solve tipping problems, use the Solve for 2D Reaction at the Supports Recipe but keep in mind one of the normal forces will be zero at the point of tipping. For instance in the image below, the reaction on the rear tire would be zero, right before the bird lands on Arnold Schwarzenegger's car, and it tumbles down the cliff!

Tipping Recipe

STEP 1: Draw FBD of the system.

STEP 2: Identify which normal force will go to zero at the point of tipping and then write equations of equilibrium.

Note: Typically you will write the moment equation about the point about which the system would try to rotate, usually that is all that is needed to solve.

STEP 3: Solve for unknowns.

EXAMPLE: TIPPING PROBLEM

Find the maximum reach distance (x) of the crane. The crane rolls freely on a rail system. The crane has a weight of 150 tons located at point G_1 and a counterweight of 40 tons at G_2. The load on the crane a point G_3 is 80 tons. The given dimensions are when the crane is in max reach mode.

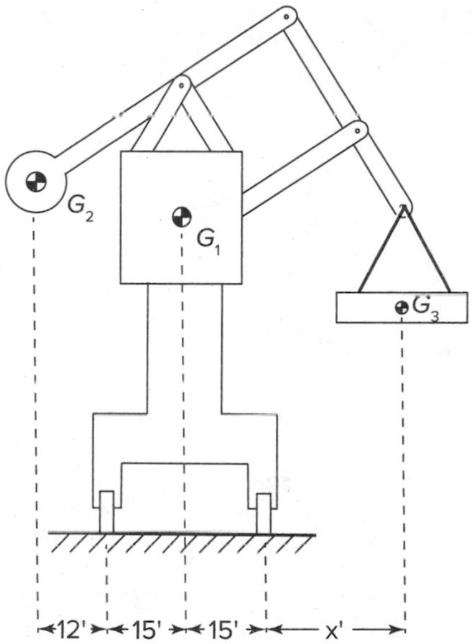

STEP 1: Draw an FBD of the crane.

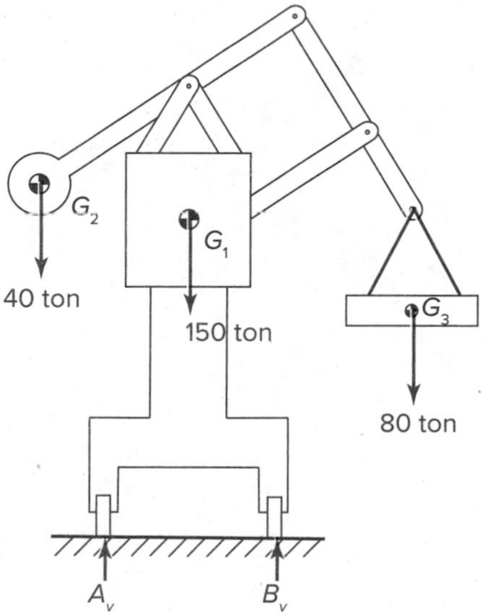

STEP 2: As the distance x increases, the crane wants to tip and rotate clockwise about point B. The instant before the crane tips, all the weight is on wheel B and the normal force at point A goes to *zero*. Knowing this information, let's take the moment about the point about which the system would try to rotate.

$$\Sigma M_b = 40(42) + 150(15) - 80(x)$$

STEP 3: Solve for distance x.

$$1{,}680 + 2{,}250 = 80(x)$$

$x = 49.12'$ (round the distance down to make sure you are safe)

If the crane tries to reach more than 49.12' from point B, it will topple.

Find the following:

1. If the lift and the man together have a weight of 9,500 #, and the box weighs 5,000 #, find the weight on the front and rear tires.
2. For the configuration of the man and the forklift, find the maximum weight of a crate that can be lifted.

Press pause on video lesson 34 once you get to the workout problem. Only press play if you get stuck.

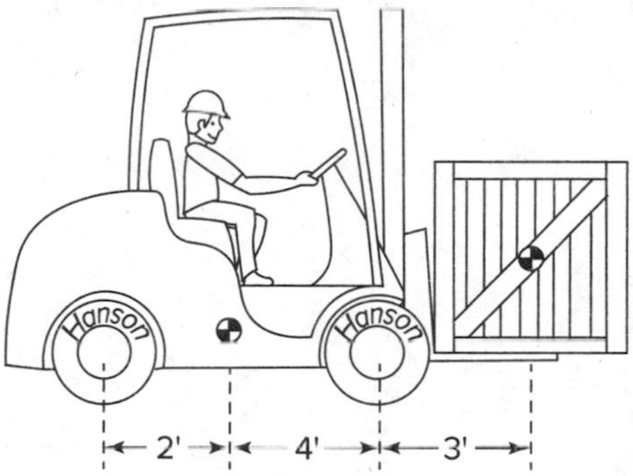

✓ TEST YOURSELF 4.4

SOLUTION TO TEST YOURSELF: Global Equilibrium Reactions

For the small mobile boom crane shown below, answer the following questions:

1. For a boom angle of $\theta = 30°$ what is the maximum load W without tipping?
2. Make a chart for the maximum load before tipping for boom angles of $\theta = 10°, 20°, 30°, 40°$, and $45°$. Does the maximum load increase or decrease as you increase the boom angle?
3. If the boom angle is 35° and the weight of the crate is 5,000 lbs, what are the reaction forces at the support A and B?

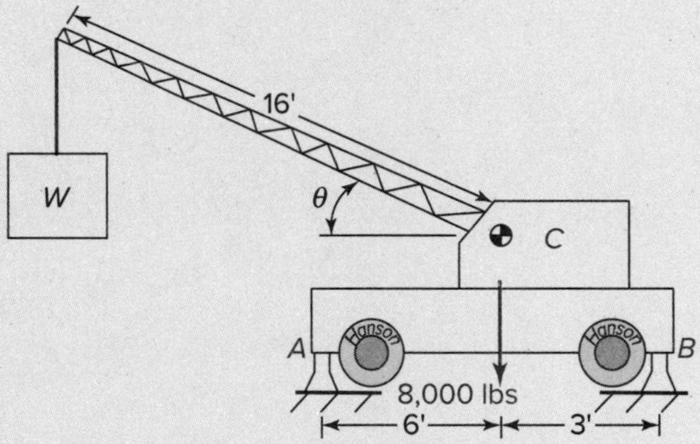

ANSWERS

1. $W = 6{,}110$ lbs
2. Increase
3. $A_y = 11{,}615$ lbs
$B_y = 1{,}385$ lbs

WATCH VIDEO

STATICS LESSON 35
3D Equilibrium of a Rigid Body, Six Equations

Solve for 3D Reactions at the Supports Recipe

STEP 1: Draw the free body diagram of the system.

Note: Do not worry about the direction of the reaction forces; simply guess them all in the positive direction.

STEP 2: Resolve any 3D forces into Cartesian components.

STEP 3: Write the three force equations of equilibrium based on FBD: $\Sigma F_x = 0$, $\Sigma F_y = 0$, and $\Sigma F_z = 0$.

STEP 4: Write the moment equations of equilibrium based on your FBD, using the origin as a "pivot point of rotation."

$\Sigma M_x = 0$ Think about "things that make me spin about the *x*-axis."

$\Sigma M_y = 0$ Think about "things that make me spin about the *y*-axis."

$\Sigma M_z = 0$ Think about "things that make me spin about the *z*-axis."

PITFALL

The biggest mistake made often with 3D reaction problems is using the wrong distance in the moment equations. When finding the distance from an axis to a particular force, use this trick: For instance, when looking for the distance from the *x*-axis to a *z* force component, the distance will always be in the remaining direction; in this case, the *y*-direction.

EXAMPLE: 3D SUPPORT REACTIONS

Your parents just installed a new motivational sign on your garage wall. The uniform density sign has a weight of 150 lbs. Find the tension in each cable and the reaction at the ball and socket joint at *A*.

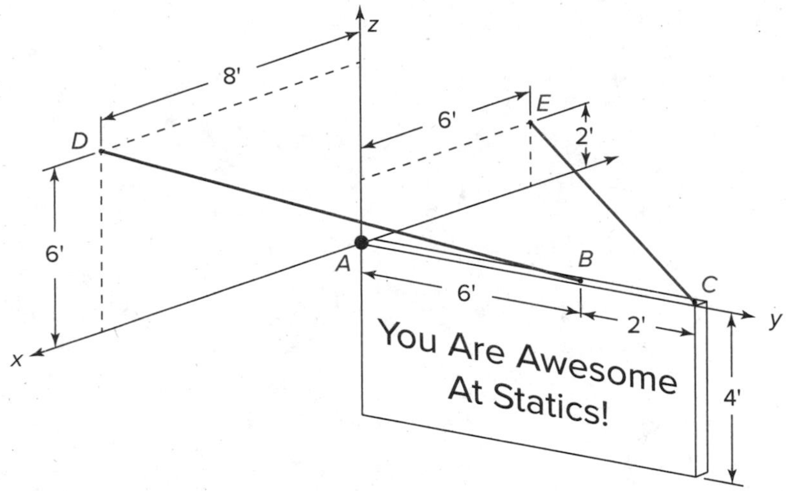

STEP 1: Draw the FBD of the system.

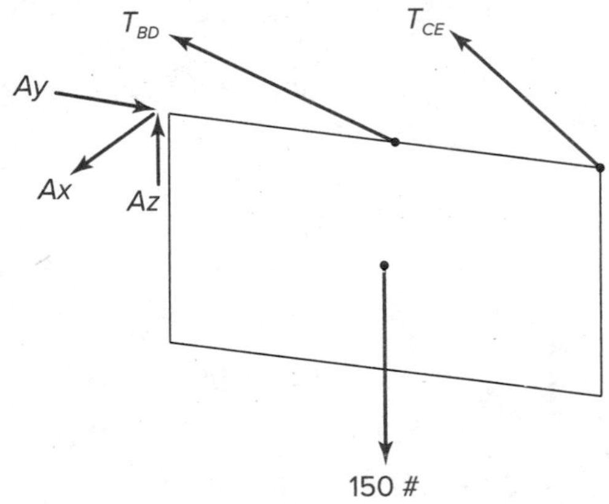

STEP 2: Write T_{CE} and T_{BD} in Cartesian vector form.

$$\vec{T}_{CE} \Rightarrow E(-6, 0, 2)$$
$$\underline{-C(0, 8, 0)}$$
$$-6\hat{\imath} - 8\hat{\jmath} + 2\hat{k}$$

$$\vec{T}_{BD} \Rightarrow D(8, 0, 6)$$
$$\underline{-B(0, 6, 0)}$$
$$8\hat{\imath} - 6\hat{\jmath} + 6\hat{k}$$

$$|\vec{T}_{CE}| = \sqrt{(-6)^2 + (-8)^2 + 2^2} = 10.2$$
$$\hat{\lambda}_{CE} = -0.588\hat{\imath} - 0.784\hat{\jmath} + 0.196\hat{k}$$
$$\vec{T}_{CE} = -0.588T_{CE}\hat{\imath} - 0.784T_{CE}\hat{\jmath} + 0.196T_{CE}\hat{k}$$

$$|\vec{T}_{BD}| = \sqrt{8^2 + (-6)^2 + 6^2} = 11.66$$
$$\hat{\lambda}_{BD} = 0.686\hat{\imath} - 0.515\hat{\jmath} + 0.515\hat{k}$$
$$\vec{T}_{BD} = 0.686T_{BD}\hat{\imath} - 0.515T_{BD}\hat{\jmath} + 0.515T_{BD}\hat{k}$$

STEP 3: Write the equations of equilibrium.

$$\Sigma F_x = 0 = A_x - 0.588T_{CE} + 0.686T_{BD}$$
$$\Sigma F_y = 0 = A_y - 0.784T_{CE} - 0.515T_{BD}$$
$$\Sigma F_z = 0 = A_z + 0.196T_{CE} + 0.515T_{BD} - 150$$
$$\Sigma M_x = 0 = 0.196T_{CE}(8) + 0.515T_{BD}(6) - 150(4)$$
$$\Sigma M_y = 0$$
$$\Sigma M_z = 0 = 0.588T_{CE}(8) - 0.686T_{BD}(6)$$

STEP 4: Solve.

From ΣM_z: $0.588(8)T_{CE} = 0.686(6)T_{BD}$

$$4.704T_{CE} = 4.116T_{BD}$$
$$T_{BD} = 1.14T_{CE}$$

From ΣM_x: $0.196(8)T_{CE} + 0.515(6)T_{BD} - 600 = 0$

$$1.568T_{CE} + 3.09T_{BD} - 600 = 0$$

Substituting for T_{BD}:

$$1.568T_{CE} + 3.09(1.14T_{CE}) = 600$$
$$5.09T_{CE} = 600$$
$$T_{CE} = 117.86 \text{ lbs}$$
$$T_{BD} = 134.37 \text{ lbs}$$

From ΣF_x: $A_x = 0.588T_{CE} - 0.686T_{BD}$

$$A_x = 69.3 - 92.18$$
$$A_x = -22.87 \text{ lbs}$$

Note: Assumed wrong direction on FBD!

From A_y: $A_y = 0.784T_{CE} + 0.515T_{BD}$

$$A_y = 92.4 + 69.2$$
$$A_y = 161.6 \text{ lbs}$$

From A_z: $A_z = -0.196T_{CE} - 0.515T_{BD} + 150$

$$A_z = -23.1 - 69.2 + 150$$
$$A_z = 57.7 \text{ lbs}$$

Find the force acting on each of the three wheels *A*, *B*, and *C*.

Press pause on video lesson 35 once you get to the workout problem. Only press play if you get stuck.

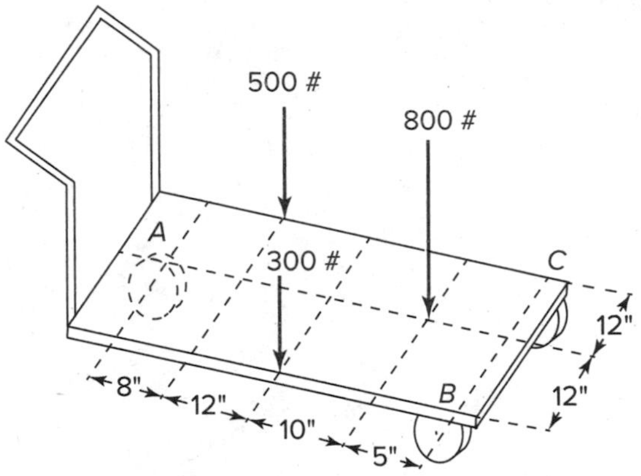

PRO TIP

Remember that the axes given in a problem are arbitrary. Sometimes a problem can be simplified if you relocate the axes so that they pass through unknowns or unknown reaction forces. This is especially true when it comes to writing the moment equations.

WATCH VIDEO

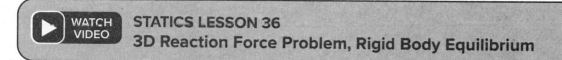

STATICS LESSON 36
3D Reaction Force Problem, Rigid Body Equilibrium

Find the reaction at *A* and find the tension in *BC*. The connection at *A* is a smooth collar on a square shaft.

Press pause on the video once you get to the workout problem. Only press play if you get stuck.

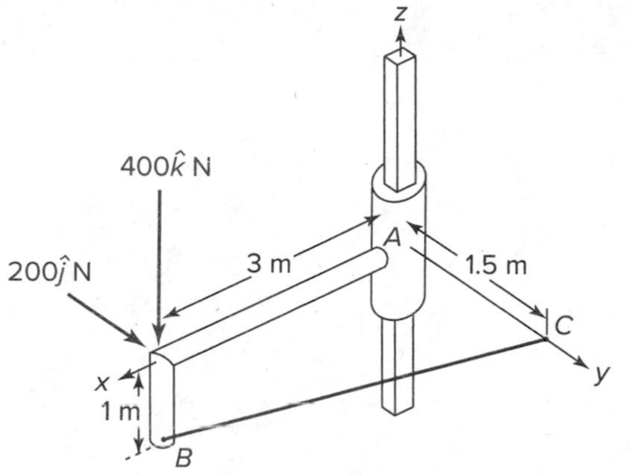

Over-Constrained Systems

To absolutely assure that there is no motion or rotation, six reactions are required: $F_x = 0$, $F_y = 0$, $F_z = 0$, $M_x = 0$, $M_y = 0$, and $M_z = 0$.

In 3D problems, it is very easy to exceed six reactions (so there are more unknowns than equations). For instance, a system with two or more pillow blocks or smooth journal bearings (four reactions each) can easily become over-constrained.

Note: The lockbars on the container in the photo are secured by smooth journal bearings, and there are more present than necessary to prevent motion.

For over-constrained problems, ignore the reaction moments. But why? Because we can't even solve this problem with what we have learned thus far. We have to make an assumption. In this case, as has been historically done in previous textbooks, assume that the journal bearings are very complaint such that the journal bearings have negligible moment reactions. I call it "slop" in the bearing.

Over-Constrained Systems Recipe

STEP 1: Draw FBD. Break the external forces down into components. It is not always possible to know the directions of the reactions, so completely guess. The math will tell if the sign is correct. If the system has more than six unknowns, it is over-constrained. If over-constrained, then ignore the reaction moments.

STEP 2: Write the equations of equilibrium:

$$\Sigma F_x = 0 \qquad \Sigma M_x = 0$$
$$\Sigma F_y = 0 \qquad \Sigma M_y = 0$$
$$\Sigma F_z = 0 \qquad \Sigma M_z = 0$$

STEP 3: Solve. This requires some high-powered algebra skills as most calculators with system solvers will not crunch this many unknowns. As a suggestion, write each equation simplified with an unknown isolated in each and then simply substitute. This is not the fastest technique but is one way that works.

EXAMPLE: OVER-CONSTRAINED SYSTEMS

Find the reactions at *A*, *B*, and *C*. Assume that the journal bearings have negligible moment reactions.

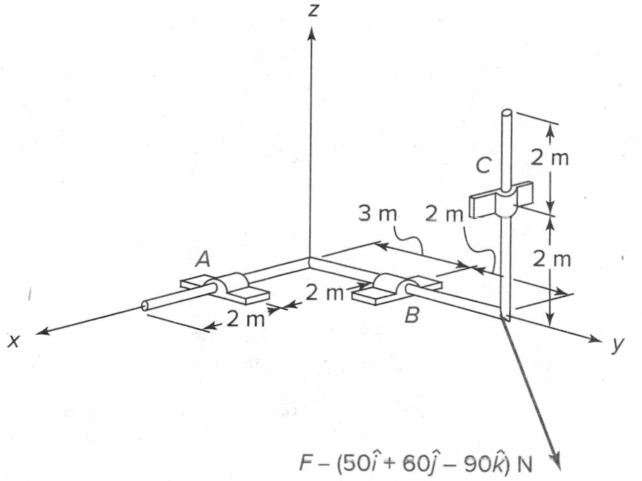

STEP 1: Draw FBD. Break the external force down into components.

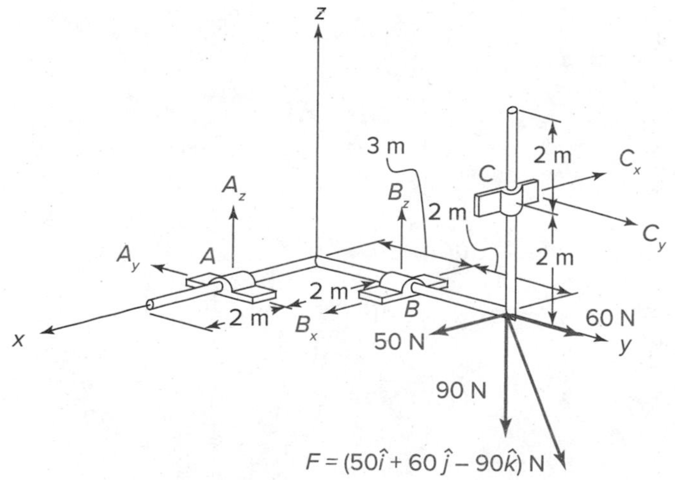

Reactions at A are (A_z), (A_y), (M_y), and (M_z).
Reactions at B are (B_x), (B_z), (M_x), and (M_z).
Reactions at C are (C_x), (C_y), (M_x), and (M_y).

System has twelve reactions and only six equations, so it is over-constrained. This means it has more reactions than necessary to ensure there is no motion. Ignore the reaction moments.

Note: The assumed directions are completely a guess, and the math will tell if the sign is correct.

STEP 2: Construct the equations of equilibrium.

Note: It is easiest if you don't worry about which one to write first; just write them in order.

$$\Sigma F_x = 0 = B_x - C_x + 50$$
$$\Sigma F_y = 0 = C_y - A_y + 60$$
$$\Sigma F_z = 0 = A_z + B_z - 90$$
$$\Sigma M_x = 0 = (3)B_z - (5)90 - (2)C_y$$
$$\Sigma M_y = 0 = -(2)C_x - (2)A_z$$
$$\Sigma M_z = 0 = -(2)A_y - (3)B_x - (5)50 + (5)C_x$$

STEP 3: Solve.

Note: Requires some high-powered algebra skills as most calculators with system solvers will not crunch this many unknowns.

Write each equation simplified with an unknown isolated in each and then simply substitute. Note this might not be the fastest technique, but this is one way that works!

1. $C_x = B_x + 50$ ← Combine with Equation 6
2. $C_y = A_y - 60$
3. $A_z = 90 - B_z$
4. $B_z = \frac{2}{3}(C_y) + 150$
5. $C_x = -A_z$
6. $A_y = 2.5(C_x) - 125 - 1.5(B_x) \Rightarrow A_y = 2.5(B_x + 50) - 125 - 1.5(B_x)$
   Therefore $A_y = B_x$

$B_z = \frac{2}{3}(A_y) + 110$

$C_x = B_z - 90$

Three new equations after three substitutions:

$B_z = \frac{2}{3}(A_y) + 110$ ← $B_z = \frac{2}{3}(B_x) + 110$

$C_x = B_z - 90$ ← Combine with Equation 1

$A_y = B_x$ ← $B_z = B_x + 140$

$B_x =$	-90 N
$A_y =$	-90 N
$C_y =$	-150 N
$C_x =$	-40 N
$A_z =$	40 N
$B_z =$	50 N

$B_z = \frac{2}{3}(B_x) + 110 \longleftrightarrow B_z = B_x + 140$

$B_x + 140 = \frac{2}{3}(B_x) + 110$

$\frac{1}{3}B_x = -30$

$B_x = -90$

From the assumed directions, the following forces were assumed in the incorrect direction: B_x, A_y, C_y, and C_x.

Challenge: Can you find a better way to do an easier substitution?

✓ TEST YOURSELF 4.5

SOLUTION TO TEST YOURSELF: Global Equilibrium Reactions

For the over-constrained rigid pipe shown below, determine the reaction forces at the three journal bearings. The 12 kip load is in the *xz* plane (that is, it has no *y* component); the force *F* is expressed as a vector (see below). Assume that the journal bearings have negligible moment reactions.

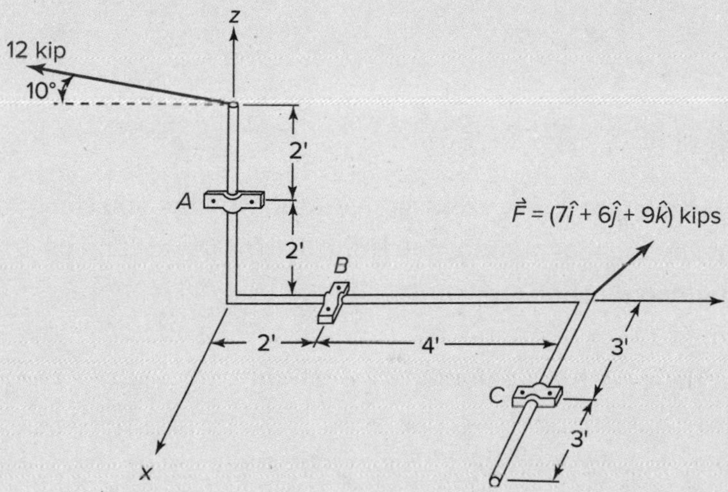

ANSWERS

$A_x = -64.6$ kips

$A_y = -46.6$ kips

$B_x = 57.6$ kips

$B_z = 32.0$ kips

$C_y = 52.4$ kips

$C_z = -43.1$ kips

KEY TAKEAWAYS

Reaction forces are points where the rigid body is connected to the "world" and prevent motion or rotation of a system.

Main takeaways from Level 4 include how to calculate:

- 2D reactions
  - Tipping problems (subpart of 2D)
- 3D reactions

2D Reactions Summary

- Remember to determine the type of reaction forces generated: ask yourself whether you can move or rotate. If no, the reaction is preventing that motion; therefore, there is a force and or moment reaction (refer to table for common reaction types).
- To be in equilibrium $\Sigma F_x = 0$, $\Sigma F_y = 0$, and $\Sigma M_z = 0$ (up stuff has to equal down stuff and right stuff has to equal left stuff). Since there are three equations, we can solve up to three unknowns.
- If you know the direction of the reaction, then put it on the FBD correctly and use it as an extra sanity check. However, you can always just guess a direction for the reaction force, and if it is opposite to what you assumed, the calculation will result in a negative value.
- A subset of 2D reaction problems are those that involve tipping. To solve these problems you must have a physical understanding of the equilibrium of the system. Right before the tip happens (the onset of tipping), one of the normal forces goes to zero.

KEY TAKEAWAYS

3D Reactions Summary

- Remember to determine the type of reaction forces generated: ask yourself whether you can move or rotate. If no, the reaction is preventing that motion; therefore, there is a force and/or moment reaction (refer to table for common reaction types).
- To be in equilibrium $\Sigma F_x = 0$, $\Sigma F_y = 0$, $\Sigma F_z = 0$, $\Sigma M_x = 0$, $\Sigma M_y = 0$, and $\Sigma M_z = 0$ (up stuff has to equal down stuff and right stuff has to equal left stuff). Since there are six equations, we can solve up to six unknowns.
- Over-constrained problems have more than six reaction forces (typically seen with pillow blocks or smooth journal bearings). Only in these special cases ignore the reaction moments since we are assuming these beams don't bend!
- It is too hard to identify the directions for 3D reactions, so always simply guess positive and let the math tell you if the direction you chose is correct (positive answer means you assumed the correct direction).

RECIPES

- **Solve for 2D Reaction at the Supports Recipe**
- **Tipping Recipe**
- **Solve for 3D Reactions at the Supports Recipe**
- **Over-Constrained Systems Recipe**

PRO TIPS

3D Reaction Tips

- In 3D, it is almost impossible to determine the direction of the reaction forces, so save some time and just assume the positive direction for all forces and moments, and let the math tell if you assumed correctly or incorrectly. An incorrect assumption will result in a negative answer.

2D Reaction Tips

- To solve tipping problems, use the Solve for 2D Reaction at the Supports Recipe but keep in mind one of the normal forces will be zero at the point of tipping. For instance in the image below, the reaction on the rear tire would be zero, right before the bird lands on Arnold Schwarzenegger's car, and it tumbles down the cliff!

2D and 3D Reaction Tips

- Remember that the axes given in a problem are arbitrary. Sometimes a problem can be simplified if you relocate the axes so that they pass through unknowns or unknown reaction forces. This is especially true when it comes to writing the moment equations.

PITFALLS

2D and 3D Reaction Tips

- One of the most common mistakes made in statics is leaving the reaction moment off of FBDs of fixed or cantilevered supports.

3D Reaction Tips

- The biggest mistake made often with 3D reaction problems is using the wrong distance in the moment equations. When finding the distance from an axis to a particular force, use this trick: For instance, when looking for the distance from the *x*-axis to a *z* force component, the distance will always be in the remaining direction; in this case, the *y*-direction.

Level 5

Centroids

Properties of Geometric Shapes and Areas

Properties of Geometric Shapes and Areas					
Shape		**$\bar{x}$**	**$\bar{y}$**	**Area**	**Length**
Triangular Area	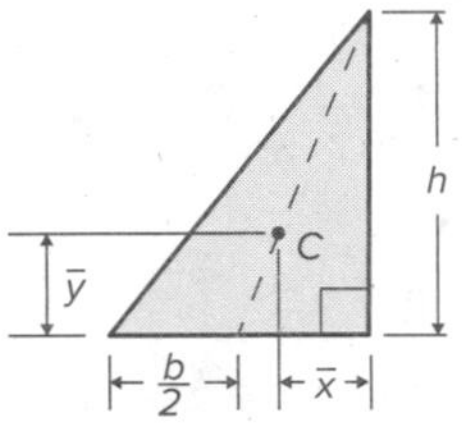	$\frac{b}{3}$	$\frac{h}{3}$	$\frac{bh}{2}$	
Quarter-circular Arc	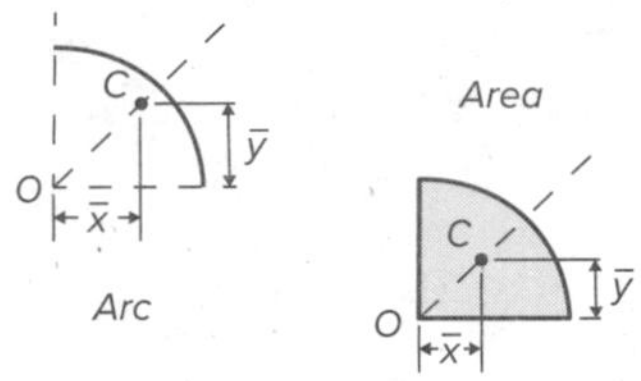	$\frac{2r}{\pi}$	$\frac{2r}{\pi}$		$\frac{\pi r}{2}$
Quarter-circular Area		$\frac{4r}{3\pi}$	$\frac{4r}{3\pi}$	$\frac{\pi r^2}{4}$	
Semicircular Arc	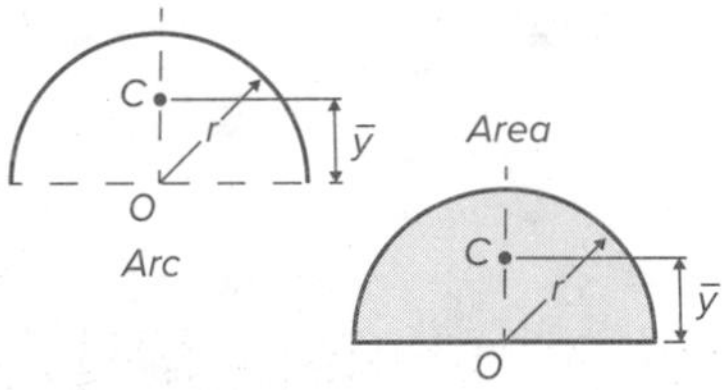	0	$\frac{2r}{\pi}$		πr
Semicircular Area		0	$\frac{4r}{3\pi}$	$\frac{\pi r^2}{2}$	
Semiparabolic Area	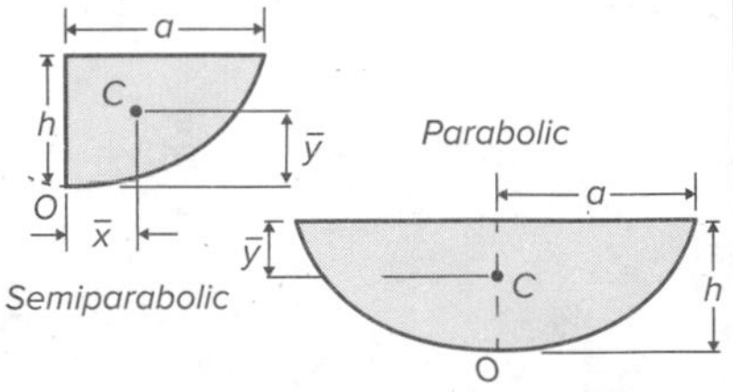	$\frac{3a}{8}$	$\frac{3h}{5}$	$\frac{2ah}{3}$	
Parabolic Area		0	$\frac{3h}{5}$	$\frac{4ah}{3}$	
Parabolic Spandrel	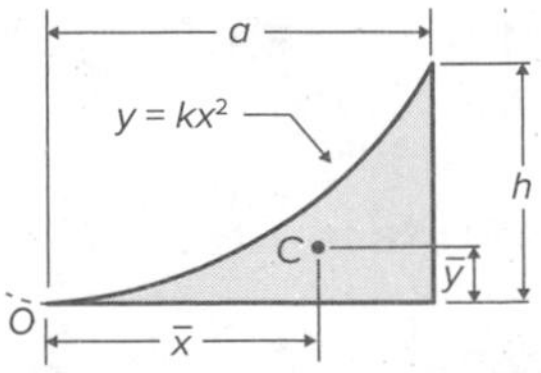	$\frac{3a}{4}$	$\frac{3h}{10}$	$\frac{ah}{3}$	
Arc of Circle	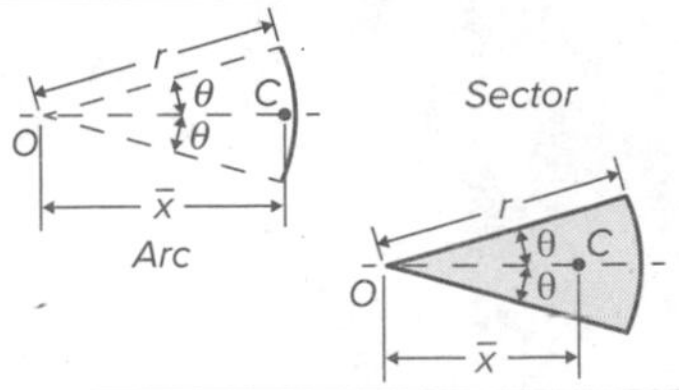	$\frac{r \sin \theta}{\theta}$	0		$2\theta r$
Circular Sector		$\frac{2r \sin \theta}{3\theta}$	0	θr^2	

WATCH VIDEO

STATICS LESSON 37
Introduction to Centroids, Where Is the Center of Texas?

Centroid is the location in space $(\bar{x}, \bar{y}, \bar{z})$ that will "balance" the weight, mass, area, and so forth at a given point.

Dr. Hanson is balancing the textbook with his hand located at the centroid.

CHALLENGE QUESTION

Where is the centroid for this statue?

ANSWER

Fisherman's heels since the system is balanced.

Composite Shape Method

Key equations for computing the location of the centroids for different properties:

For Areas:	For Volume:	For Lengths:	For Mass:
$\bar{x} = \dfrac{\Sigma x_i A_i}{\Sigma A_i}$	$\bar{x} = \dfrac{\Sigma x_i V_i}{\Sigma V_i}$	$\bar{x} = \dfrac{\Sigma x_i L_i}{\Sigma L_i}$	$\bar{x} = \dfrac{\Sigma x_i M_i}{\Sigma M_i}$
$\bar{y} = \dfrac{\Sigma y_i A_i}{\Sigma A_i}$	$\bar{y} = \dfrac{\Sigma y_i V_i}{\Sigma V_i}$	$\bar{y} = \dfrac{\Sigma y_i L_i}{\Sigma L_i}$	$\bar{y} = \dfrac{\Sigma y_i M_i}{\Sigma M_i}$
$\bar{z} = \dfrac{\Sigma z_i A_i}{\Sigma A_i}$	$\bar{z} = \dfrac{\Sigma z_i V_i}{\Sigma V_i}$	$\bar{z} = \dfrac{\Sigma z_i L_i}{\Sigma L_i}$	$\bar{z} = \dfrac{\Sigma z_i M_i}{\Sigma M_i}$

Note: Here "*i*" represents a particular piece of the entire object you are analyzing. The *x*, *y*, and *z* are the distance to the center of that particular piece as measured from the datum. Thus, you are essentially adding up all of the pieces. If a piece is a void, don't forget to subtract it on both the numerator and the dominator!

Calculus Method

The calculus method can also be used:

$$\bar{x} = \frac{\int x\,dA}{\int dA} \qquad \bar{y} = \frac{\int y\,dA}{\int dA} \qquad \bar{z} = \frac{\int z\,dA}{\int dA}$$

Use these equations if you have complex shapes, typically described by mathematical equations (discussed later in this level).

PITFALL

The number one mistake made when calculating centroids is to forget that locations are *always* referenced from the origin (i.e., datum).

Area Cutout Method for Determining Centroids

- Make a cutout of the area of interest making sure the thickness is uniform.
- Add several suspension points.
- Draw a straight line downward from each suspension.
- Centroid is where all the lines intercept.

Where is the center of Texas? Mercury, Texas.

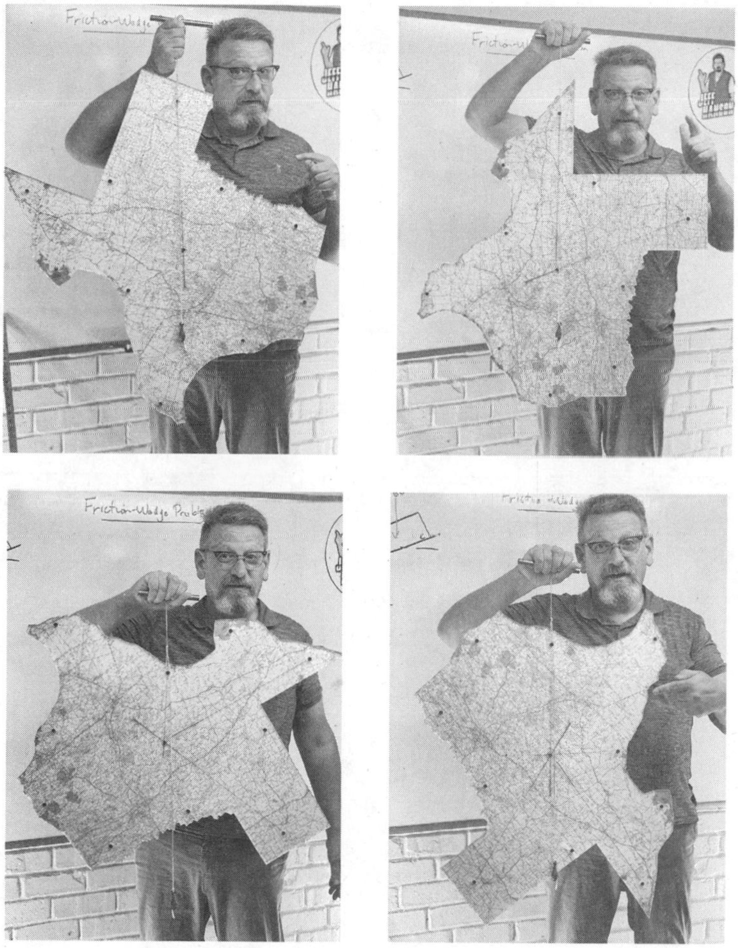

Where is the center of your region?

WATCH VIDEO **STATICS LESSON 38**
Introduction to Centroids by Composite Shapes, Table Method

Composite Shapes Method for Determining Centroids

PRO TIPS

- **Use the table found in the beginning of this level for geometric properties of areas (i.e., centroids) and envision the total shape as being composed of simple shapes with properties listed in the table.**

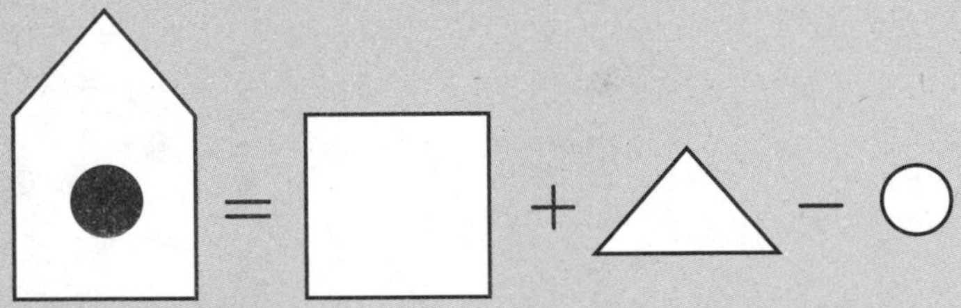

- **Use symmetry and observation (intuition) to identify the centroid when possible.**

Composite Shapes Centroid Recipe

STEP 1: Identify the correct equation based on the type of problem given:

- Area equation for simple composite area shapes
- Volume equation for composition of volume shapes (***Note:*** Each piece of the composition must be of the same density, i.e., homogeneous material)
- Length equation for slender bodies such as a wire or wire frame (***Note:*** Homogeneous material, with constant density and diameter)
- Mass equation for volume compositions of differing densities. Must convert each component into a mass: $m = \rho V$ where ρ is density and V is volume.

Composite Shapes Centroid Recipe (continued)

STEP 2: Divide the composite shape into components and place a dot where you think the centroid is located for each piece's part.

STEP 3: Expand the centroid equation to fit your problem. Remember to use the table found in the beginning of this level for equations of centroids of common shapes such as triangles, semi- and quarter circles, and semi-parabolic shapes.

STEP 4: Fill in your equations such that the distances are all measured from the origin (i.e., datum). Remember, if you put garbage in your equations, you get garbage out!

STEP 5: Solve.

STEP 6: Sanity check your answer. You should have some idea, using your intuition, where the centroid of the given shape should approximately be located. Ask yourself, does my answer make sense?

EXAMPLE: COMPOSITE SHAPE METHOD

Find $\bar{x}$ and $\bar{y}$ for the given shape as measured from the coordinate axes.

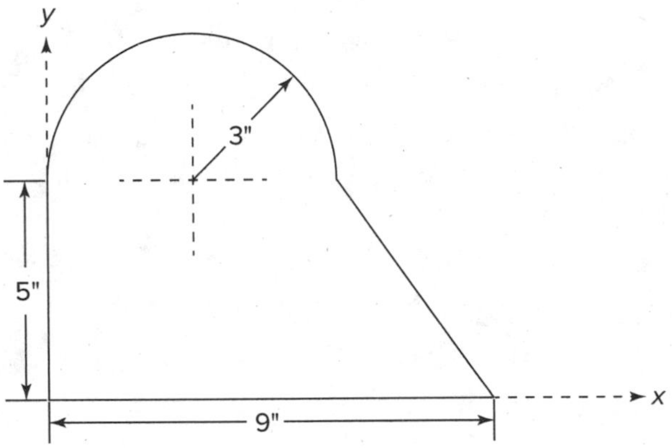

STEP 1: Identify what version of the composite equation is relevant to the problem. Use an area equation since this is an area.

$$\bar{x} = \frac{\Sigma x_i A_i}{\Sigma A_i}, \ \bar{y} = \frac{\Sigma y_i A_i}{\Sigma A_i}$$

STEP 2: Divide the composite shape into components: a rectangle, a semi-circle, and a triangle.

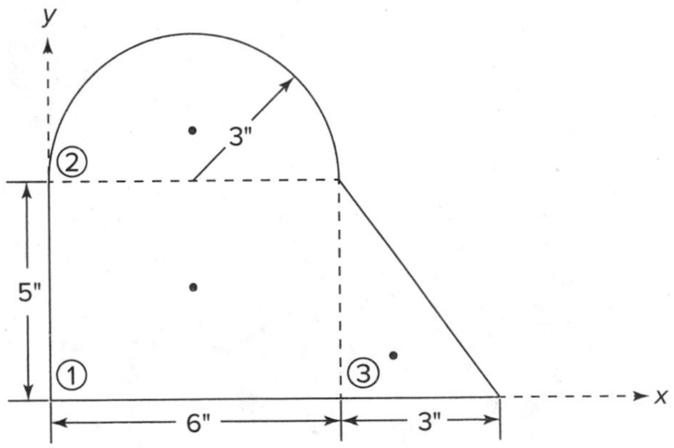

STEP 3: Expand the centroid equation.

$$\bar{x} = \frac{\Sigma x_i A_i}{\Sigma A_i} \text{ becomes } \bar{x} = \frac{\bar{x}_1 A_1 + \bar{x}_2 A_2 + \bar{x}_3 A_3}{A_1 + A_2 + A_3}$$

$$\bar{y} = \frac{\Sigma y_i A_i}{\Sigma A_i} \text{ becomes } \bar{y} = \frac{\bar{y}_1 A_1 + \bar{y}_2 A_2 + \bar{y}_3 A_3}{A_1 + A_2 + A_3}$$

STEP 4: Fill in your equations.

$$\bar{x} = \frac{(3)(30) + (3)(\frac{\pi(3)^2}{2}) + (7)(\frac{1}{2})(3)(5)}{30 + \frac{\pi(3)^2}{2} + \frac{1}{2}(15)}$$

$$\bar{y} = \frac{(2.5)(30) + (5 + \frac{4(3)}{3\pi})(\frac{\pi(3)^2}{2}) + (\frac{5}{3})(\frac{1}{2})(3)(5)}{30 + \frac{\pi(3)^2}{2} + \frac{1}{2}(15)}$$

STEP 5: Solve.

$$\bar{x} = \frac{90 + 42.41 + 52.5}{30 + 14.14 + 7.5} = \frac{184.9}{51.64} = 3.58''$$

$$\bar{y} = \frac{75 + 88.69 + 12.5}{30 + 14.14 + 7.5} = \frac{176.19}{51.64} = 3.41''$$

STEP 6: Sanity check your answer. *Yes* it does make sense!

CHALLENGE QUESTION

Does the centroid always have to be located on the body? Hint: What is Homer Simpson's favorite food?

ANSWER

No. Things such as boomerangs, horseshoes, and donuts have centroids that are not on their bodies.

Find $\bar{x}$ and $\bar{y}$ for the given shapes as measured from the coordinate axis.

Press pause on video lesson 38 once you get to the workout problem. Only press play if you get stuck.

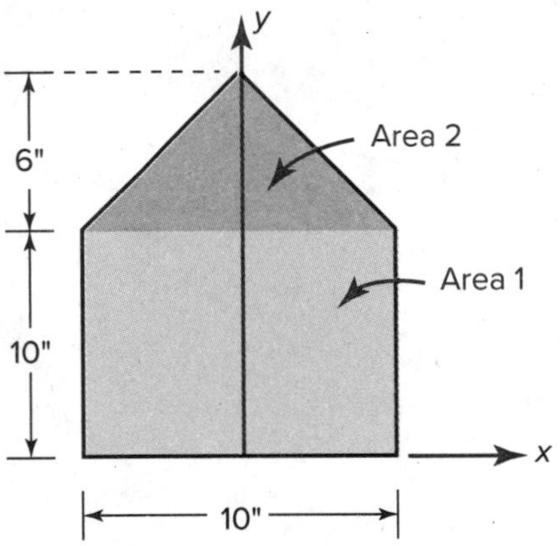

EXAMPLE: TABLE METHOD

Exact same type of problems *except* in Step 4 of the Composite Shapes Centroid Recipe build a table to organize data instead of writing a long equation.

Here is the previous example problem using the table method.

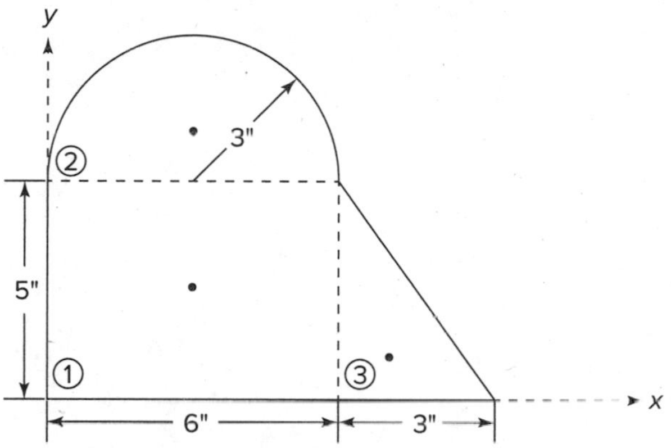

Table would be as follows:

Area	x_i	y_i	A_i	x_iA_i	y_iA_i
1	3	2.5	30	90	75
2	3	6.27	14.14	42.42	88.66
3	7	1.67	7.5	52.5	12.53

$\Sigma A_i = 51.64$, $\Sigma x_iA_i = 184.9$, $\Sigma y_iA_i = 176.2$

$$\bar{x} = \frac{\Sigma x_iA_i}{\Sigma A_i} = \frac{184.9}{51.64} = 3.58"$$

$$\bar{y} = \frac{\Sigma y_iA_i}{\Sigma A_i} = \frac{176.2}{51.64} = 3.41"$$

PRO TIPS

The table method is especially helpful when solving composite problems with many parts.

Find the centroid of the shape composed of a square hole (area 1) and quarter-circle (area 2) using the table method.

Press pause on video lesson 38 once you get to the workout problem. Only press play if you get stuck.

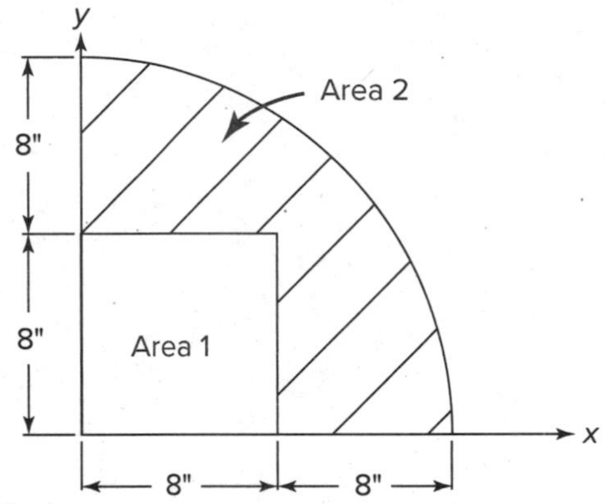

WATCH VIDEO

STATICS LESSON 39
Centroid Using Composite Shapes, Center of Area

Find the centroid $(\bar{x}, \bar{y})$ of the shape shown using the table method.
Press pause on the video once you get to the workout problem. Only press play if you get stuck.

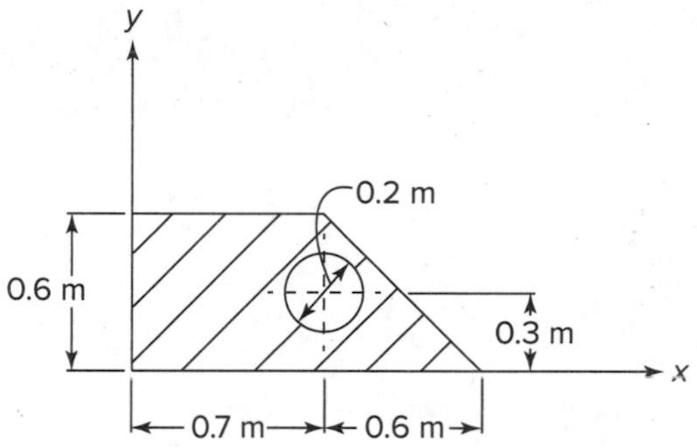

TEST YOURSELF 5.1

SOLUTION TO TEST YOURSELF: Centroids

Compute the centroid of the shape shown below.

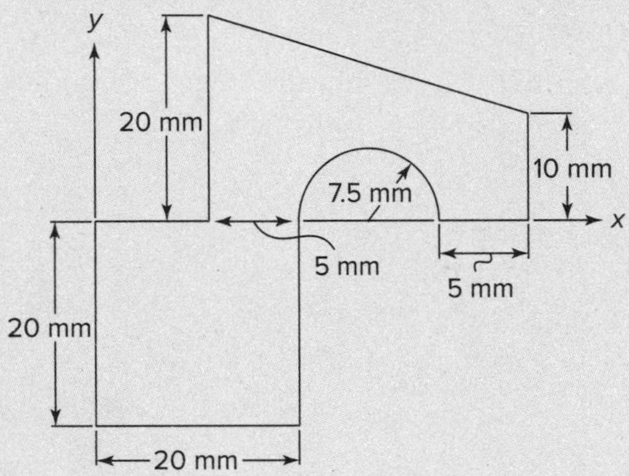

ANSWERS

$\bar{x} = 16.55$ mm

$\bar{y} = -1.99$ mm

WATCH VIDEO

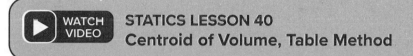

STATICS LESSON 40
Centroid of Volume, Table Method

Find the centroid of the volume $(\bar{x}, \bar{y}, \bar{z})$ of the shape show below using the table method.

Press pause on the video once you get to the workout problem. Only press play if you get stuck.

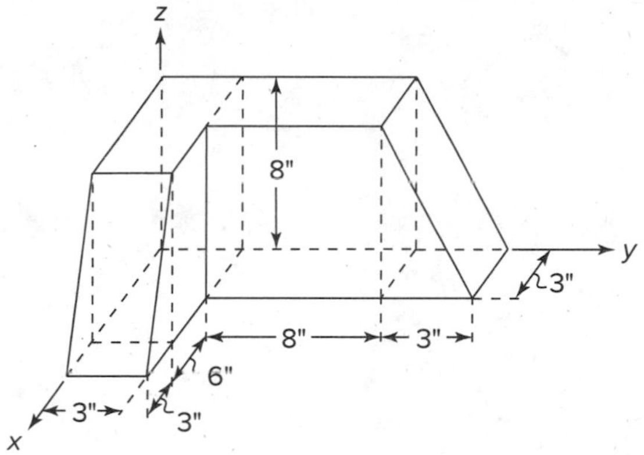

WATCH VIDEO

STATICS LESSON 41
Centroid of Mass, Body with Different Densities

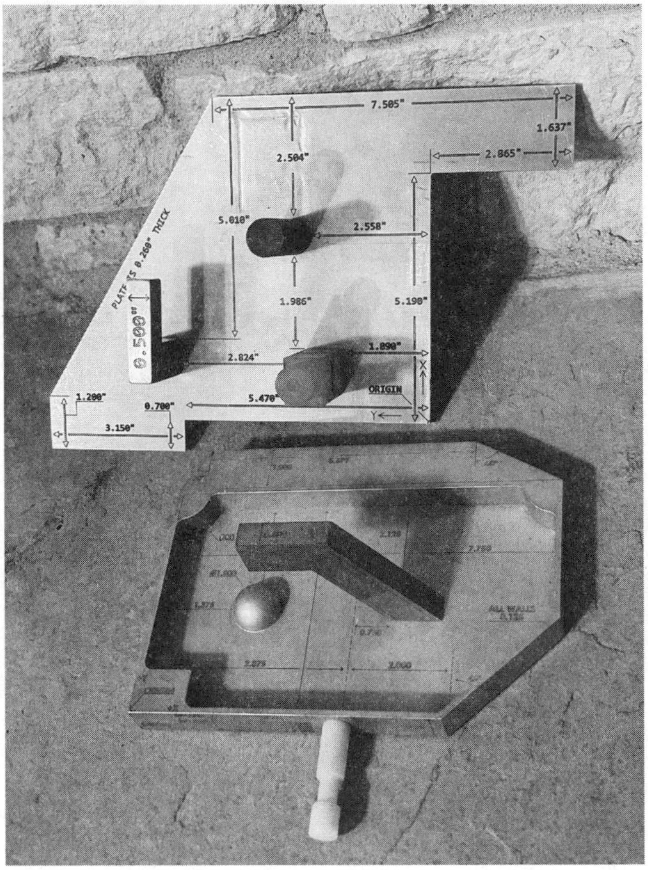

These are examples of strange assemblies of components of different densities such as aluminum, brass, and nylon. As a challenge, I have students calculate the centroids of each. Do you think you could do that?

Find the centroid of the mass $(\bar{x}, \bar{y}, \bar{z})$ of the shape shown below using the table method.

Press pause on video lesson 41 once you get to the workout problem. Only press play if you get stuck.

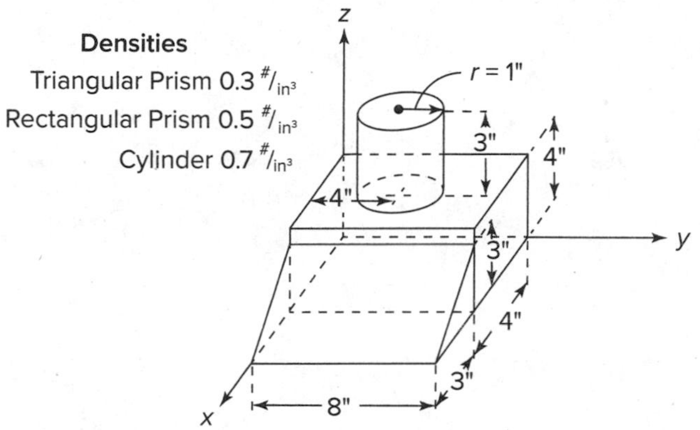

PRO TIP

Make sure to use the mass and not only the volume to find the center of mass (i.e., don't forget to multiply volume by density!).

WATCH VIDEO **STATICS LESSON 42**
Introduction to Centroid by Calculus Method, Flip-the-Strip Method

Calculus Method of Centroids

Recall from calculus: Summation (Σ) in calculus is $\int$, which I like to think of as a capital "S," meaning sum.

$$\bar{x} = \frac{\int x\, dA}{\int dA}, \quad \bar{y} = \frac{\int y\, dA}{\int dA}$$

Centroid by Calculus (Flip-the-Strip) Recipe

STEP 1: Draw a single differential element on the shape. When calculating $\bar{x}$, the width of the strip should be dx. When calculating $\bar{y}$, the height of the strip should be dy.

STEP 2: To calculate $\bar{x}$, determine the "height" of each "slice" at any position x.

STEP 3: At position x write the expression for the differential area of width dx and height y. Make sure to write the function so that dA is a function of x only (i.e., replace y with an expression in terms of x such that $y = f(x)$):

$$A = \int dA = \int y dx = \int f(x)\, dx$$

The integral symbol is just a fancy way of writing a capital "S" for summation. Note we already used sigma when summing discrete things, and so now we use the fancy "S" for summing infinitesimally thin things.

Centroid by Calculus (Flip-the-Strip) Recipe (continued)

STEP 4: Compute the area by integrating the differential area between the two limits along the x-axis. This becomes the denominator in the $\bar{x}$ equation.

Note: When determining the limits for integration, remember to look at the direction that the "slices" need to be added, then use the limits from the beginning to the end of that direction. (Like adding up a stack of books: the books go from where to where? Those are your limits of integration.)

STEP 5: To determine the numerator in the $\bar{x}$ equation, compute $\int x\, dA$ (read as x times unintegrated dA) between the same two limits along the x-axis (use dA from above).

STEP 6: Now, compute the x-centroid of the area along the x-axis using the equation $\bar{x} = \dfrac{\int x\, dA}{\int dA}$.

STEP 7: Flip the strip. Determine the "width" of each "slice" at any position y.

STEP 8: At position y write the expression for the differential area: $dA = x\, dy$. Make sure to substitute the function so that dA is a function of y only (i.e., replace x with an expression in terms of y such that $x = f(y)$):

$A = \int dA = \int x\, dy = \int f(y)\, dy$

STEP 9: Compute the area by integrating the differential area between the two limits along the y-axis. This becomes the denominator in the $\bar{y}$ equation.

Note: You should get the exact same area as previously! If not, review your work carefully to find your error.

STEP 10: Compute $\int y\, dA$ between the two limits along the y-axis.

STEP 11: Now, compute the y-centroid of the area along the y-axis using the equation $\bar{y} = \dfrac{\int y\, dA}{\int dA}$.

EXAMPLE: CALCULUS METHOD

Find the centroid $(\bar{x}, \bar{y})$ of the triangle shown using the flip-the-strip method that is based on calculus:

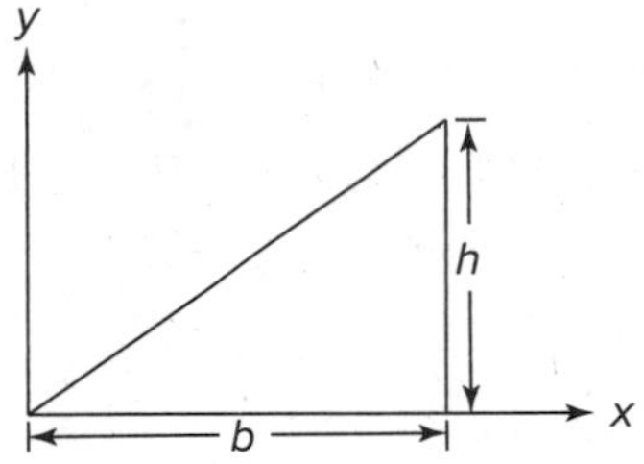

$$\bar{x} = \frac{\int x\,dA}{\int dA}, \quad \bar{y} = \frac{\int y\,dA}{\int dA}$$

STEP 1: Sketch a rectangular slice of width dx and height y on the shape (see figure below). The area of this "slice" is a rectangle (height times width) that is cross-hatched, and it is given as $y \times dx$ and represents *one* differential area.

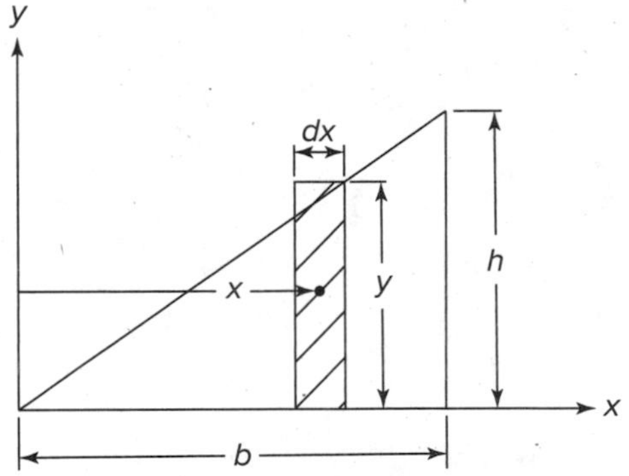

STEP 2: For this problem, calculate the "height" of each "slice" at any position x via determining the function for the straight line representing the top of the triangle in the form $y = mx + b$. (***Note:*** b here is the y-intercept of the equation, not the base of the triangle in our figure—sorry for the confusion!)

- To compute the y-intercept (b), determine the value of y when x is equal to zero. In this case, $b = 0$.
- To compute the slope (m), think of the "rise" over the "run." The rise is the height (h), and the run is b. Therefore the slope (m) is: $m = \frac{h}{b}$.

  In this case: $y = mx + b = \left(\frac{h}{b}\right)x + 0 = \left(\frac{h}{b}\right)x$

STEP 3: At position x write the expression for the differential area of width dx and height y:

height: $y = \frac{h}{b}x$ width: dx area of one slice: $\left(\frac{h}{b}x\right)dx$

STEP 4: Compute the total area by integrating between the two limits along the y-axis (x goes from zero to b): $A = \int dA = \int_0^b \left(\frac{h}{b}x\right)dx = \left(\frac{h}{2b}x^2\right)\Big|_0^b = \frac{bh}{2}$.

STEP 5: Compute $\int x\,dA$ between the same two limits $x = 0 \rightarrow b$ (x goes from zero to b):

$\int x\,dA = \int_0^b \left(\frac{h}{b}(x)(x)\right)dx = \left(\frac{h}{3b}x^3\right)\Big|_0^b = \frac{1}{3}h\,b^2$.

STEP 6: Compute the x-centroid: $\bar{x} = \frac{\int x\,dA}{\int dA} = \frac{\frac{1}{3}h\,b^2}{\frac{bh}{2}} = \left(\frac{2}{3}\right)b$.

STEP 7: Flip the strip. Determine the "width" of each "slice" at any position y.

From step 2, $y = \frac{h}{b}x \Rightarrow x = \frac{b}{h}y$.

Notice how the width of the strip doesn't start at the y-axis so we need to think about what is the width really.

$x = b - y\left(\frac{b}{h}\right)$

STEP 8: At position y write the expression for the differential area of height dy and width x:

width: $x = b - y\left(\frac{b}{h}\right)$ height: dy area of one slice: $\left(b - y\left(\frac{b}{h}\right)\right)dy$

STEP 9: Compute the area by integrating between the two limits along the y-axis:

$$A = \int dA = \int_0^h \left(b - \left(\frac{b}{h}\right) y\right) dy = \left(by - \frac{by^2}{2h}\right)\Bigg|_0^h = b\,h - b\frac{h^2}{(2\,h)} = \frac{1}{2}(b\,h).$$

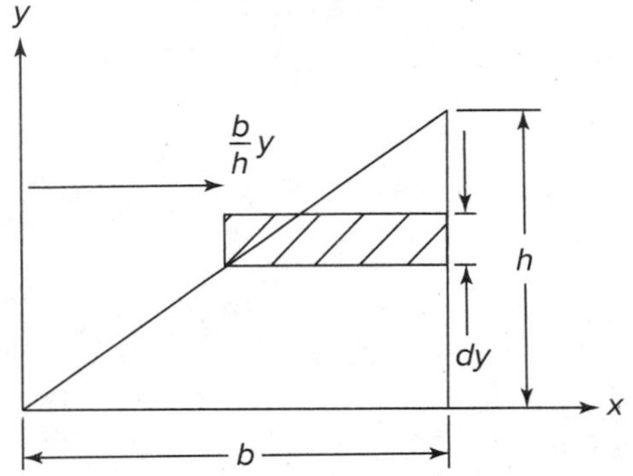

STEP 10: Compute the numerator of the centroid equation between the two limits along the y-axis:

$$\int y\,dA = \int_0^h y\left(b - \left(\frac{b}{h}\right) y\right) dy = \left(\frac{by^2}{2} - \frac{by^3}{3h}\right)\Bigg|_0^b = b\left(\frac{h^2}{2}\right) - \frac{b}{h}\left(\frac{h^3}{3}\right) = \frac{bh^2}{2} - \frac{bh^2}{3} = \frac{3bh^2}{6} - \frac{2bh^2}{6} = \frac{bh^2}{6}.$$

STEP 11: Compute the y-centroid: $\bar{y} = \dfrac{\int y\,dA}{\int dA} = \dfrac{\frac{bh^2}{6}}{\frac{bh}{2}} = \dfrac{h}{3}.$

WATCH VIDEO

STATICS LESSON 43
Centroids by Calculus Example Problem

Find the $\bar{x}$ centroid of the shape shown below bounded by two lines.

Press pause on the video once you get to the workout problem. Only press play if you get stuck.

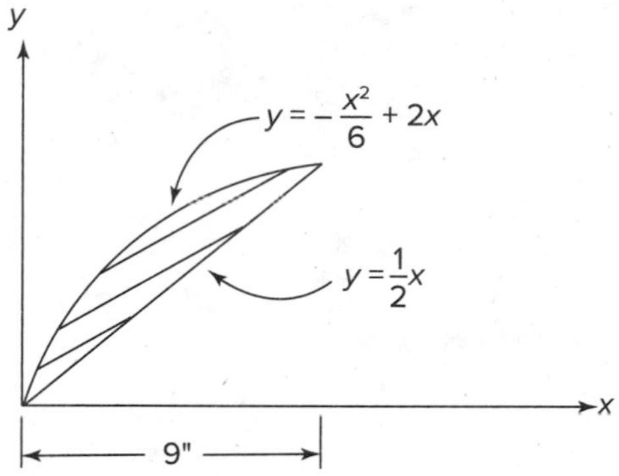

PRO TIP

When computing the centroid $\bar{x} = \frac{\int x\,dA}{\int dA}$, it is easiest to start with the denominator (i.e., the area), and then compute the numerator.

PITFALL

When computing the height of the differential area, don't forget that it is the total height. If the shape you are trying to find the centroid of lives entirely above the *x*-axis, then take the upper *y* function, and subtract the lower *y* function. However, if it straddles the *x*-axis, then you will need to add the upper *y* function to the lower *y* function.

TEST YOURSELF 5.2

SOLUTION TO TEST YOURSELF: Centroids

Find the centroid of the area bound by the straight line and the parabola shown below (dimensions in inches).

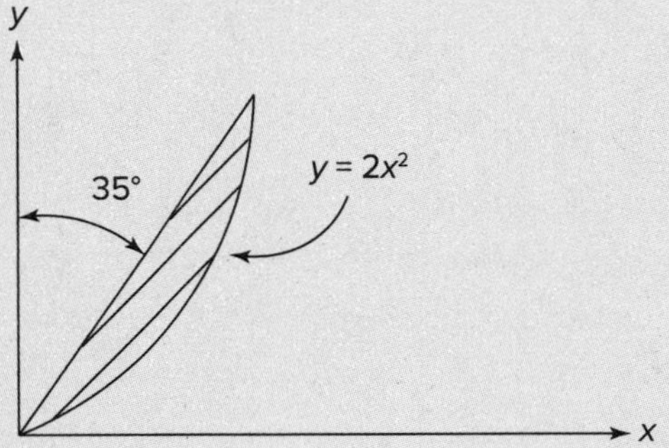

ANSWERS

$\bar{x} = 0.357$

$\bar{y} = 0.4078$

WATCH VIDEO

STATICS LESSON 44
Very Challenging Centroids by Calculus Problem

Can you use the flip-the-strip method every time? Yes, but the math gets super-duper complicated, and you have to remember chain rules, etc. It is possible to find the centroid without flipping the strip using calculus.

PRO TIP

Two clues to find when you shouldn't use flip-the-strip and instead use integration:

- **Equation that has an operator in it such as $y = 4x^2 + 7x$ (it's difficult to isolate x)**
- **When you have more than one function defining a shape (top part has a function and bottom part has a different function such that it requires integration by parts)**

Determination of Centroids by Integration (Non-Flip-the-Strip) Recipe

STEP 1: Find $\bar{x}$ using Flip-the-Strip Recipe Steps 1–5.

STEP 2: $\bar{y} = \dfrac{\int y\,dA}{\int dA}$. Since we computed $\int dA$ already, we know the denominator (the area doesn't change).

STEP 3: Compute $\int y\,dA$. Multiply unintegrated dA by $\bar{y}$ for one strip. Since dA is the height of the strip and the width of the strip, $\bar{y}$ must just be the height of the strip divided by 2 referenced from the origin. Use the same limits as we did when calculating $\bar{x}$.

STEP 4: Solve for $\bar{y}$.

CENTROIDS

EXAMPLE: NON–FLIP-THE-STRIP

Find $\bar{x}$ and $\bar{y}$ for the give shape.

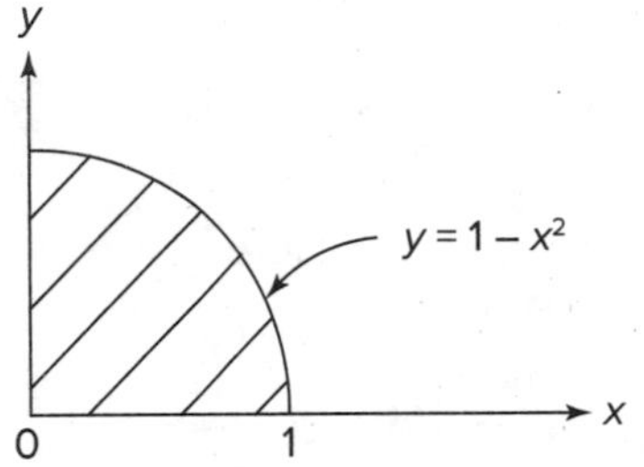

STEP 1: Visualize the summation of rectangular strips by shading one of them on your diagram.

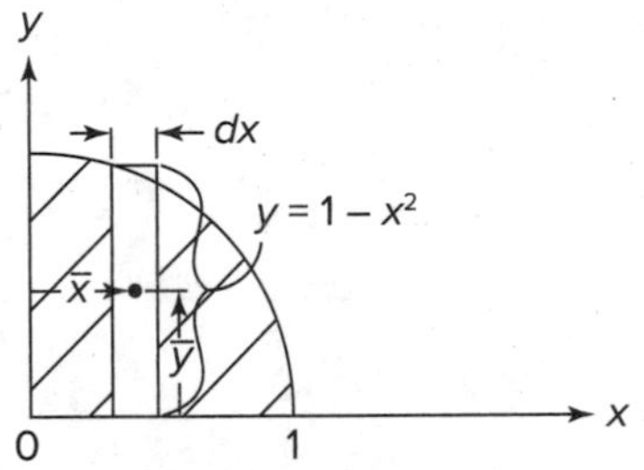

STEP 2: Solve for $\bar{x}$ the same way as we learned earlier. First find the denominator of the $\bar{x}$ equation ($\int dA$), then find the numerator and simplify.

Begin with calculating the denominator:

$$\int dA = \int_0^1 \underbrace{(1-x^2)}_{\text{height}} \underbrace{dx}_{\text{width}}$$

$$= 1x - \frac{x^3}{3}\Big|_0^1$$

$$= 1 - \frac{1}{3} = \frac{2}{3}$$

Note: 2/3 is the total area of the shape.

Complete the centroid calculation by calculating the numerator:

$$\int xdA = \int_0^1 x(1-x^2)dx$$

$$= \int_0^1 (x-x^3)dx$$

$$= \frac{x^2}{2} - \frac{x^4}{4}\Big|_0^1$$

$$= \frac{1}{2} - \frac{1}{4} = \frac{1}{4}$$

$$\bar{x} = \frac{\int xdA}{\int dA} = \frac{\frac{1}{4}}{\frac{2}{3}} = \frac{1}{4} \cdot \frac{3}{2} = \frac{3}{8}$$

STEP 3: Without "flipping the strip," let's see if we can find $\bar{y}$.

$$\bar{y} = \frac{\int ydA}{\int dA}$$

We already found $\int dA$ in Step 2, and since the area hasn't changed we can use $\int dA$ in the $\bar{y}$ equation. We need to find $\int ydA$ only. Find the location of $\bar{y}$ on our vertical strip.

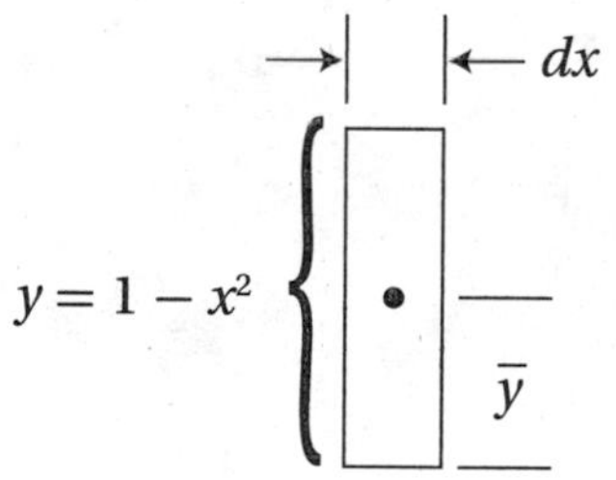

So in this case, $\bar{y}$ is located at:

$$\bar{y} = \frac{1-x^2}{2}$$

CENTROIDS

STEP 4: Find $\int y dA$ (remember to use dA from Step 2).

$$\int y dA = \int_0^1 \underbrace{\left(\frac{1-x^2}{2}\right)}_{y} \underbrace{(1-x^2)dx}_{dA}$$

$$= \int_0^1 \left(\frac{1}{2} - \frac{x^2}{2}\right)(1-x^2)dx$$

Expanding, $\int y dA = \int_0^1 \left(\frac{1}{2} - \frac{x^2}{2} - \frac{x^2}{2} + \frac{x^4}{2}\right)dx$

$$= \int_0^1 \left(\frac{1}{2} - x^2 + \frac{x^4}{2}\right)dx$$

$$= \left(\frac{x}{2} - \frac{x^3}{3} + \frac{x^5}{10}\right)\Bigg|_0^1$$

$$= \frac{1}{2} - \frac{1}{3} + \frac{1}{10} = \frac{15}{30} - \frac{10}{30} + \frac{3}{30} = \frac{8}{30} = \frac{4}{15}$$

STEP 5: Solve for $\bar{y}$.

$$\bar{y} = \frac{\int y dA}{\int dA} = \frac{\frac{4}{15}}{\frac{2}{3}} = \frac{4}{15} \cdot \frac{3}{2} = \frac{12}{30} = \frac{2}{5}$$

CENTROIDS

Below is an 8" thick plate described by the two equations. The plate density is 0.284 lbs/in^3. Determine the weight of the plate; compute the centroid; using equilibrium, determine the reactions.

Press pause on video lesson 44 once you get to the workout problem. Only press play if you get stuck.

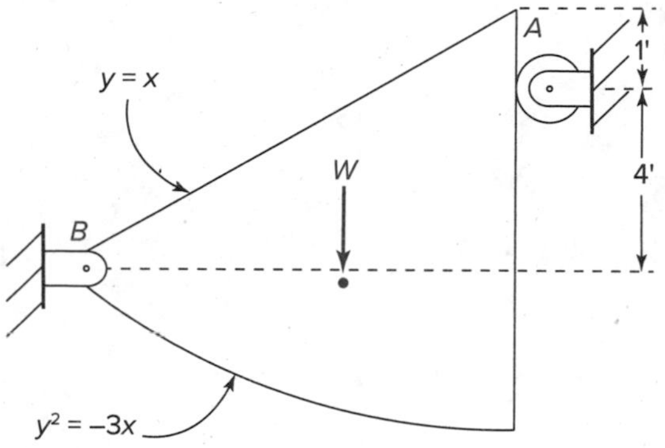

PITFALL

Don't forget the length of the strip can't be negative.

✓ TEST YOURSELF 5.3

SOLUTION TO TEST YOURSELF: Centroids

Compute the centroid of the shape shown using the non-flip-the-strip method.

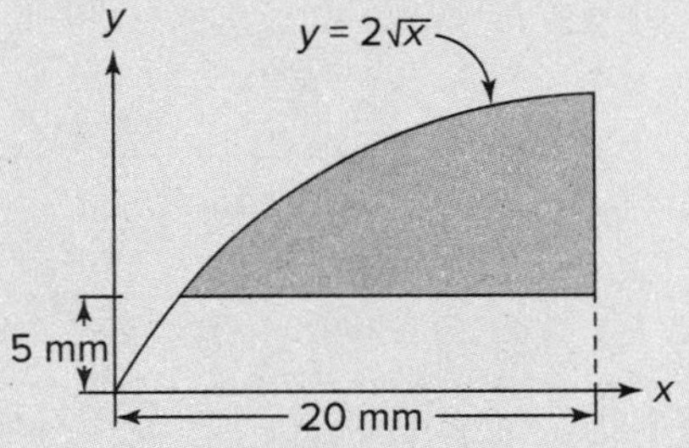

ANSWERS

$\bar{x} = 15.19$ mm

$\bar{y} = 6.37$ mm

WATCH VIDEO **STATICS LESSON 45**
Theorems of Pappus Guldinus, Volume and Surface Area

Theorems of Pappus Guldinus

The Theorems of Pappus Guldinus have to do with calculating volume and surface areas of "axis symmetric" shapes. Also called *bodies of revolution* as they are revolved around some axis.

CHALLENGE QUESTION

If you went to the machine shop and wanted to make an axis symmetric part, which machine would you use?

First Theorem (Surface Area)

$SA = \theta \bar{r} L$

Second Theorem (Volume)

$V = \theta \bar{r} A$

where:

θ = the rotation angle (radians, between zero and 2π)

$\bar{r}$ = distance from the axis of rotation to either the centroid of the line or the centroid of the area (for rotation around the x-axis, use $\bar{y}$, for rotation around the y-axis, use $\bar{x}$)

A = area of shape

L = length of line

PITFALL

The centroid of a line is different than the centroid of a shape! A common mistake is to use ⅓ the distance for an angled line because it sweeps out a triangle, when, in fact, the centroid of the line is in the middle of the line.

ANSWER

A lathe would be a good choice.

Pappus Guldinus Recipe

STEP 1: Sketch the generating shape that, when swept around the axis of rotation, creates the final volume.

This is absolutely the hardest part! Imagine taking the part and slicing it in half length-wise, and then take half of that section view. What would that cross section look like? That is the "generating shape."

Part *Length-wise* *Generating Shape*

STEP 2: For computing surface area: draw the generating shape again, this time paying attention to lines not areas (i.e., divide into parts). Put a dot for where you think the centroid of each line is located.

STEP 3: Compute the surface area: $SA = \theta \bar{r} L$.

STEP 4: For computing volume: modify the sketch to show the bounding lines that create areas of the generating shape (i.e., divide into parts). Put a dot for where you think the centroid of each area is located.

STEP 5: Compute the volume for each part using the equation: $V = \theta \bar{r} A$.

EXAMPLE: PAPPUS GULDINUS

Find the surface area and volume of the shape show using the theorems of Pappus Guldinus.

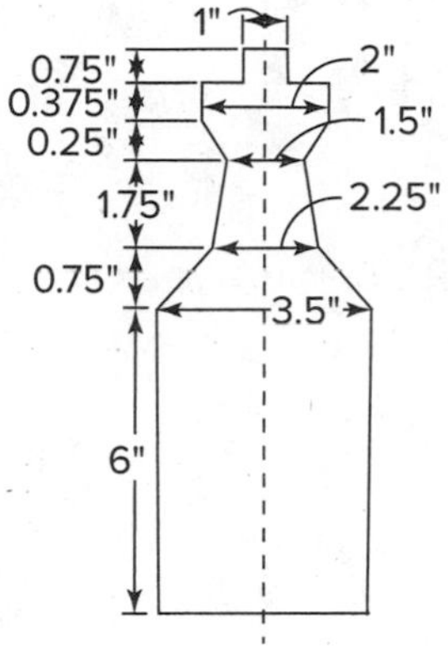

STEP 1: Draw the generating shape.

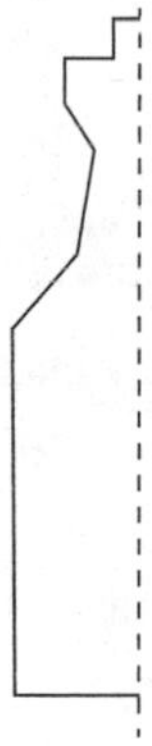

STEP 2: Divide into line parts and add dots at centroids of the lines.

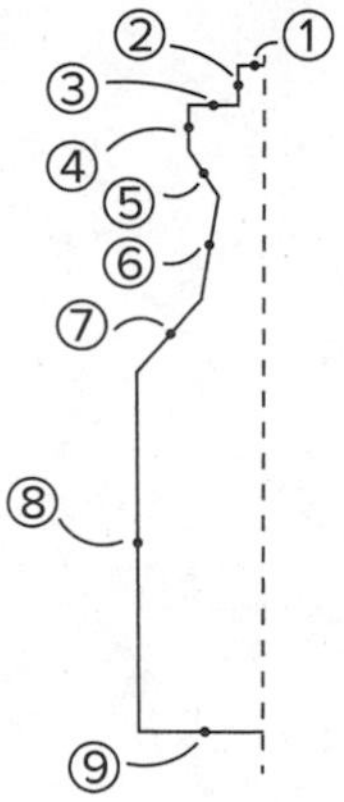

STEP 3: Compute the surface area for each part using equation: $SA = \theta \bar{r} L$. Since we are rotating around the y-axis, we are using $\bar{x}$ so $SA = \theta \bar{x} L$.

$$SA = 2\pi[x_1L_1 + x_2L_2 + x_3L_3 + x_4L_4 + x_5L_5 + x_6L_6 + x_7L_7 + x_8L_8 + x_9L_9]$$

$$SA = 2\pi[(0.25)(0.5) + (0.5)(0.75) + (0.75)(0.5) + (1)(0.375) + (0.875)\sqrt{0.25^2 + 0.25^2} + (0.9375)\sqrt{0.375^2 + 1.75^2} + (1.4375)\sqrt{0.625^2 + 0.75^2} + (1.75)(6) + (0.875)(1.75)]$$

$$SA = 2\pi[0.125 + 0.375 + 0.375 + 0.375 + 0.309 + 1.678 + 1.403 + 10.500 + 1.531] = 104.8 \text{ in}^2$$

STEP 4: Divide into area parts and add dots at the centroids of the areas.

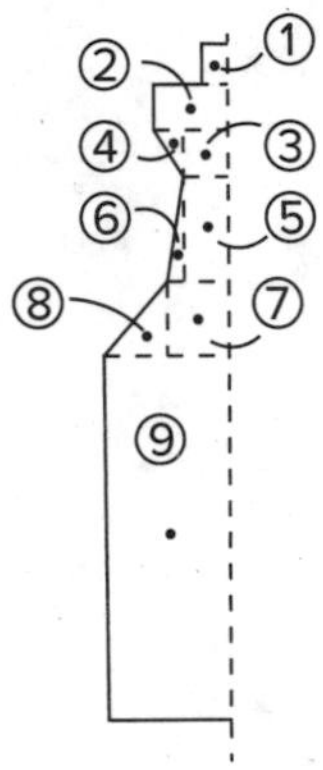

STEP 5: Compute the volume for each part using the equation: $V = \theta \bar{r} A$. Since we are rotating around the y-axis, we are using $\bar{x}$ so $V = \theta \bar{x} A$.

$$V = 2\pi[x_1A_1 + x_2A_2 + x_3A_3 + x_4A_4 + x_5A_5 + x_6A_6 + x_7A_7 + x_8A_8 + x_9A_9]$$

$$V = 2\pi[(0.25)(0.5)(0.75) + (0.5)(0.375)(1) + (0.375)(0.25)(0.75) + (0.833)(\tfrac{1}{2})(0.25)(0.25) + (0.375)(0.75)(1.75) + (0.875)(\tfrac{1}{2})(0.375)(1.75) + 0.5625(1.125)(0.75) + (1.33)(\tfrac{1}{2})(0.625)(0.75) + 0.875(6)(1.75)]$$

$$V = 2\pi[0.09375 + 0.1875 + 0.0703 + 0.026 + 0.492 + 0.287 + 0.475 + 0.312 + 9.188] = 69.94 \text{in}^3$$

Determine the surface area and volume for the shape.

Press pause on video lesson 45 once you get to the workout problem. Only press play if you get stuck.

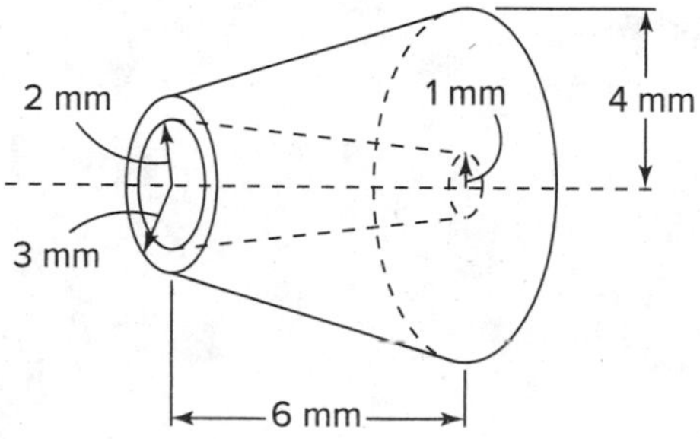

TEST YOURSELF 5.4

SOLUTION TO TEST YOURSELF: Centroids

For the special cylindrical materials handling container shown below, compute the surface area and volume.

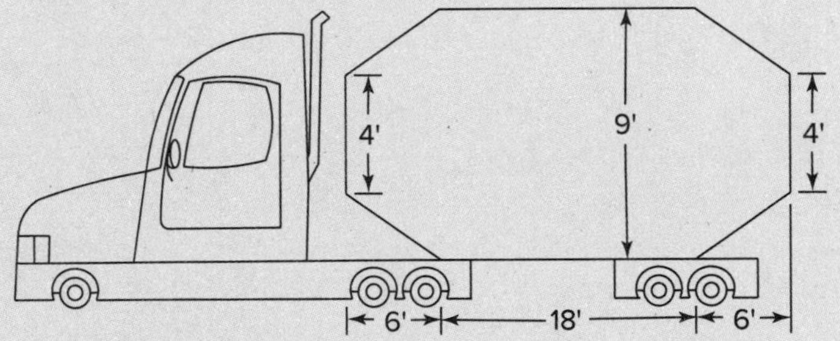

ANSWERS

Surface area = 789.3 ft^2

Volume = 1,562.9 ft^3

WATCH VIDEO **STATICS LESSON 46**
Distributed Loads Using Centroids

Calculating Distributed Loads

What Is a Distributed Load?

- Weight (force) that is spread over a surface
- Not a concentrated load (i.e., a point load)
- Example: snow load on a roof or a wind load on a building
- For 2D problems, typically given as the force/length, e.g., lb/ft or kN/m

Converting a Distributed Load into an Equivalent Point Load Recipe

STEP 1: Calculate the total *magnitude* (i.e., the total force) of a distributed load by computing the "area" under the distributed load. This is the equivalent point load.

STEP 2: Find *where* the equivalent point load acts by determining the centroid of the area of the distributed load and apply that point load found in Step 1 at the centroid of that shape.

STEP 3: Solve as a regular statics problem.

PRO TIP

To compute global equilibrium, convert each distributed load to a point load and place each point load at the centroid of its distributed load.

PITFALL

Remember to pay attention to the direction of the distributed load.

EXAMPLE: DISTRIBUTED LOADS

Find the reactions at the supports.

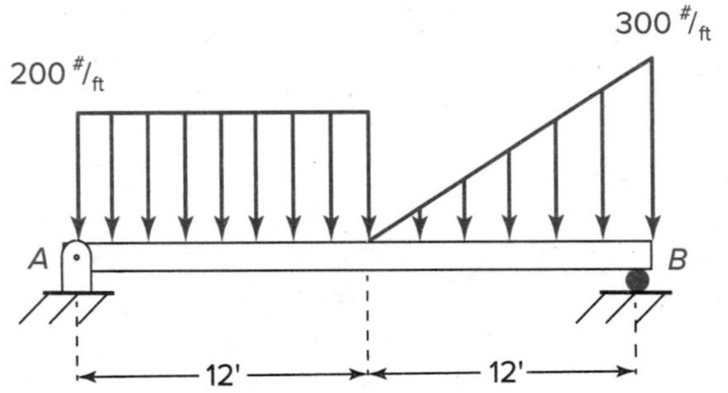

STEP 1: Convert the two distributed loads into concentrated loads by finding the "area" of each shape.

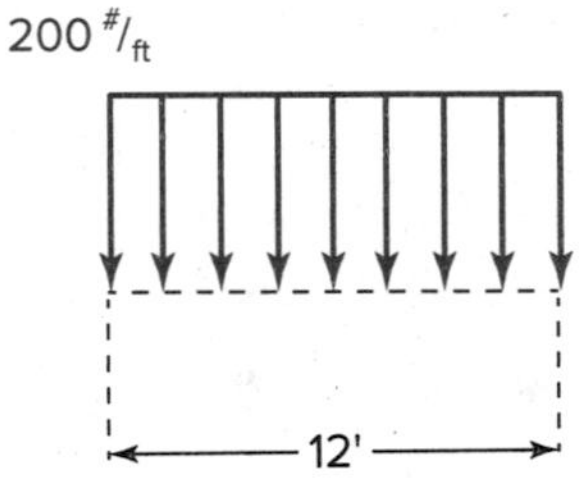

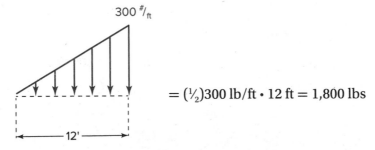

$= 200\ \text{lb/ft} \cdot 12\ \text{ft} = 2{,}400\ \text{lbs}$

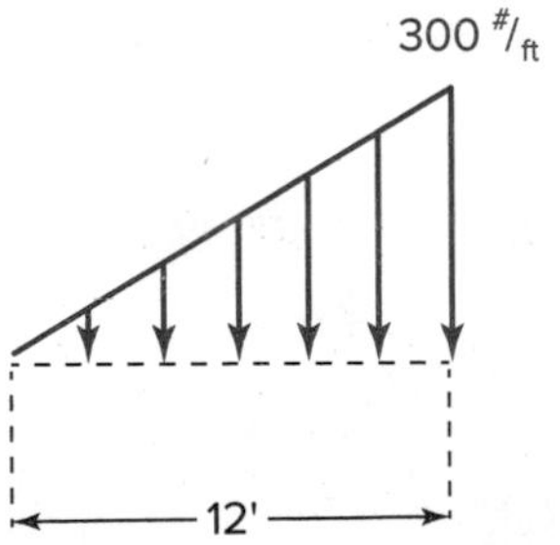

$= (\tfrac{1}{2})300\ \text{lb/ft} \cdot 12\ \text{ft} = 1{,}800\ \text{lbs}$

STEP 2: Apply the concentrated loads at the centroids of the respective areas (centered for the rectangle and at ⅔ of the base for the triangle) and construct an FBD.

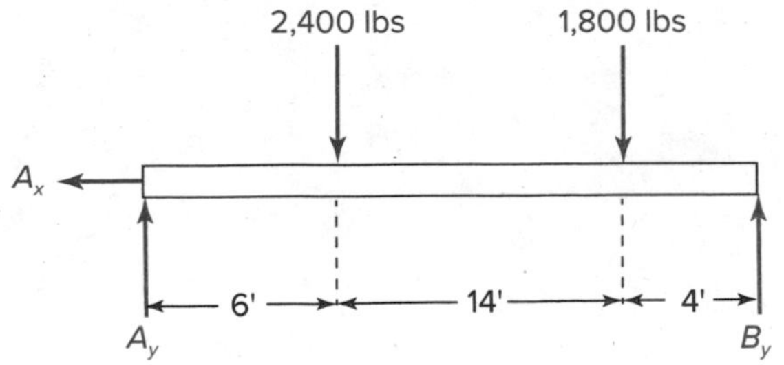

STEP 3: Solve for A_y, A_x, and B_y like normal (using equilibrium equations).

$\Sigma F_x = 0 = A_x$

$\Sigma F_y = 0 = A_y - 2{,}400 - 1{,}800 + B_y$

$A_y + B_y = 4{,}200$

$\Sigma M_A = 0 = -2{,}400(6) - 1{,}800(20) + B_y(24)$

$B_y = 2{,}100$ lbs

$A_y = 2{,}100$ lbs

$A_x = 0$

For the three distributed loads on the trailer shown below, compute the magnitude of each load and the location on the trailer. Once this is completed, figuring the reaction forces should be a snap!

Press pause on video lesson 46 once you get to the workout problems. Only press play if you get stuck.

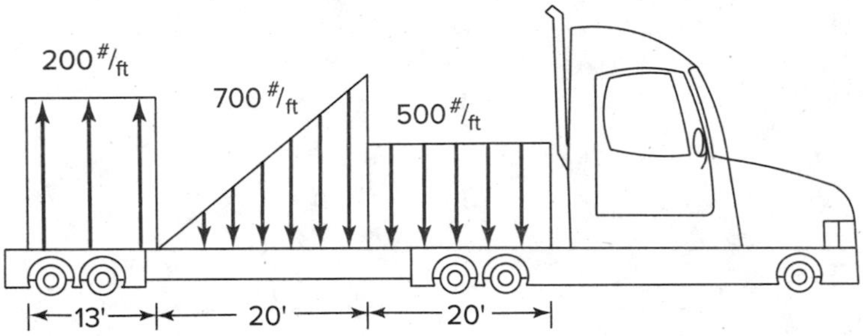

Find the reaction forces for the beam below.

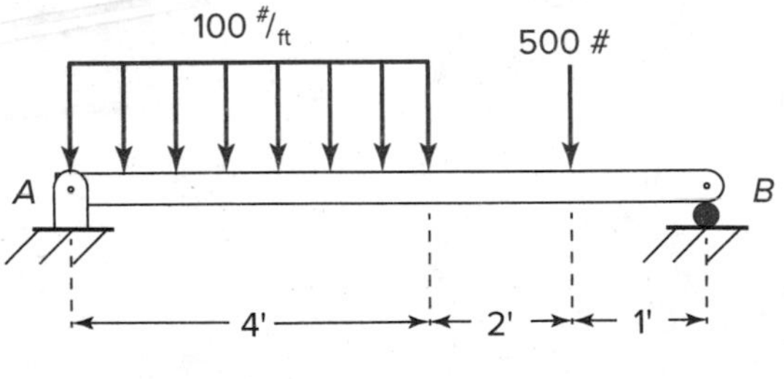

✓ TEST YOURSELF 5.5

SOLUTION TO TEST YOURSELF: Centroids

For the loading on the beam as shown below, compute the reaction forces at the two supports.

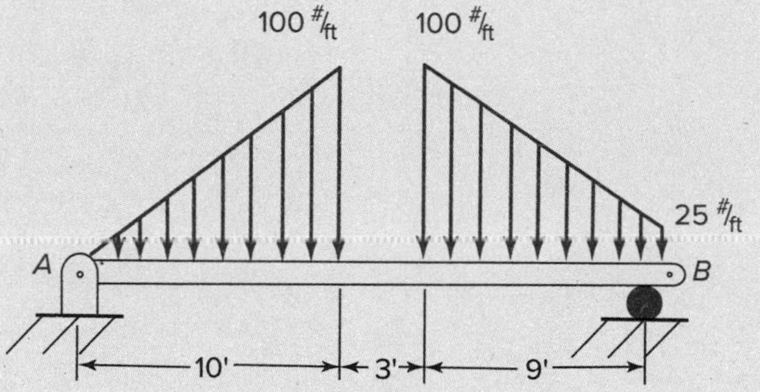

ANSWERS

$A_x = 0$

$A_y = 486.5$ #

$B_y = 576$ #

KEY TAKEAWAYS

Centroids are points where you can perfectly balance the object. The main takeaways from Level 5 include how to calculate:

- Composite shapes
  - Areas
  - Volumes
  - Masses (pieces with different density)
  - Lengths
- Centroids by calculus
- Theorems of Pappus Guldinus
- Converting distributed loads into a concentrated load

Composite Shapes Summary

- Decide what you are dealing with and which of the following equations you should use.

For Areas:	For Volume:	For Lengths:	For Mass:
$\bar{x} = \frac{\Sigma x_i A_i}{\Sigma A_i}$	$\bar{x} = \frac{\Sigma x_i V_i}{\Sigma V_i}$	$\bar{x} = \frac{\Sigma x_i L_i}{\Sigma L_i}$	$\bar{x} = \frac{\Sigma x_i M_i}{\Sigma M_i}$
$\bar{y} = \frac{\Sigma y_i A_i}{\Sigma A_i}$	$\bar{y} = \frac{\Sigma y_i V_i}{\Sigma V_i}$	$\bar{y} = \frac{\Sigma y_i L_i}{\Sigma L_i}$	$\bar{y} = \frac{\Sigma y_i M_i}{\Sigma M_i}$
$\bar{z} = \frac{\Sigma z_i A_i}{\Sigma A_i}$	$\bar{z} = \frac{\Sigma z_i V_i}{\Sigma V_i}$	$\bar{z} = \frac{\Sigma z_i L_i}{\Sigma L_i}$	$\bar{z} = \frac{\Sigma z_i M_i}{\Sigma M_i}$

- Always make sure that all your dimensions are referenced from the origin!
- Don't forget to utilize the Properties of Geometric Shapes and Areas table (found at the beginning of this level).

KEY TAKEAWAYS

Centroids by Calculus Summary

- Use this method for complicated shapes that aren't in the table:

$$\bar{x} = \frac{\int x\,dA}{\int dA} \qquad \bar{y} = \frac{\int y\,dA}{\int dA} \qquad \bar{z} = \frac{\int z\,dA}{\int dA}$$

- Flip-the-strip method
  - Easiest to use but can only be used when the equation of the function is simple to solve for the variable (i.e., it doesn't contain an operator symbol: +, –, *, /, etc.)
- Non-flip-the-strip method
  - Helps prevent having to use the chain rule or "u" substitutions, but it does require a bit of algebra, so be careful with your terms (several will typically drop out)

Theorems of Pappus Guldinus

- First theorem: $SA = \theta \bar{r} L$
- Second theorem: $V = \theta \bar{r} A$
- Don't forget theta is in radians so if you have a full revolution it is 2π
- $\bar{r}$ is the distance from the axis of rotation to the centroid of the shape
  - If rotating about the x-axis, use $\bar{y}$
  - If rotating about the y-axis, use $\bar{x}$

Distributed Loads

- These don't work well in our statics equations, so we have to convert them into concentrated loads (i.e., point loads).
- Distributed loads are calculated by calculating the "area" of the distributed load shape.
- The magnitude of this load is applied at the centroid of this shape.

RECIPES

- Composite Shapes Centroid Recipe
- Centroid by Calculus (Flip-the-Strip) Recipe
- Determination of Centroids by Integration (Non–Flip-the-Strip) Recipe
- Pappus Guldinus Recipe
- Converting a Distributed Load into an Equivalent Point Load Recipe

PRO TIPS

Composite Shapes

- Use the table found in the beginning of this level for geometric properties of areas (i.e., centroids) and the total shape as being composed of simple shapes with properties listed in the table.
- Use symmetry and observation (intuition) to identify the centroid when possible.
- The table method is especially helpful when solving composite problems with many parts.
- Make sure to use the mass and not only the volume to find the center of mass (i.e., don't forget to multiply volume by density!).

Calculus Method

- When computing the centroid $\bar{x} = \dfrac{\int x\,dA}{\int dA}$, it is easiest to start with the denominator (i.e., the area), and then compute the numerator.
- Two clues to find when you shouldn't use flip-the-strip and instead use integration:
  - Equation that has an operator in it such as $y = 4x^2 + 7x$ (it's difficult to isolate x)
  - When you have more than one function defining a shape (top part has a function and bottom part has a different function such that it requires integration by parts)

Distributed Loads

- To compute global equilibrium, convert each distributed load to a point load and place each point load at the centroid of its distributed load.

PITFALLS

Composite Shapes

- The number one mistake made when calculating centroids is to forget that locations are *always* referenced from the origin (i.e., datum).

Calculus Method

- When computing the height of the differential area, don't forget that it is the total height. If the shape you are trying to find the centroid of lives entirely above the *x*-axis, then take the upper *y* function, and subtract the lower *y* function. However, if it straddles the *x*-axis, then you will need to add the upper *y* function to the lower *y* function.
- Don't forget the length of the strip can't be negative.

Pappus Guldinus

- The centroid of a line is different than the centroid of a shape! A common mistake is to use ⅓ the distance for an angled line because it sweeps out a triangle, when, in fact, the centroid of the line is in the middle of the line.

Distributed Loads

- Remember to pay attention to the direction of the distributed load.

Level 6

Trusses, Frames, and Machines

WATCH VIDEO

STATICS LESSON 47
Introduction to Trusses, Frames, and Machines

This entire topic is about finding the internal force(s) in connections or members in trusses, frames, and machines.

TRUSS

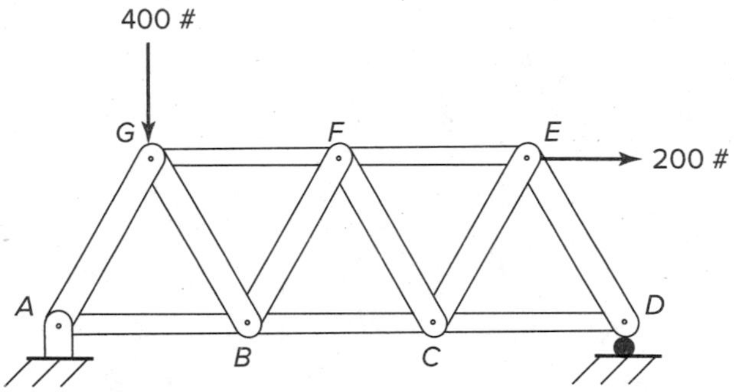

- Trusses are made entirely of two force members.
  - All members are assumed to be pin connected at their ends.
  - Weights of the members are assumed negligible.
  - Loaded only at the joints. No forces applied at the middle of members!
- Methods for solving include:
  - Method of Joints
  - Method of Sections

FRAME

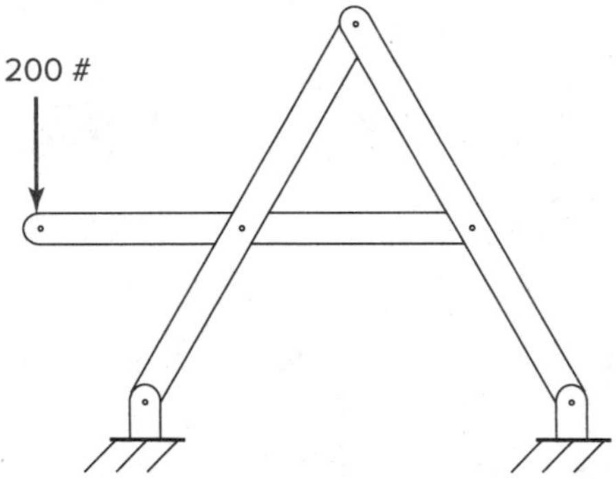

- The difference between truss and a frame is that a frame can contain multiforce members and two force members not only loaded at the pins.
  - ***Reminder:*** A multiforce member has more than two forces or moments (can be loaded along its length and not only at ends). The frame above has three members (continuous) with pins along their length; each member is a multiforce member.
- Frames do not have to be pin-connected.

MACHINE

- Machines have moving parts.
- A machine can contain the same characteristics as a frame except it does not necessarily require a reaction force with the ground (i.e., it is not hooked to the world, so you don't need to find global equilibrium).

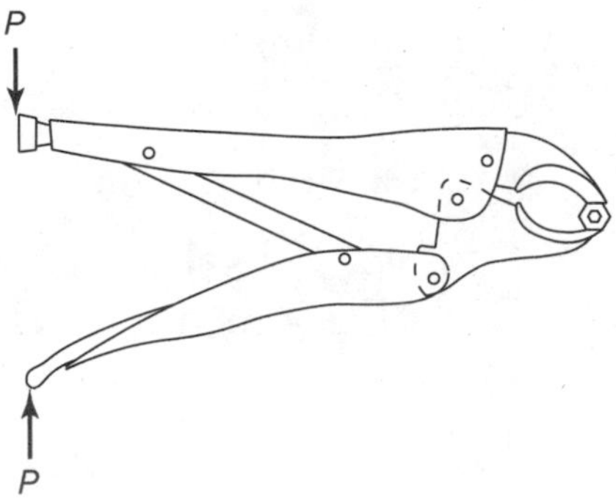

TEST YOURSELF 6.1

SOLUTION TO TEST YOURSELF: Trusses, Frames, and Machines

Label the images as either a truss, machine, or frame.

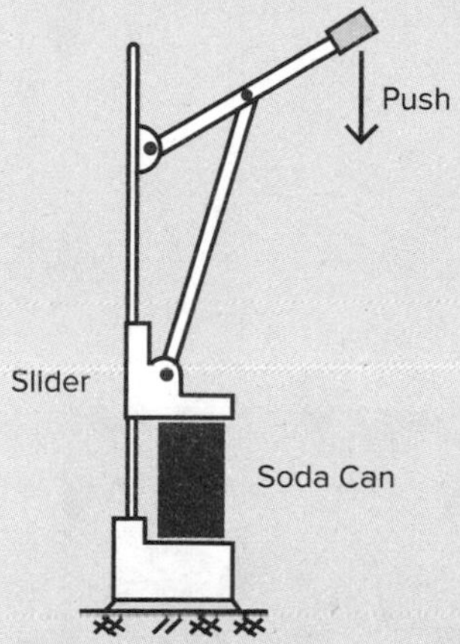

1.

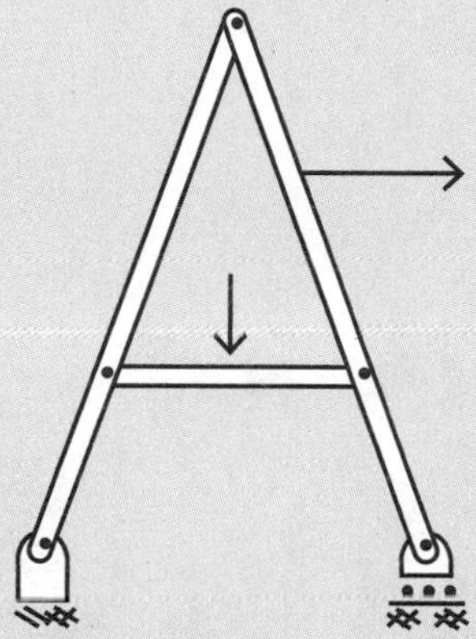

2.

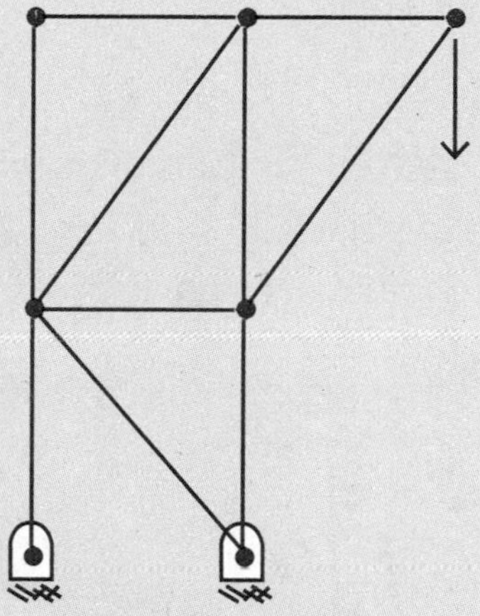

3.

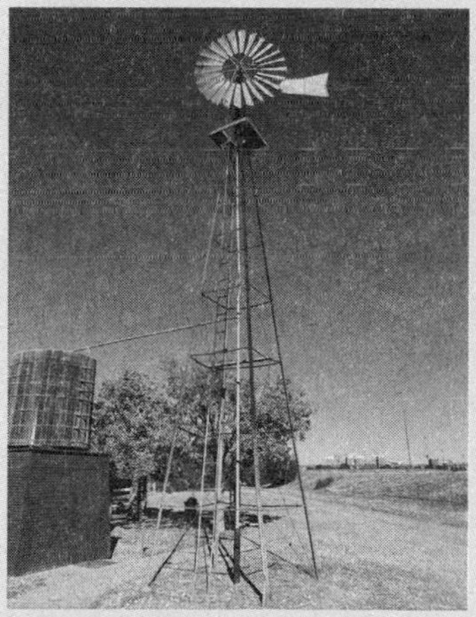

4.

5.

6.

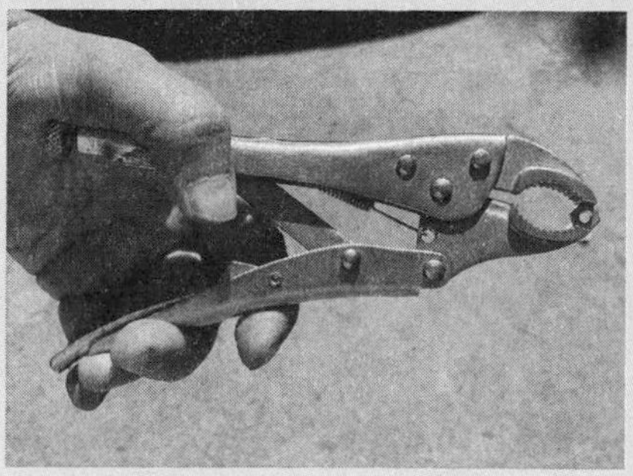

7.

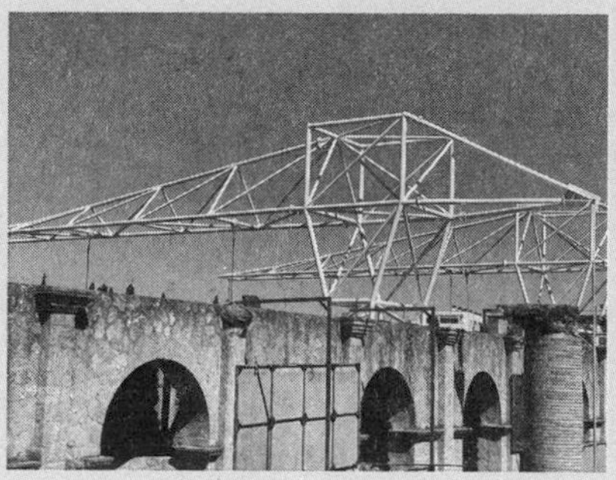

8.

9.

ANSWERS

1. Machine; 2. frame; 3. truss; 4. truss; 5. machine; 6. machine; 7. machine; 8. truss; 9. could be argued to be a truss, frame, or a machine

TRUSSES

Why Do I Need to Know About Trusses?

When engineers design bridges, in order to correctly size the members of the truss, they have to know the forces carried in each member of the truss. Some common questions that engineers might want to answer are:

- Is the bridge safe under given loading?
- Might it fail?
- Does it need repair?

COMPONENTS OF A TRUSS

The joints are the location where different members of the truss are joined together typically via welding, riveting, or bolting them to a gusset plate. In a truss we often call gusset plates a joint.

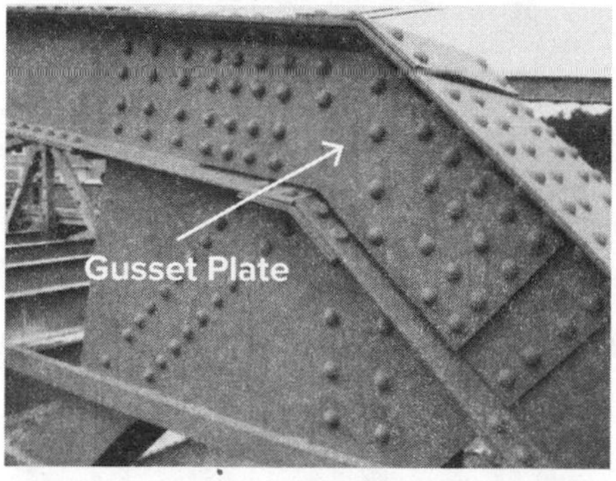

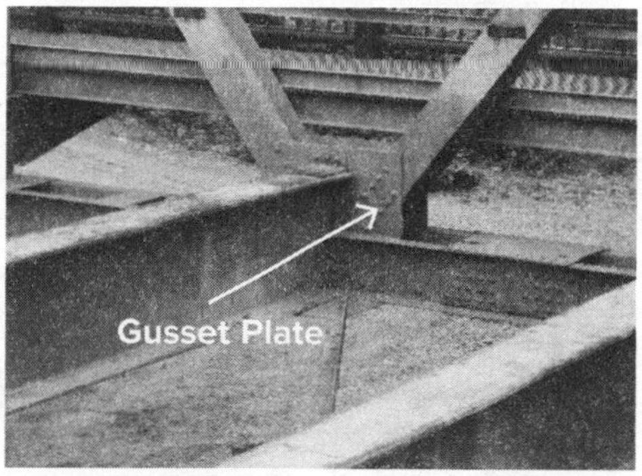

To simplify gusset plates (joints) between members we:

- Ignore the weight of the members
- Assume that the loads are concentrated at the ends (so pin-connected).

Pins and rollers are common reaction forces at supports as it allows for expansion and contraction of the bridge due to temperature changes.

ROLLER OR ROCKER

One reaction force

- Allows the bridge to expand and contract easily without breaking
- Restricts vertical movement only (R_y) (it allows movement in the horizontal direction)
- When replaced with a reaction force, the force only exists in vertical direction

PIN

Two reaction forces

- Common assumption is that the pin is frictionless
- Restricts horizontal (R_x) and vertical movement (R_y)

TRUSSES, FRAMES, AND MACHINES

Need a Refresher?

WATCH VIDEO **STATICS LESSON 31**
System Equilibrium, 2D Reactions at the Supports

Sign Convention for Truss Structures: Tension or Compression

Members in a truss are either in tension (*pulling apart*) or compression (*pushing together*). If this wasn't the case and both arrows pointed in the same direction, then the entire member would be moving in the direction of the arrows.

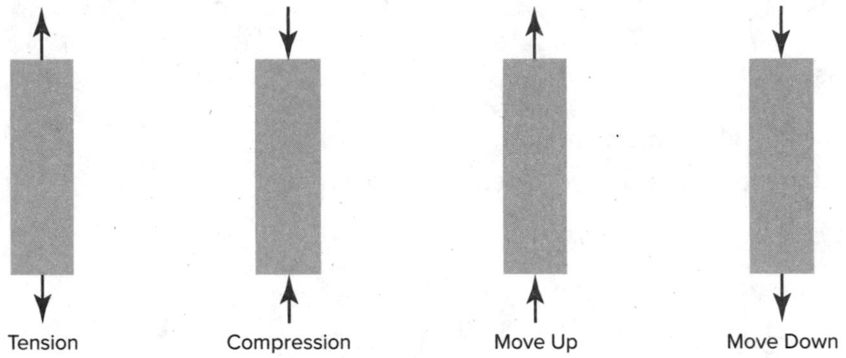

Whether a member is in tension or compression is important because:

- Members can fail differently when experiencing tension or compression.
- Some materials behave differently when experiencing tension or compression.

According to Newton's Third Law (for every action there is an equal and opposite reaction):

- If you assume the forces are pointed away from a joint, then you are assuming the member is in tension.
- If you assume the forces are pointed toward a joint, then you are assuming the member is in compression.

MEMBER *AB* IN TENSION

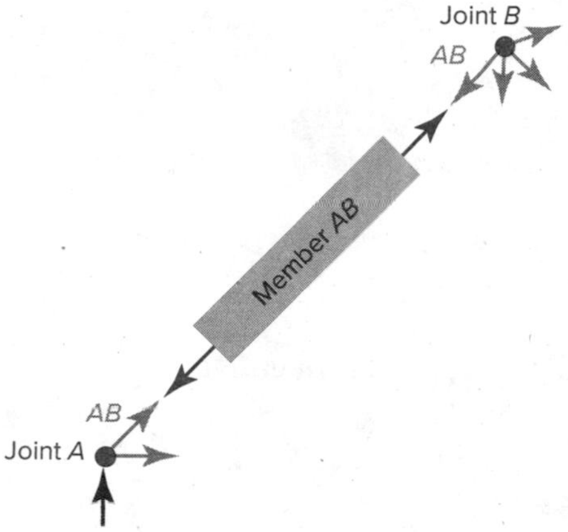

MEMBER *AB* IN COMPRESSION

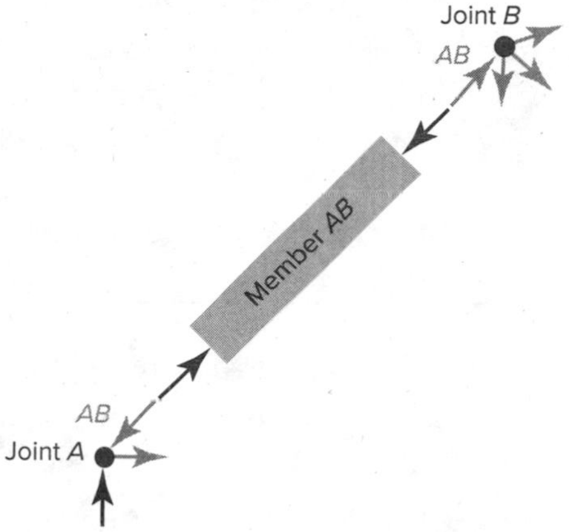

PRO TIP

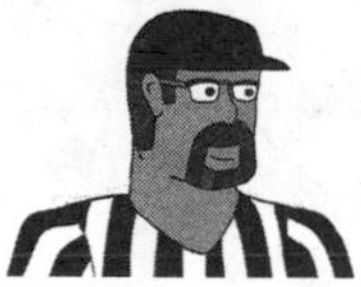

When solving problems, if you get a negative sign when solving for *AB*, then it is opposite to what you assumed! Sometimes we won't know which direction the force acts, so you have to assume a direction. If you get a negative sign when solving, then it is opposite to what you assumed.

✓ TEST YOURSELF 6.2

SOLUTION TO TEST YOURSELF: Trusses, Frames, and Machines

If you drew your FBD as shown in the images below, answer each question.

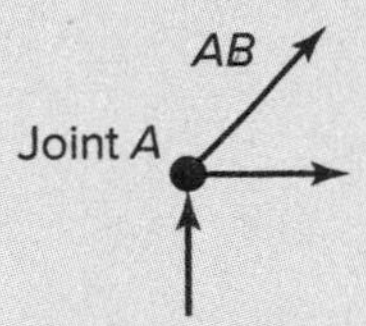

1. If you drew your FBD like this and you find *AB* is positive, then member *AB* is in:

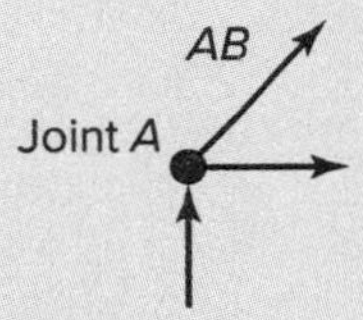

2. If you drew your FBD like this and you find *AB* is negative, then member *AB* is in:

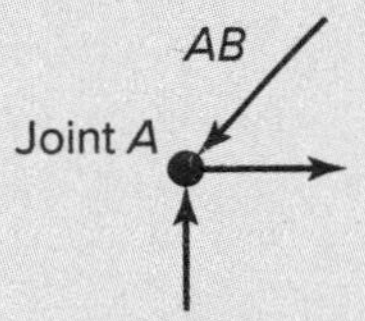

3. If you drew your FBD like this and you find *AB* is positive, then member *AB* is in:

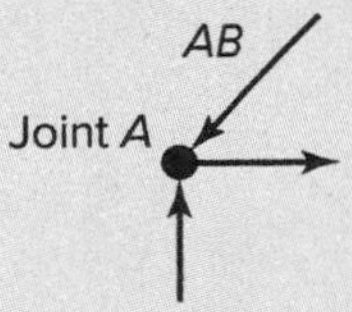

4. If you drew your FBD like this and you find *AB* is negative, then member *AB* is in:

ANSWERS

1. Tension; 2. Compression; 3. Compression; 4. Tension

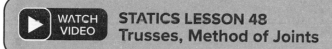

WATCH VIDEO **STATICS LESSON 48**
Trusses, Method of Joints

Method of Joints

- Will work for *all* truss problems, but they take a lot of time to solve using this method.
- Use this method when the problem says to find the force in every single method.
- Uses only $\Sigma F_x = 0$ and $\Sigma F_y = 0$ to solve for forces in truss members since all forces on the FBD of a joint pass through the same point (i.e., since there is no distance, you cannot use the moment equations to find the internal force in the beam). ***Note:*** You might need the moment equation to solve for the reaction forces (i.e., global equilibrium).
- When selecting a joint to solve for unknown forces in members of the truss, we must not select a joint with more than two unknowns since we only have two equations available ($\Sigma F_x = 0$ and $\Sigma F_y = 0$).

Method of Joints Recipe

STEP 1: Find global equilibrium (i.e., find the reactions at the supports).

STEP 2: Select a joint.

Note: Do not pick a joint with more than two unknowns.

STEP 3: Draw a free body diagram of that joint.

STEP 4: Resolve into Cartesian components.

STEP 5: Solve by taking sum of forces.

STEP 6: Repeat.

Note: Make sure you keep consistent with your assumption for whether a member was in tension or compression as you go from joint to joint.

PRO TIP

Always look for a symmetric truss with a symmetric load. You can then quickly add up all of the downward forces and divide by 2 to immediately obtain the global equilibrium.

EXAMPLE: METHOD OF JOINTS

Solve for the forces in every member.

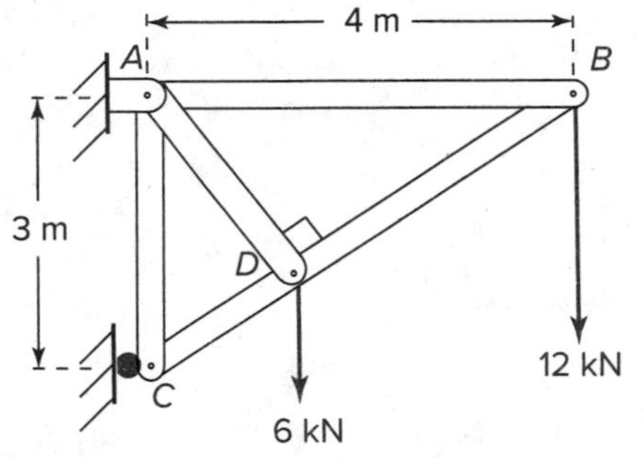

STEP 1: Find global equilibrium.

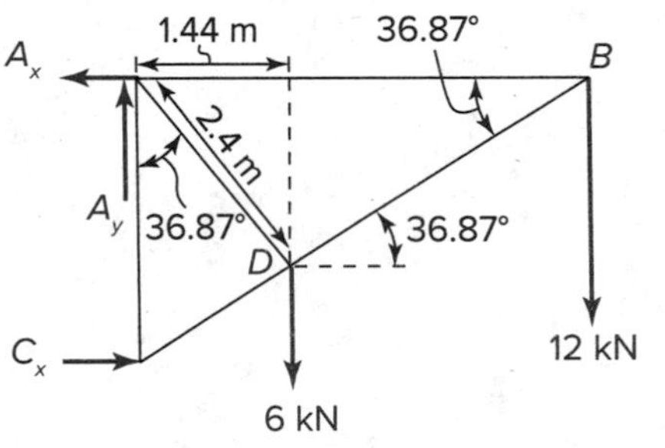

$\tan^{-1}\left(\frac{3}{4}\right) = 36.87°$

$\measuredangle ACD = 180° - 90° - 36.87° = 53.13°$

$\measuredangle CAD = 180° - 90° - 53.13° = 36.87°$

$\measuredangle BAD = 180° - 90° - 36.87° = 53.13°$

$3cos(36.87°) = 2.4\text{ m}$

$2.4cos(53.13°) = 1.44\text{ m}$

$$\Sigma M_A = 0 = C_x(3) - 12(4) - 6(1.44)$$
$$C_x = 18.88\text{ kN}$$
$$\Sigma F_x = 0 = -A_x + C_x$$
$$A_x = 18.88\text{ kN}$$
$$\Sigma F_y = 0 = A_y - 6 - 12$$
$$A_y = 18\text{ kN}$$

STEPS 2 and 3: Select a joint that has no more than two unknowns. Draw a FBD of that joint.

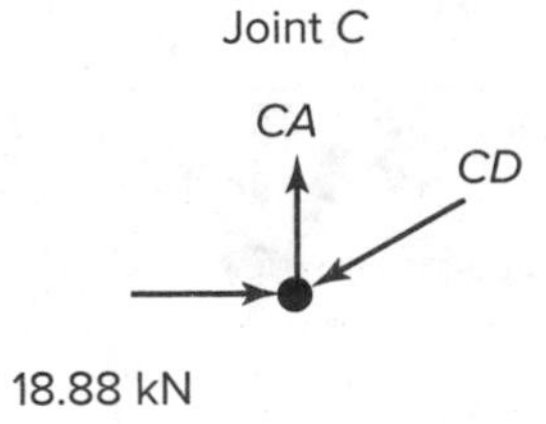

STEP 4: Resolve into Cartesian components.

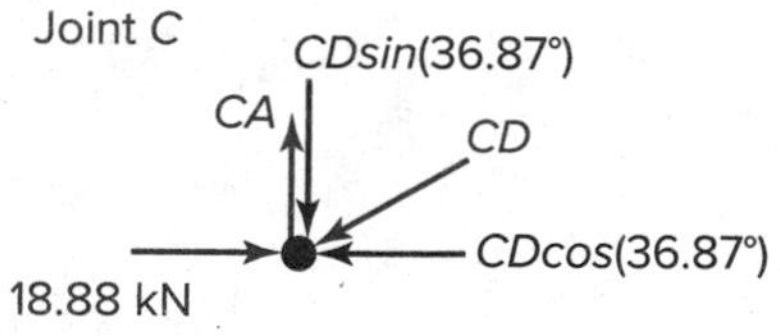

STEP 5: Solve by taking sum of forces.

Note: "C" represents compression and "T" represents tension.

$\Sigma F_x = 0 = 18.88 - CDcos(36.87°)$
$CD = 23.6$ kN (C)

Note: We assumed that *CD* was in compression and got a positive so in the direction we assumed.

$\Sigma F_y = 0 = CA - CDsin(36.87°)$
$CA = 14.16$ kN (T)

Note: We assumed *CA* was in tension and got a positive so in the direction we assumed.

STEP 6: Repeat.

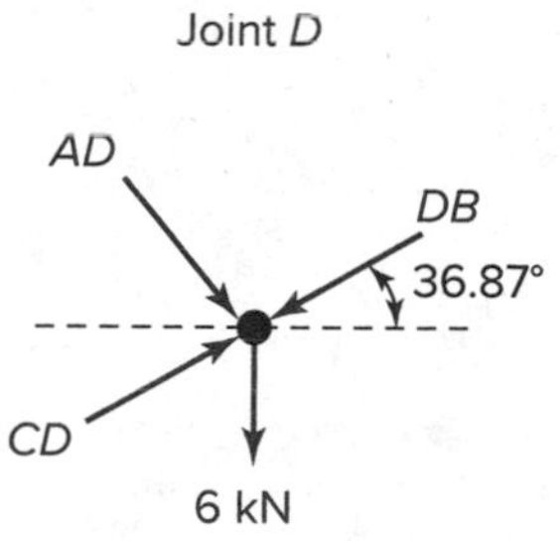

$\Sigma F_x = 0 = CDcos(36.87°) - DBcos(36.87°) + ADcos(53.13°)$
$18.88 = 0.8\ DB - 0.6\ AD$

$\Sigma F_y = 0 = CDsin(36.87°) - 6 - ADsin(53.13°) - DBsin(36.87°)$
$8.16 = 0.6\ DB + 0.8\ AD$

Using system solver:
$DB = 20$ kN (C)

Note: We assumed that *DB* was in compression and got a positive so in the direction we assumed.

$AD = -4.83$ kN $= 4.83$ kN (T)

Note: We assumed that *AD* was in compression but it is actually in tension!

We can use joint A or joint B.

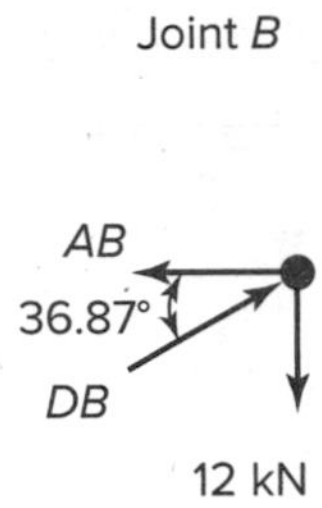

$$\Sigma F_x = 0 = DBcos(36.87°) - AB$$
$$AB = 16 \text{ kN (T)}$$

Note: We assumed that AB was in tension and got a positive so in the direction we assumed.

PRO TIPS

- **You must keep consistent with your assumption for whether a given member is in tension or compression as you go from joint to joint. For instance, if the same force vector (i.e., CD and AB) are on more than one FBD, the force vectors *must* be in opposite directions on each of the FBDs. In other words, the *CD* would go to the left on one and to the right on the other—even if the directions are simply a guess.**
- **Students often get confused about which forces are affecting global equilibrium and want to include things such as forces at pins or other internal loads. A great way to think about global equilibrium is to think of the entire system as a potato. It's just an amorphous shape with some external forces acting on it. The external forces are the reactions (see the FBD below).**

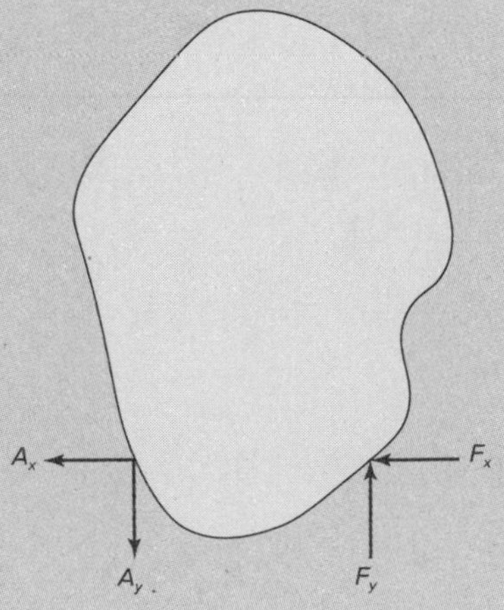

Find the force in each member of the truss and state whether it is in tension or compression.

Press pause on video lesson 48 once you get to the workout problem. Only press play if you get stuck.

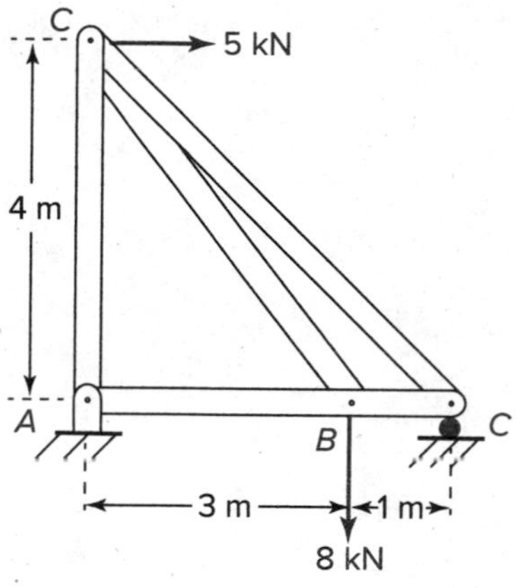

✓ TEST YOURSELF 6.3

SOLUTION TO TEST YOURSELF: Trusses, Frames, and Machines

Compute the forces in each member of truss as shown below; indicate tension, compression, or a zero force member.

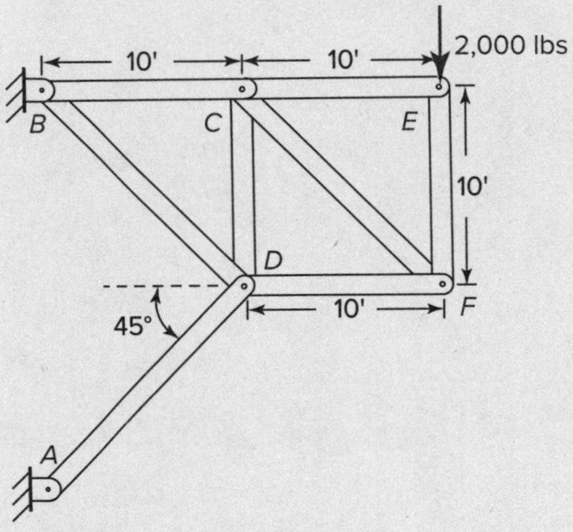

ANSWERS

CE = 0
EF = 2,000 lbs (C)
CD = 2,000 lbs (C)
BC = 2,000 lbs (T)
CF = 2,828.4 lbs (T)
AD = 2,828.4 (C)
DB = 0

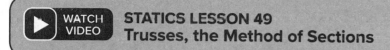

WATCH VIDEO **STATICS LESSON 49**
Trusses, the Method of Sections

Method of Sections

- The Method of Joints will work for *all* truss problems, but they take a lot of time to solve using this method. The Method of Sections will speed up this process.
- How do I know when to use the Method of Joints versus the Method of Sections?
  - Use the Method of Sections when a problem asks for the forces in specific members (not the forces in every single member).
  - If the problem says find the force in every single member, then just use Method of Joints.

  Recall: The Method of Joints uses only $\Sigma F_x = 0$ and $\Sigma F_y = 0$ to solve for forces in truss members since all forces on the FBD of a joint pass through the same point (i.e., since there is no distance, you cannot use the moment equations).

- The Method of Sections has an FBD where all forces do *not* pass through the same point. Thus, we can now use a moment equation to solve the FBD.
- Since we have the $\Sigma F_x = 0$, $\Sigma F_y = 0$, and $\Sigma M_z = 0$ available, we can find up to three unknowns. This is why we cannot cut through more than three unknowns on our FBD for these problems.

Method of Sections Recipe

STEP 1: Find global equilibrium (i.e., find the reactions at the supports).

STEP 2: Cut through members of interest.

Note: Do not cut through more than three members.

STEP 3: Draw a free body diagram of the easiest side.

STEP 4: Resolve the forces into Cartesian components.

STEP 5: Solve by taking sum of forces and/or moments.

EXAMPLE: METHOD OF SECTIONS

Find the force in members *BC, CH, HG*.

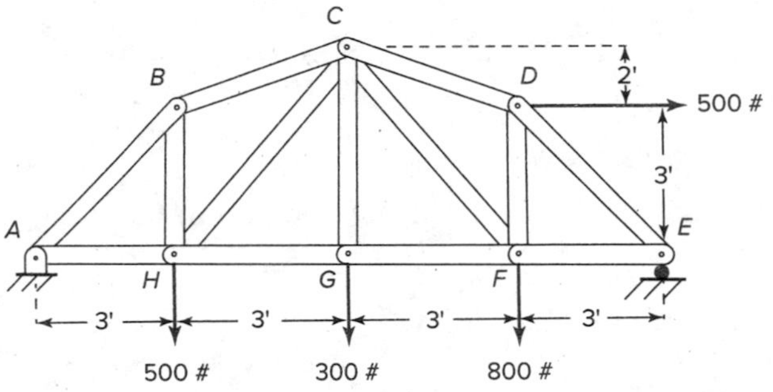

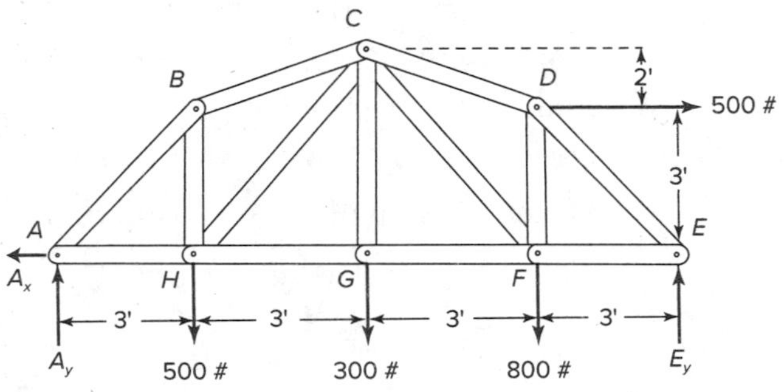

STEP 1: Find global equilibrium from the FBD shown on the right.

$\Sigma M_A = 0 = -500(3) - 300(6) - 800(9) - 500(3) + E_y(12)$

$E_y = 1{,}000\ \#$

$\Sigma F_y = 0 = A_y + E_y - 500 - 300 - 800$

$A_y = 600\ \#$

$\Sigma F_x = 0 = 500 - A_x$

$A_x = 500\ \#$

STEP 2: Cut through members of interest.

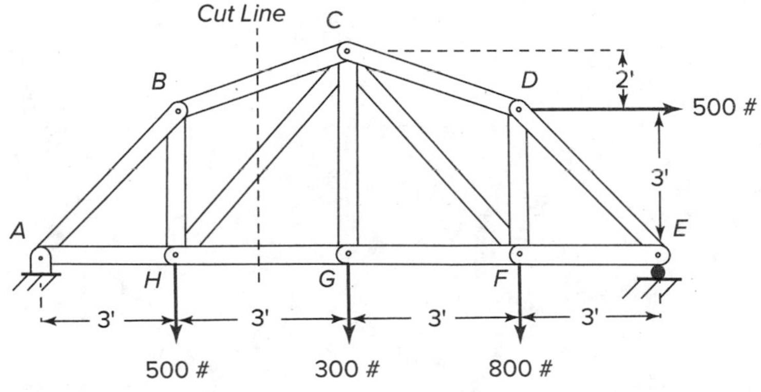

STEP 3: Draw an FBD of easiest side.

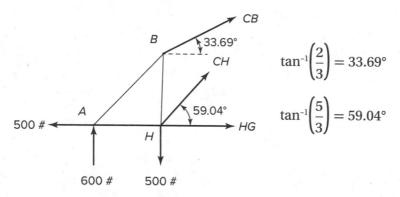

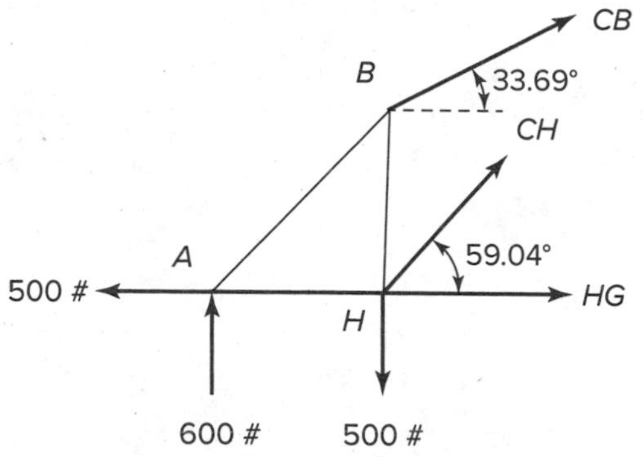

$$\tan^{-1}\left(\frac{2}{3}\right) = 33.69°$$

$$\tan^{-1}\left(\frac{5}{3}\right) = 59.04°$$

STEP 4: Resolve into Cartesian components.

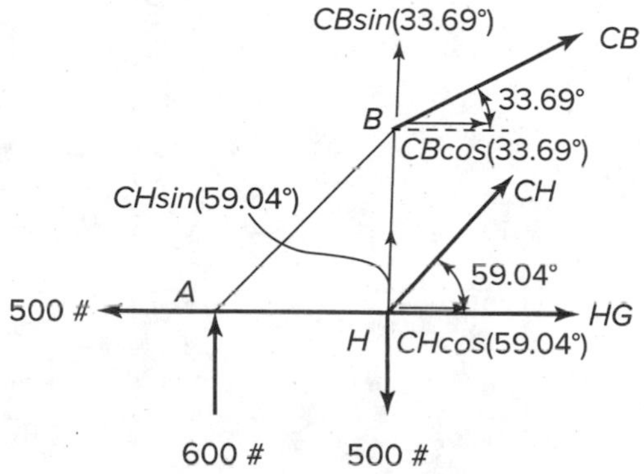

STEP 5: Solve by taking sum of forces and/or moments.

Note: We chose H to take our moment about since several of the unknowns pass through this point (i.e., no perpendicular distance). This eliminates them for being included in the moment equation making it easier to solve by hand. You can definitely take the moment about a different point; depending where you pick, it just might have two unknowns in the equation making it slightly more difficult to solve for equilibrium.

$\Sigma M_H = 0 = -600(3) - CBcos(33.69°)(3)$

$CB = -721.1\,\# = 721.1$ lbs (C)

Note: Assumed member CD in tension, actually in compression.

$\Sigma F_y = 0 = 600 - 500 + CBsin(33.69°) + CHsin(59.04°)$

$0 = 100 + (-721.1)sin(33.69°) + CHsin(59.04°)$

$300 = CHsin(59.04°)$

$CH = 350$ lbs (T)

$\Sigma F_x = 0 = HG + CHcos(59.04°) + CBcos(33.69°) - 500$

$0 = HG + 350cos(59.04°) + (-721.1)cos(33.69°) - 500$

$HG = 919.9$ lbs (T)

Find the force in members *GH, HE,* and *DE* and state whether the members are in tension or compression.

Press pause on video lesson 49 once you get to the workout problem. Only press play if you get stuck.

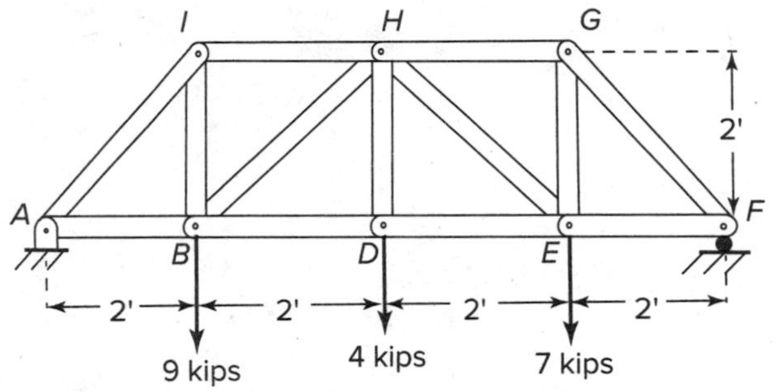

WATCH VIDEO

STATICS LESSON 50
Trusses, How to Find a Zero Force Member

Identifying zero force members will save you lots of time in solving a truss.

But don't worry if you miss one; you will still get a zero when you solve for the member of the truss by using our typical methods. What to look for:

- When an FBD has only three members and two members are on the same line of action, those two members will *always* have the same magnitude. The third member will *always* be zero, no matter what the angle.

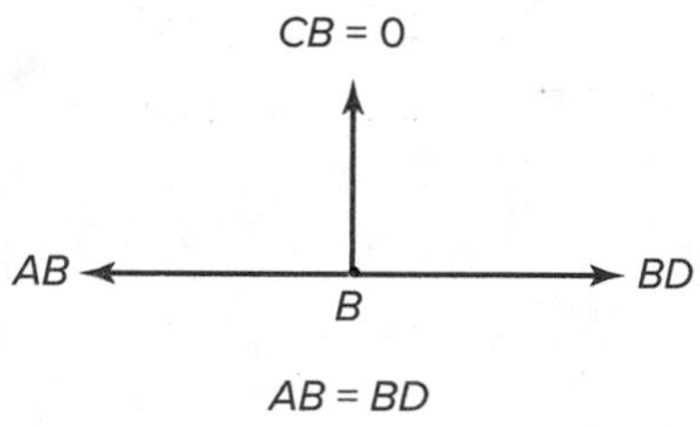

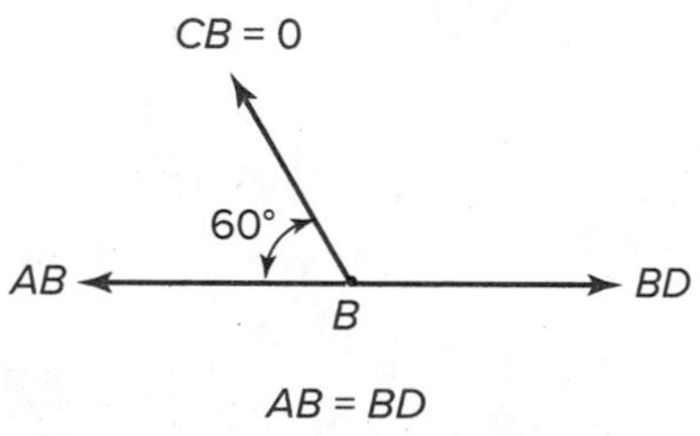

- When an FBD has only two members at any angle (other than on the same line of action), both of those members will *always* be zero.

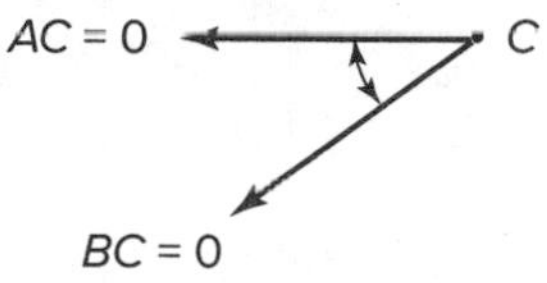

CHALLENGE QUESTION

Which of the members in the truss below are zero force members?

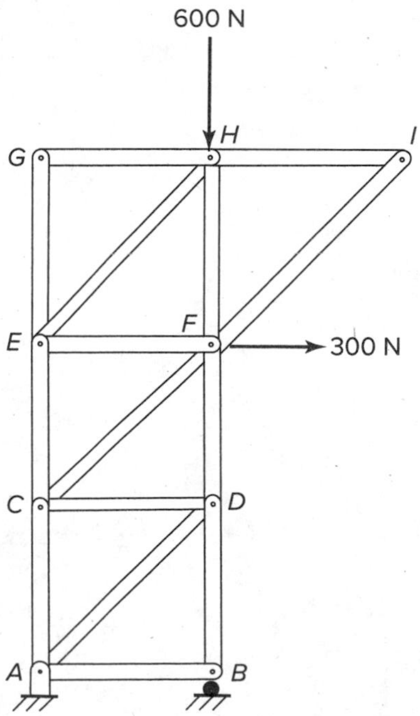

ANSWERS

BA, GH, GE, HI, FI, HE, EF, EC

Find the force in every member of the truss. State whether each member is in tension or compression. Don't forget to utilize the pro tips!

Press pause on video lesson 50 once you get to the workout problem. Only press play if you get stuck.

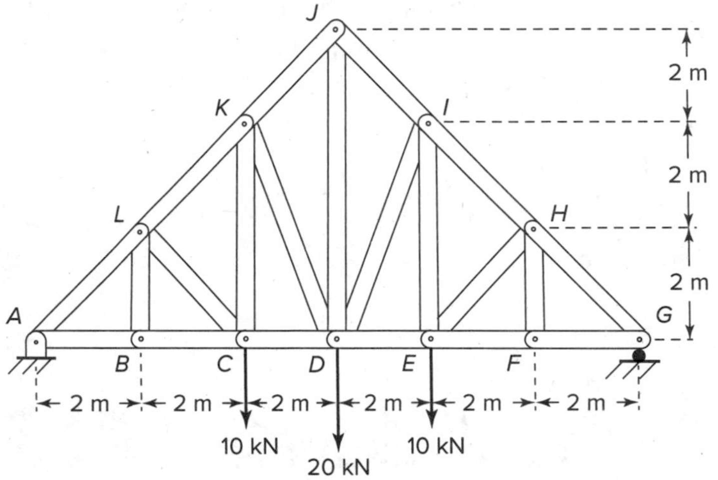

PRO TIPS

Look for zero force members:

- **Occurs when two forces are on the same line of action and they oppose each other. Then there is a third force that could be in any direction. The third force is always a zero force member.**
- **Occurs when an FBD has only two members at any angle (other than on the same line of action); both of those members will *always* be zero.**

TEST YOURSELF 6.4

SOLUTION TO TEST YOURSELF: Trusses, Frames, and Machines

Using the method of sections, compute the forces in members *DF, CB, CG,* and *BE* and indicate if they are tension, compression, or a zero force member.

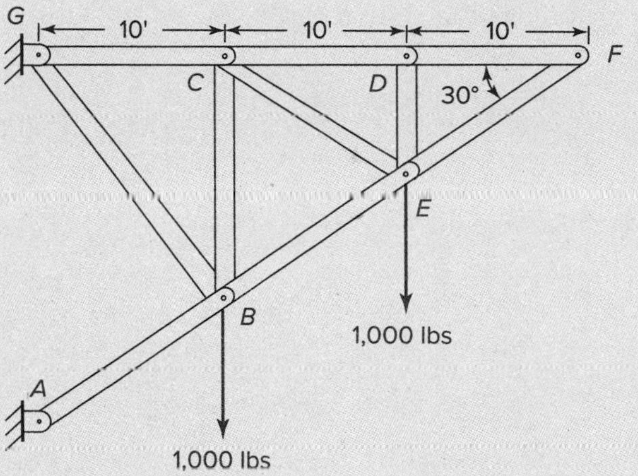

ANSWERS

$DF = 0$

$CB = 500$ lbs (C)

$BE = 1{,}000$ lbs (C)

$CG = 866$ lbs (T)

WATCH VIDEO

STATICS LESSON 51
Trusses, Method of Sections, Truss Tips and Tricks

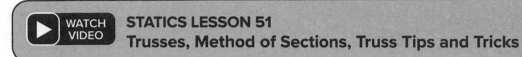

Find forces in members *BC, BE,* and *FE*. State whether each member is in tension or compression. Press pause on the video once you get to the workout problem. Only press play if you get stuck.

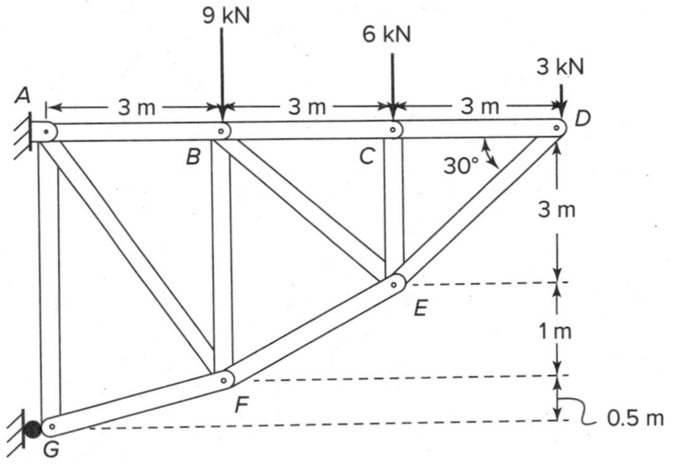

WATCH VIDEO

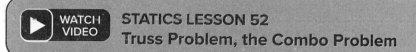

STATICS LESSON 52
Truss Problem, the Combo Problem

Find the forces in members *IJ*, *CJ*, *CB*, and *CI* below. State whether each member is in tension or compression.

Press pause on the video once you get to the workout problem. Only press play if you get stuck.

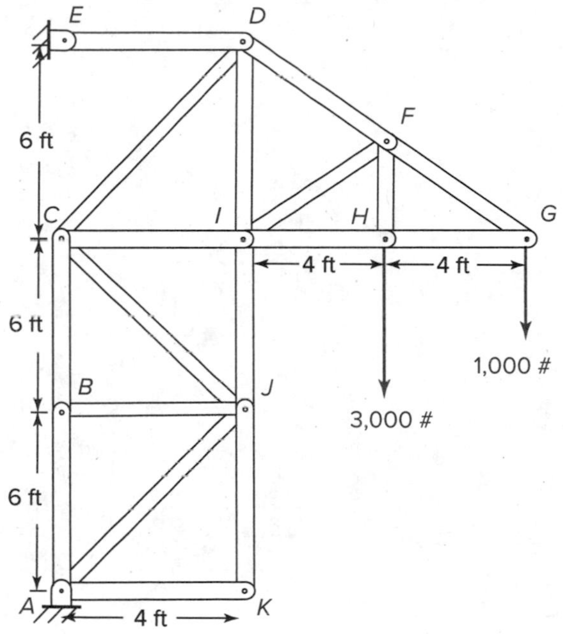

✓ TEST YOURSELF 6.5

SOLUTION TO: Trusses, Frames, and Machine

The ceiling truss supports the roof (and any loads on it). Now, the designer wants to hang a very heavy light fixture (1,000 lbs) directly at the center of the bottom member of the truss. Find the forces in the two members closest to the left support (*AB* and *AD*). The room width is 16', height of the truss is 6', and the two diagonal members meet the top members at a right angle.

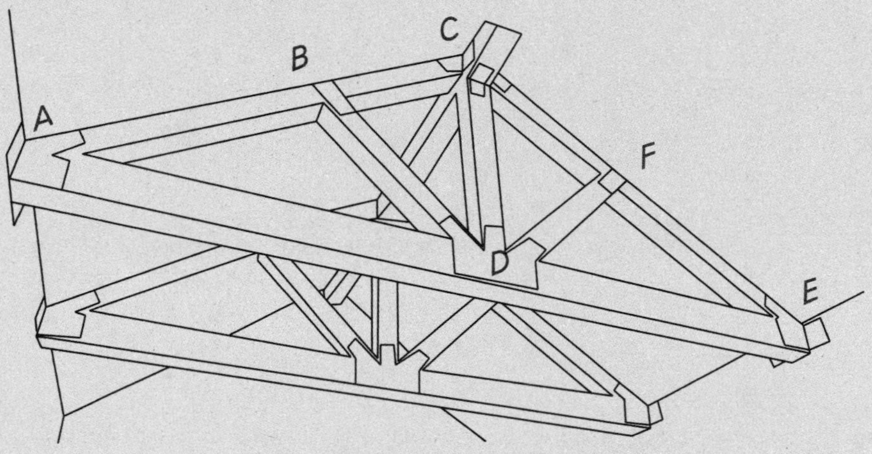

ANSWERS
AB = 833.3 lbs (C)
AD = 666.6 lbs (T)

✓ TEST YOURSELF 6.6

SOLUTION TO: Trusses, Frames, and Machine

Find the forces that develop in the members of the given truss.

Shown below is an idealization of the actual bridge. We choose this simpler version to make our analysis easier.

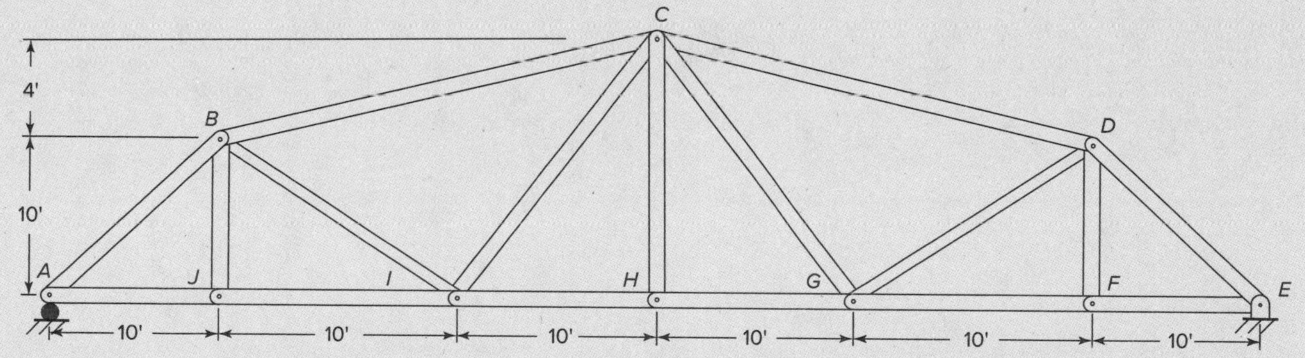

To determine the internal forces in the bridge, we have to know how bridge is loaded. Let's assume a truck to have the following loads:

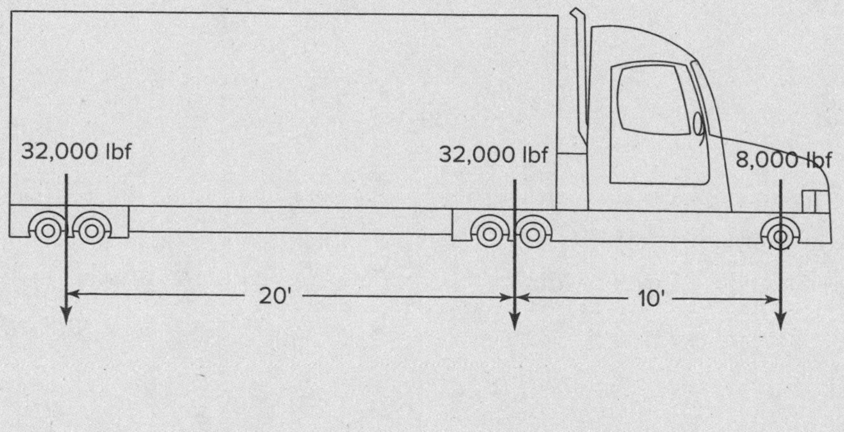

For this analysis let's assume this bridge is a single lane bridge with one truck. Because the bridge consists of two trusses (one on each side of the roadway), we are only looking at one side (left or right). Because of symmetry, each truss on the side of the roadway will take half the total load of the truck. The truck is positioned as below:

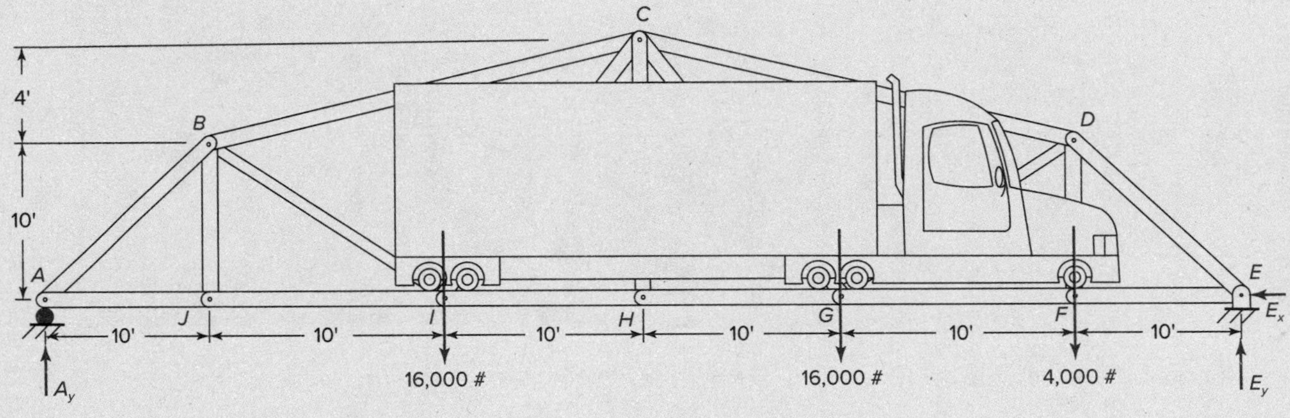

ANSWERS

$AJ = 16.7$ kips (T); $JI = 16.7$ kips (T); $IH = 24.3$ kips (T); $HG = 24.3$ kips (T); $GF = 19.3$ kips (T); $FE = 19.3$ kips (T); $ED = 27.3$ kips (C); $DC = 29.5$ kips (C); $CB = 28.3$ kips (C); $BA = 23.57$ kips (C); $BJ = 0$; BI = 15.7 kips (T); $CI = 6$ kips (T); $CH = 0$; $CG = 7.9$ kips (T); $DG = 13.5$ kips (T); $DF = 4$ kips (T)

WATCH VIDEO **STATICS LESSON 53**
Frame Problem with 2 (Two) Force Members

FRAMES

Frame Problems

- Consist of two force members as well as multiforce members
- Extremely similar to truss problems except trusses are made entirely of two force members
- Note the similarities in the recipes for frames versus trusses

Method of Solving Frame Problems Recipe

STEP 1: Look for two force members.

- What's a two force member?
  - A member that is pin connected on both ends with no other forces or moments on it.
- Why is it important to find two force members?
  - Two force members reduce the number of unknowns from two to one because the direction is always known since it acts in a direction along the length of the member.
- Should I draw an FBD of a two force member?
  - No, because all you will learn is that $F = F$.

STEP 2: Find global equilibrium where possible (i.e., find the reactions at the supports).

Even if I can only find one piece of global equilibrium, such as an x-reaction or a y-reaction but can't find anything else, that one piece of information may be important for solving the whole problem.

For some problems, global equilibrium won't be able to be found, so in those cases move on to the next recipe step.

STEP 3: Blow the frame apart. Draw the FBD of every part of the frame except for the two force members as they won't help.

STEP 4: Solve.

PITFALL

Remember, to turn a uniformly distributed load into an equivalent point force, take the magnitude of the distributed load and multiply it by the length it acts over to find the value of the point force load. Then apply that load at the centroid.

EXAMPLE: FRAME PROBLEM

Find the forces in member *ABC*.

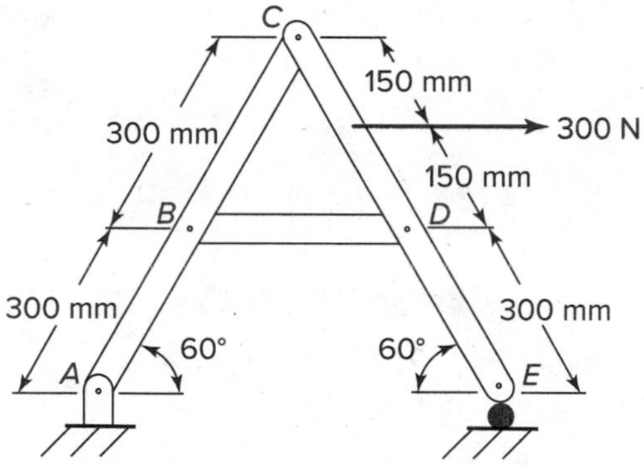

STEP 1: Draw the FBD of the entire system and solve for global equilibrium.

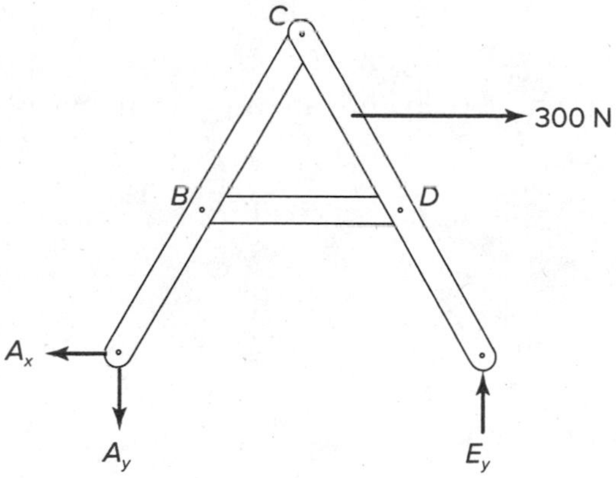

$$450sin(60°) = 389.7$$

$$\Sigma M_A = 0 = E_y(600) - 300(389.7)$$
$$E_y = 194.85\text{ N}$$
$$\Sigma F_y = 0 = E_y - A_y$$
$$A_y = 194.85\text{ N}$$
$$\Sigma F_x = 0 = 300 - A_x$$
$$A_x = 300\text{ N}$$

STEP 2: Identify any two force members. Recall, we do not draw FBDs of two force members because we can't learn anything from them. However, they are important because they take us from two unknowns to use one unknown. It is also handy if you can identify if they are in tension or compression. If you cannot, it's okay to just guess.

In this case, member *BD* is a two force member as it is pinned connected at both ends and has no forces or moments applied along its length.

STEP 3: Explode the frame and draw an FBD of each member.

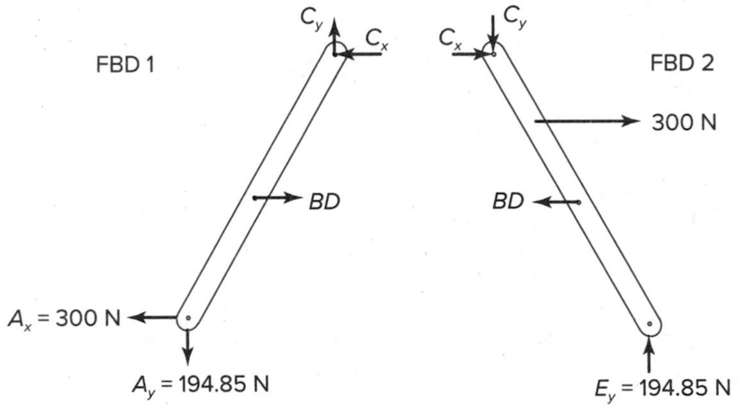

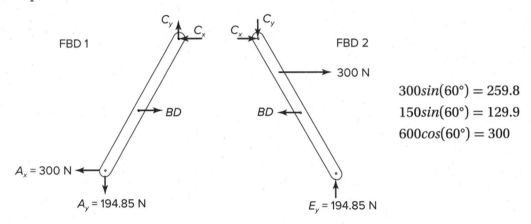

$300sin(60°) = 259.8$

$150sin(60°) = 129.9$

$600cos(60°) = 300$

STEP 4: Solve the FBDs for all needed information. We can find one value from one of the free bodies to help solve the second free body.

From FBD 1:

$\Sigma F_y = 0 = C_y - A_y$

$C_y = 194.85$ N

From FBD 2:

$\Sigma M_C = 0 = 300(129.9) - BD(259.8) + 194.85(300)$

$$BD = \frac{38{,}970 + 58{,}455}{259.8} = 375 \text{ N}$$

Finally FBD 2:

$\Sigma F_x = 0 = 300 - 375 + C_x$

$C_x = 75$ N

So for member *ABC*:

For the final answer lets look at beam *ABC*. Report the force as being negative if pointing to the left and positive if pointing to the right. Report the force as being positive if upward and downward if negative.

$A_y = -194.85$ N

$A_x = -300$ N

$BD = 375$ N (T)

$C_x = -75$ N

$C_y = 194.85$ N

Find the forces on bar *BCD* and the backboard.

Press pause on video lesson 53 once you get to the workout problem. Only press play if you get stuck.

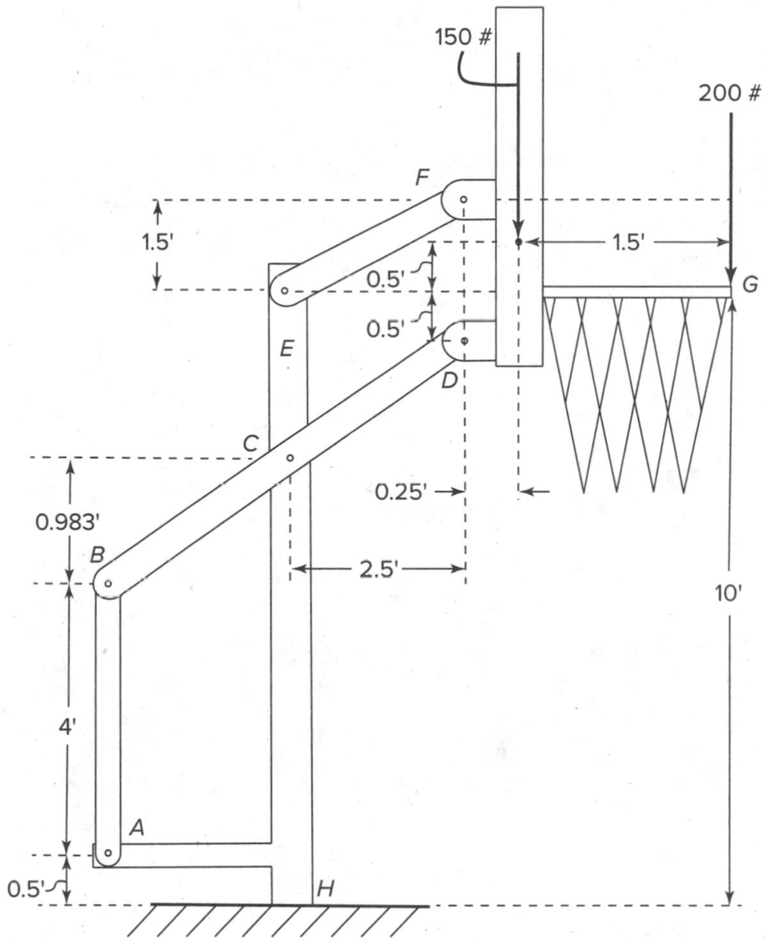

WATCH VIDEO

STATICS LESSON 54
More Difficult Frame Problem

Find all forces acting on member *DEF* of the billboard. ***Hint:*** You can't solve for all of the reaction forces.

Press pause on the video once you get to the workout problem. Only press play if you get stuck.

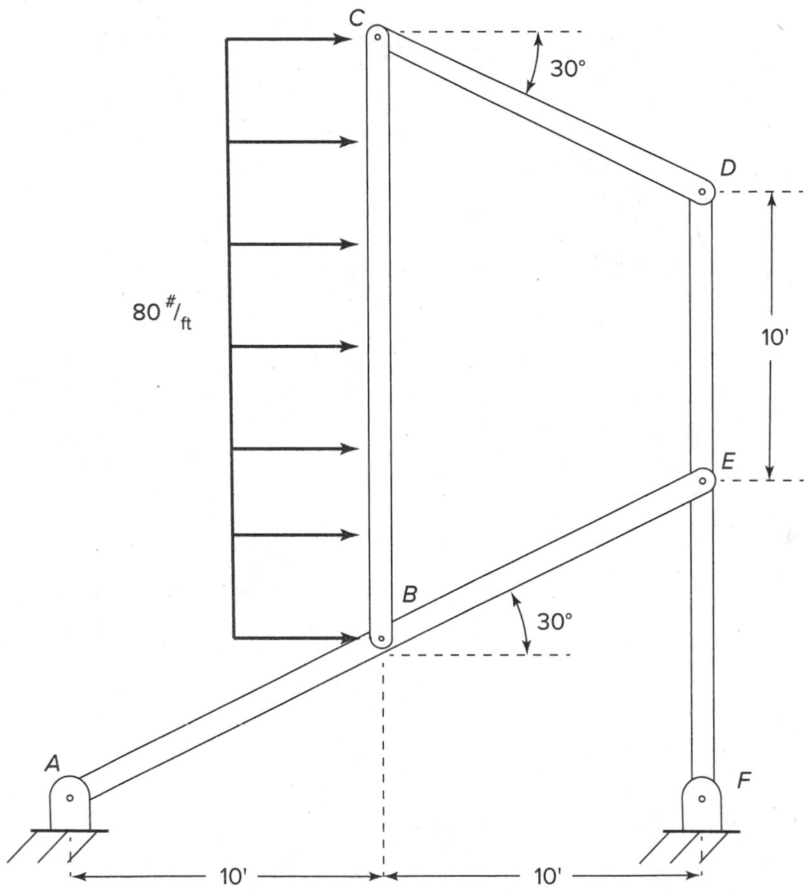

✓ TEST YOURSELF 6.7

SOLUTION TO TEST YOURSELF: Trusses, Frames, and Machines

For the frame shown below, determine the support forces at the wall.

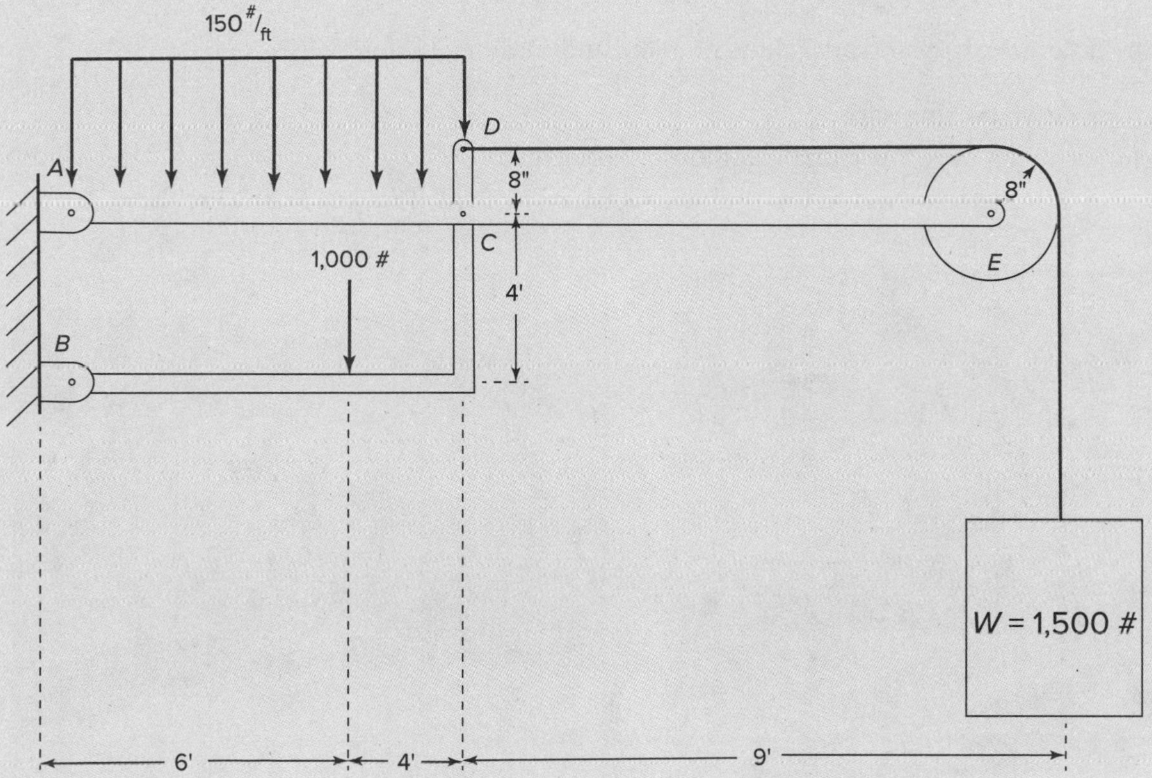

ANSWERS

A_x = 10,500 lbs acting left, A_y = 500 lbs downward; B_x = 10,500 lbs acting right, B_y = 4,500 lbs upward

WATCH VIDEO **STATICS LESSON 55**
Machine Problem, You Must Know How to Do This!

MACHINES

What's the difference between a machine problem and a frame problem?

- Machines have moving parts.
- Machines do not have any global equilibrium to solve for.
- Solved in the exact same way as frame problems, so see Method of Solving Frame Problems Recipe.

Method of Solving Machine Problems Recipe

STEP 1: This recipe is almost identical to frame problems, but the only difference is that machine problems do not have global equilibrium. Therefore, Step 1 will be to look for two force members.

STEP 2: Blow the machine apart. Draw the FBD of every part of the machine except for the two force members as they won't help.

STEP 3: Solve.

EXAMPLE: MACHINE PROBLEM

If the bucket holds 3,000 N of dirt, find the forces on member *BEG*. Ignore the weight of the bucket. Note the dirt's centroid position is shown below.

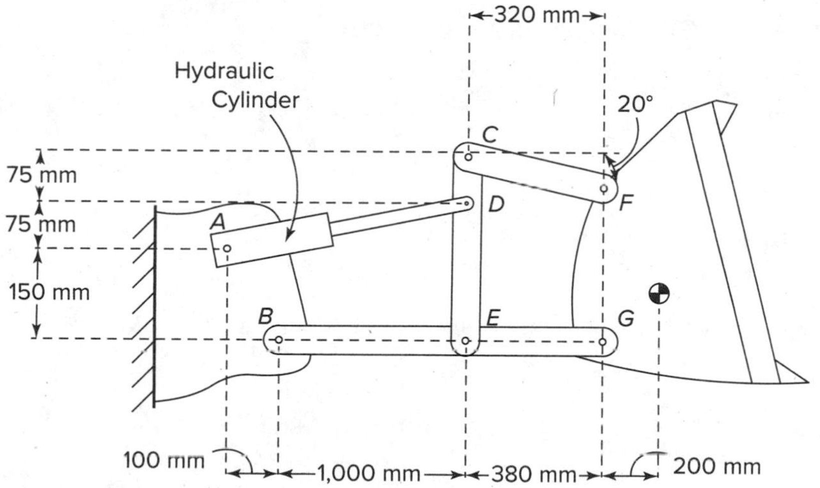

STEP 1: Identify any and all two force members. In this system, two force members are *CF* and *AD*. Note hydraulic cylinders are always two force members and can be either in tension or compression. Based off the bucket's position, members *CF* and *AD* are in tension.

STEP 2: Explode the system and draw an FBD of the remaining parts (do not draw FBD of two force members).

$$\tan^{-1}\left(\frac{75}{1{,}100}\right) = 3.9°$$

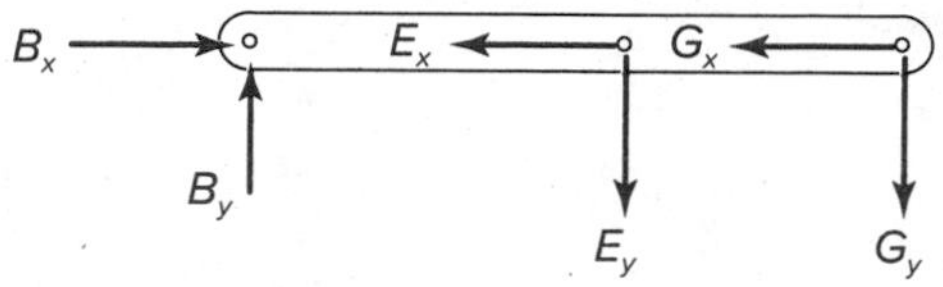

FBD 1

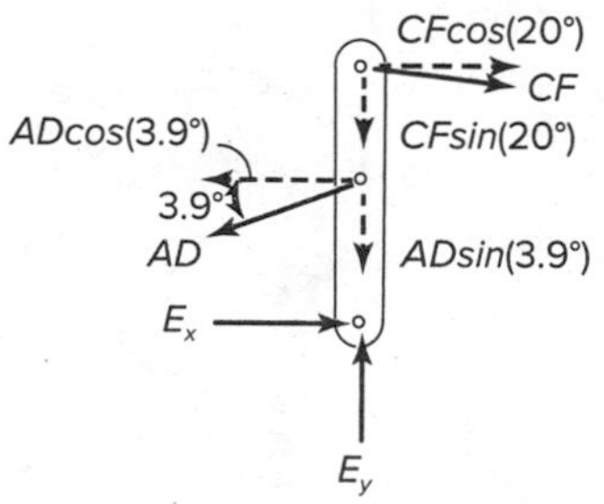

FBD 2

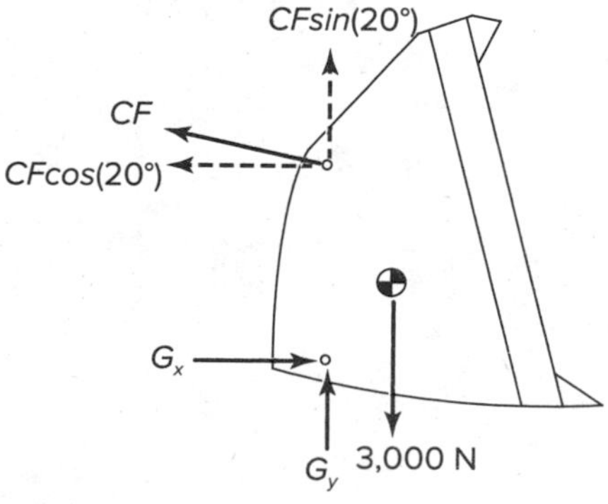

FBD 3

STEP 3: Solve FBDs using information from one free body to solve the next until you have everything needed. Let's start with FBD 3.

$\Sigma M_G = 0 = CFcos(20°)(300 - 320tan(20°)) - 3{,}000(200)$
$CF = 3{,}479\ \text{N}$

$\Sigma F_x = 0 = G_x - CFcos(20°) = G_x - 3{,}479cos(20°)$
$G_x = 3{,}269\ \text{N}$

$\Sigma F_y = 0 = G_y - 3{,}000 + CFsin(20°) = G_y - 3{,}000 + 3{,}479sin(20°)$
$G_y = 1{,}810\ \text{N}$

Next solve for FBD 2:

$\Sigma M_E = 0 = -CFcos(20°)(300) + ADcos(3.9°)(225)$
$= -(3{,}479)cos(20°)(300) + ADcos(3.9°)(225)$
$AD = 4{,}369\ \text{N}$

$\Sigma F_x = 0 = CFcos(20°) - ADcos(3.9°) + E_x = 3{,}479cos(20°) - 4{,}369cos(3.9°) + E_x$
$E_x = 1{,}090\ \text{N}$

$\Sigma F_y = 0 = -CFsin(20°) - ADsin(3.9°) + E_y = -(3{,}479)sin(20°) - 4{,}369sin(3.9°) + E_y$
$E_y = 1{,}487\ \text{N}$

Finally, solving FBD 1 we can easily find B_x and B_y, which are the last two remaining unknowns.

$\Sigma F_x = 0 = B_x - E_x - G_x = B_x - 1{,}090 - 3{,}269$
$B_x = 4{,}359\ \text{N}$

$\Sigma F_y = 0 = B_y - E_y - G_y = B_y - 1{,}487 - 1{,}810$
$B_y = 3{,}297\ \text{N}$

TRUSSES, FRAMES, AND MACHINES

Find the force that the pliers exert on the bolt.

Press pause on video lesson 55 once you get to the workout problem. Only press play if you get stuck.

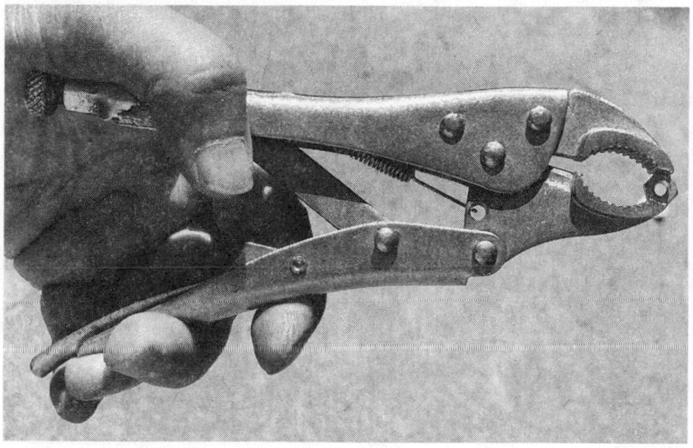

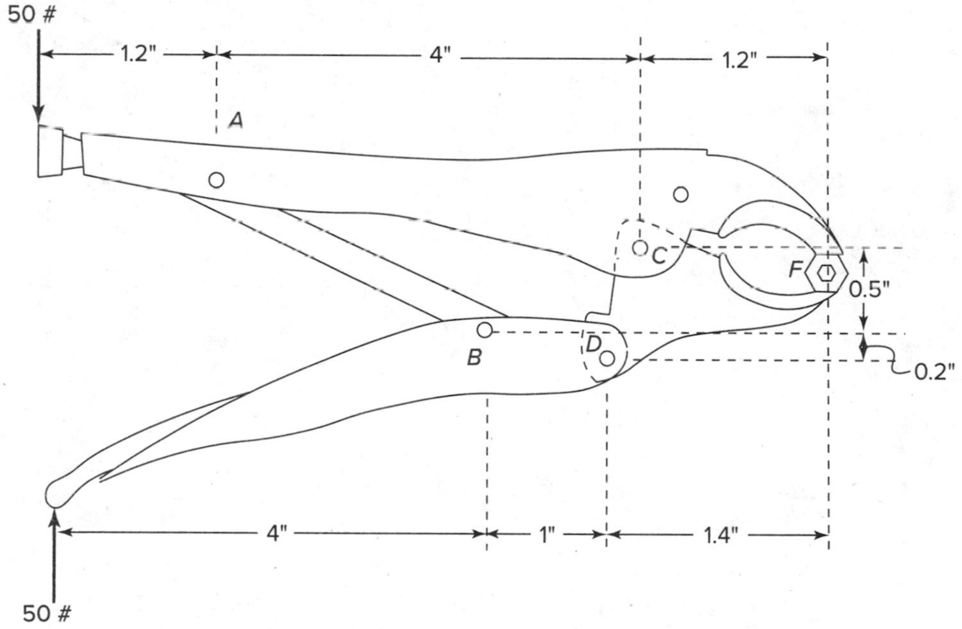

✓ TEST YOURSELF 6.8

SOLUTION TO TEST YOURSELF: Trusses, Frames, and Machines

For the backhoe pictured below (left) and idealized sketch (right), compute the forces at point *A, B, C,* and the forces in the cylinders at *HD* and *CH.*

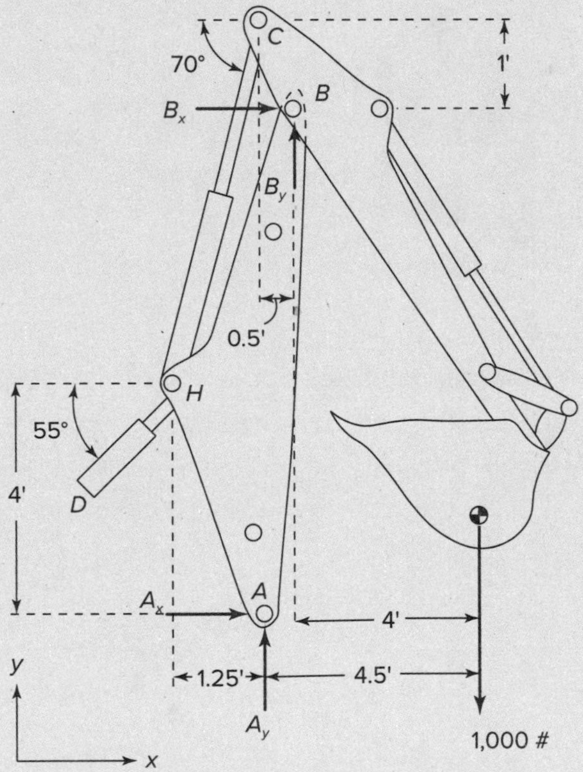

ANSWERS

At *HD*: 1,356 lbs along the cylinder; *CH* = 4,927 lbs, A_x = 777.7 lbs to the right, A_y = 2,110 lbs upward, B_x = 1,685 lbs to the right, B_y = 5,630 lbs upward

WATCH VIDEO

STATICS LESSON 56
Pulley Palooza Problems

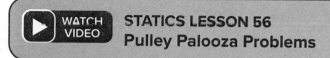

Pulleys are simple machines, and thus, pulley problems are generally included with frames and machines in statics.

All pulleys primarily consist of a wheel with a rope around it. Depending on the pulley arrangement, a pulley may either allow:

- A heavy object to be lifted with less force, called mechanical advantage
- Simply redirect the force applied in a different direction, or
- A combination of both of the above

The different types of pulley systems include:

- **Fixed.** Pulley attached to a supporting structure while rope is free.
- **Movable.** Pulley is not attached to supporting structure whereas rope is attached.
- **Compound.** A combination of fixed and movable.
- **Block and tackle.** Different from compound in that you have mechanical advantage, whereas compound only changes the direction of the force.

Types of Pulleys

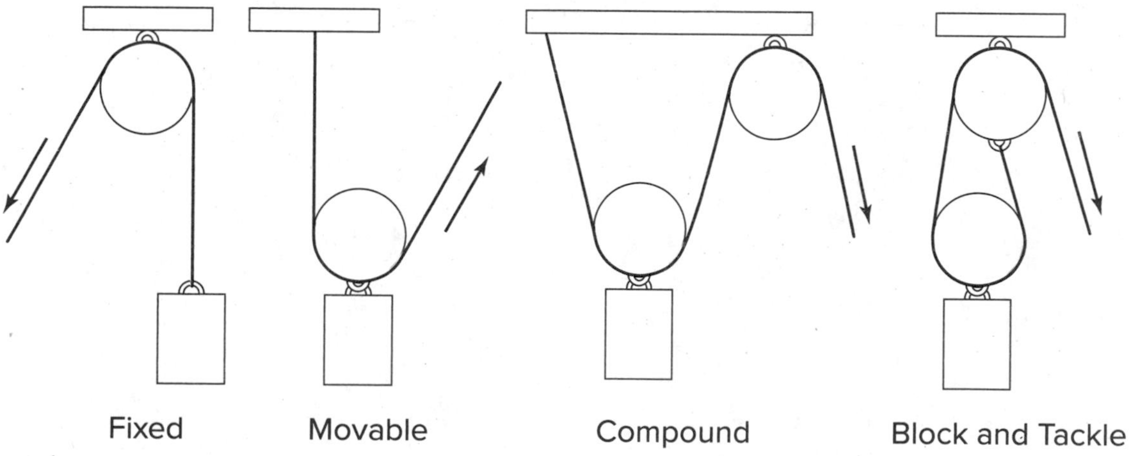

Method of Solving Pulley Problems Recipe

STEP 1: Draw FBD by cutting through ropes until the pulley hits the floor. The key is you can't cut through more than one rope at a time. If you do, then you will have two unknowns. (A key assumption is that the pulley is frictionless.)

STEP 2: If there are multiple pulleys, then start with one. Write your equilibrium equation. Go to the next pulley, writing equilibrium equations again including information learned from the previous pulley.

Note: For each free body diagram, you should be able to write $\Sigma F_y = 0$. Once you have written that equation for each of your FBDs, then you should be able to form a system of equations and use your system solver in your calculator.

STEP 3: Solve.

PRO TIP

Unless otherwise stated, assume pulleys are frictionless. When you have a rope going around a pulley or you can follow a rope through a system of pulleys, the tension in the rope will be the same at all points.

EXAMPLE: PULLEY PROBLEM

For the given system, find the weight of W_1 and W_2 to maintain equilibrium.

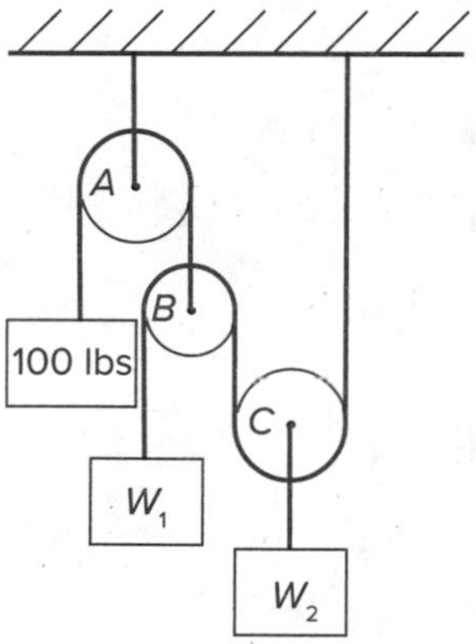

STEP 1: Isolate pulley *A*. Draw an FBD and solve it. Remember you cannot cut through more than one rope at a time.

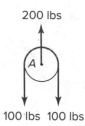

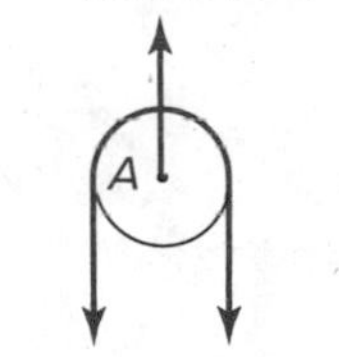

The rope of the pulley *has* to have the same tension on both sides of the pulley. Since this is the case, then the force at point *A* is equal to 200 #.

STEP 2: Continue isolating pulleys and working your way across the pulley system. Note, nothing other than the sum of the forces is usually required to solve these problems.

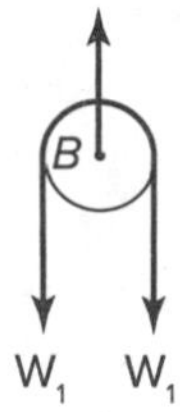

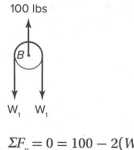

$\Sigma F_y = 0 = 100 - 2(W_1)$

$W_1 = 50\ \#$

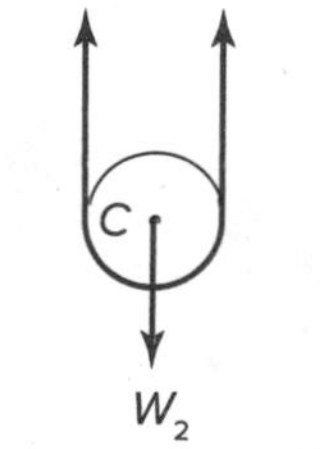

$\Sigma F_y = 0 = 2(50) - W_2$

$W_2 = 100\ \#$

TRUSSES, FRAMES, AND MACHINES

Find the tension in the rope for the following pulleys. ***Note:*** When a rope passes through the centroid of the pulley (shown with a dot), that means that it is rigidly attached to the rope.

Press pause on video lesson 56 once you get to the workout problem. Only press play if you get stuck.

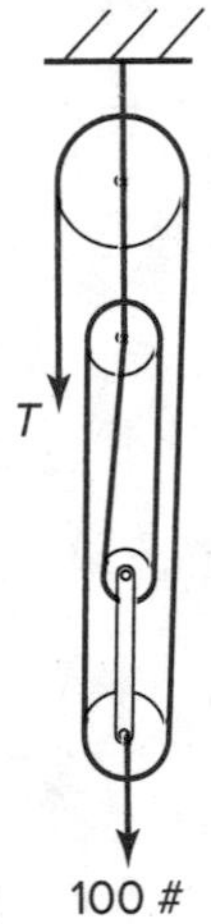

a)

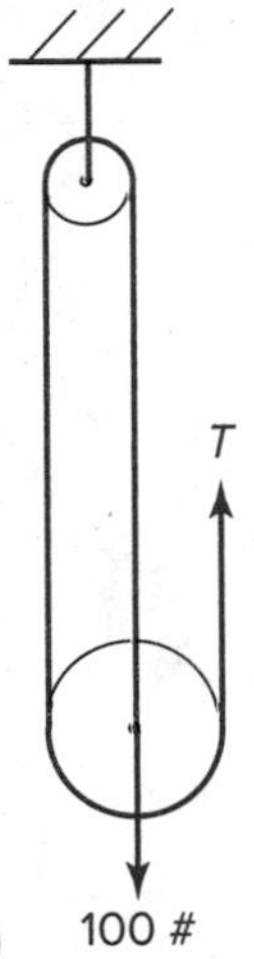

b)

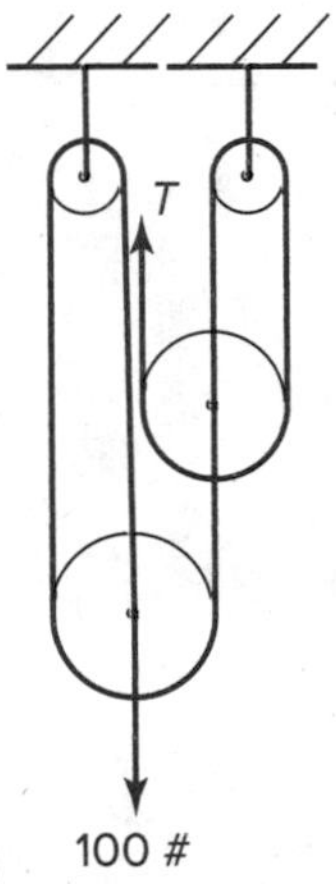

c)

Find the force in the rope and the force of the man on the platform if the man is 90 kg and the platform is 10 kg.

If $T_{max} = 1{,}500$ N, find W_{max}.

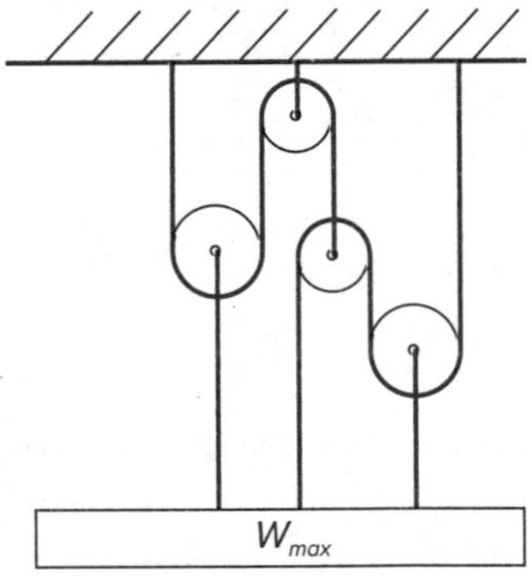

TEST YOURSELF 6.9

SOLUTION TO TEST YOURSELF: Trusses, Frames, and Machines

If W_1 is 100 lbs, what is W_2 to keep the system in equilibrium?

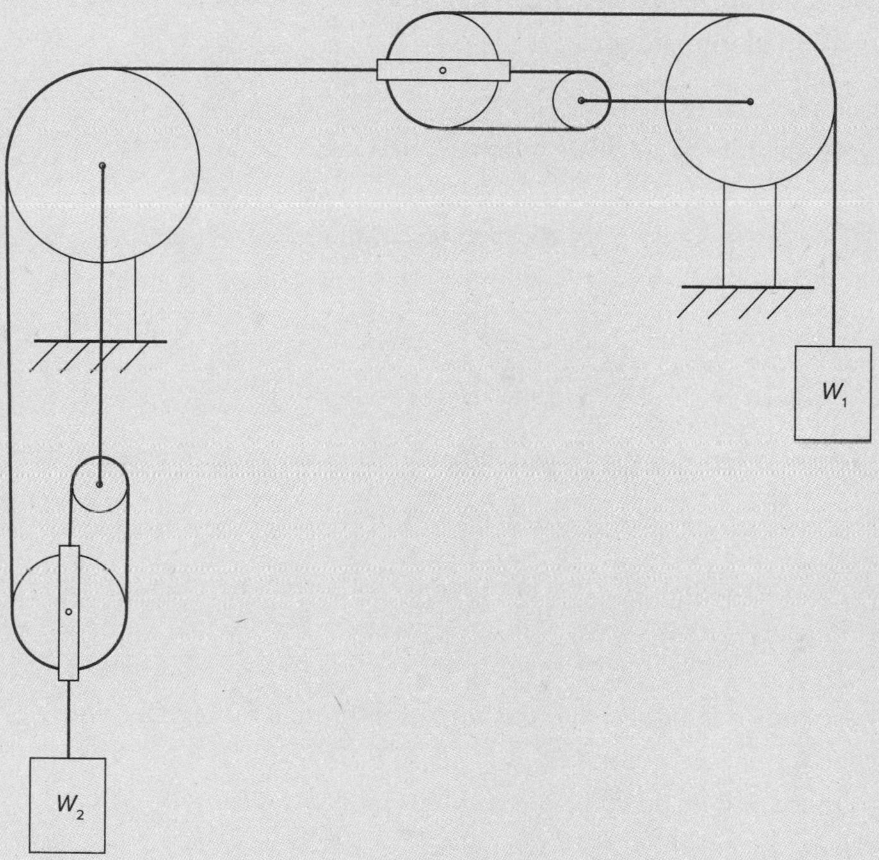

ANSWER
900 lbs

- **Trusses.** Composed entirely of two force members and always have global equilibrium.
- **Frames.** Composed of two force members and multiforce members and have global equilibrium, but you can't always solve for it immediately.
- **Machines.** Have moving parts and are composed of two force members and multiforce members; do not have global equilibrium to solve for.

All of these systems absolutely rely on the ability to draw FBD.

Main takeaways from Level 6 includes how to calculate internal forces for:

- Trusses
- Frames
- Machines

Trusses Summary

- Two force members are pin connected at the ends with no other forces or moments applied at the middle of the beam.
- Method of Joints
  - Analyze one joint of the truss at a time, and that joint can never have more than two unknowns; solve one joint and work your way across the truss.
- Method of Sections
  - Make a cut through the member of interest; be sure not to cut through more than three unknowns as you only have three equations ($\Sigma F_x = 0$, $\Sigma F_y = 0$, $\Sigma M_z = 0$).
- Compression members point to the joint and tension members point away from the joint.

Frames Summary

- It is important to identify two force members so that it takes you from two unknowns to one unknown (as you know the direction). (***Note:*** Drawing FBDs of two force members isn't very helpful as you simply prove that $F = F$.)
- If you try to solve these problems and you have too many unknowns, so you probably missed a two force member.

Machine Summary

- Machine problems have moving parts such as backhoe buckets, tractor loaders, pliers, bolt cutters, locking pliers, rose bush clippers, etc.
- It is not necessary or helpful to solve for the global equilibrium (machines are not connected to the world, so this is helpful when identifying machines); otherwise solve exactly like a frame problem.

RECIPES

- Method of Joints Recipe
- Method of Sections Recipe
- Method of Solving Frame Problems Recipe
- Method of Solving Machine Problems Recipe
- Method of Solving Pulley Problems Recipe

PRO TIPS

Trusses

- When solving problems, if you get a negative sign when solving for *AB*, then it is opposite to what you assumed! Sometimes we won't know which direction the force acts, so you have to assume a direction. If you get a negative sign when solving, then it is opposite to what you assumed.
- Always look for a symmetric truss with a symmetric load. You can then quickly add up all of the downward forces and divide by 2 to immediately obtain the global equilibrium.
- You must keep consistent with your assumption for whether a given member is in tension or compression as you go from joint to joint. For instance, if the same force vector (i.e., CD and AB) are on more than one FBD, the force vectors *must* be in opposite directions on each of the FBDs. In other words, the *CD* would go to the left on one and to the right on the other—even if the directions are simply a guess.

(continued on next page)

PRO TIPS

Trusses *(continued)*

- Students often get confused about which forces are affecting global equilibrium and want to include things such as forces at pins or other internal loads. A great way to think about global equilibrium is to think of the entire system as a potato. It's just an amorphous shape with some external forces acting on it. The external forces are the reactions.
- Look for zero force members:
  - Occurs when two forces are on the same line of action and they oppose each other. Then there is a third force that could be in any direction. The third force is always a zero force member.
  - Occurs when an FBD has only two members at any angle (other than on the same line of action); both of those members will *always* be zero.

Pulleys

- Unless otherwise stated, assume pulleys are frictionless. When you have a rope going around a pulley or you can follow a rope through a system of pulleys, the tension in the rope will be the same at all points.

PITFALL

Frames

- Remember, to turn a uniformly distributed load into an equivalent point force, take the magnitude of the distributed load and multiply it by the length it acts over to find the value of the point force load. Then apply that load at the centroid.

Level 7

Internal Forces in Beams

WATCH VIDEO

STATICS LESSON 57
Introduction to Internal Forces *M*, *N*, *V*

In statics, there are three types of loads:

1. **External loads or applied loads.** Concentrated forces, distributed loads, or concentrated moments applied to the system.
2. **Reaction forces or moments.** Reactions at supports caused by the external loads or moments.
3. **Internal forces and moments.** Axial, shear, and/or bending *within* beams and other rigid bodies caused by external forces.

To analyze these internal loads, section (i.e., cut through) the beam.

Three internal forces:

- ***M* – Bending Moment.** Causes beam to bend with curvature.
- ***N* – Normal Force.** Causes the beam to extend or compress along the length (similar to what we saw with trusses).
- ***V* – Shear Force.** Induces a tearing or shearing in the beam.

Bending moment (*M*)

Pulling force (*N*)

Shearing force (*V*)

PRO TIP

When cutting a beam, it generates three components: *M*, *N*, and *V*. Otherwise, it can be treated as a regular equilibrium problem. You can remember with my handy (silly) jingle . . .

♫♫♩♩♩ Every time you cut the beam, you must have an *M*, *N*, *V*!

The positive sign convention for M, V, and N is given below for either side of the cut:

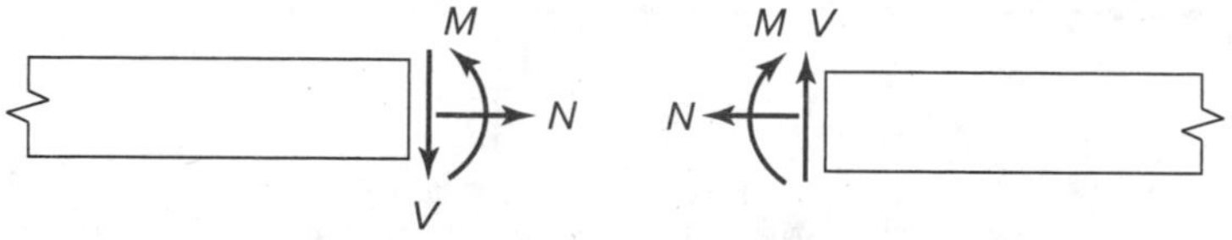

PRO TIP

When cutting a beam and using the positive sign convention above, feel free to use either free body diagram (right or left side) as you will get the same answer. *Note:* All answers to internal force questions should have signs in accordance with the positive sign convention.

PITFALL

It is easy to get the positive convention for the shear force, *V*, in the wrong direction. When using the left-side free body diagram, a trick for remembering the sign convention is "Don't get *left* at the store or you'll feel *down*." Silly, right? But I bet you'll remember it!

Solving Internal Forces at a Given Point Recipe

STEP 1: Find global equilibrium. Draw a free body diagram of the beam and solve for the reactions.

STEP 2: Section (cut) the beam at the point of interest and draw the FBD of either side of the beam. Be sure to add M, N, and V to the cut surface of your FBD.

Note: If there is a concentrated force at the point of interest, you will have to cut the beam "one molecule" to the right or left of that force. This will always be clearly stated in the problem.

STEP 3: Use either free body diagram (either one should give you the exact same answers) and solve for M, N, and V (the internal forces).

EXAMPLE: INTERNAL FORCES

For the given loaded beam, find the internal forces on the beam at a point just to the left of the 600 lbs concentrated load.

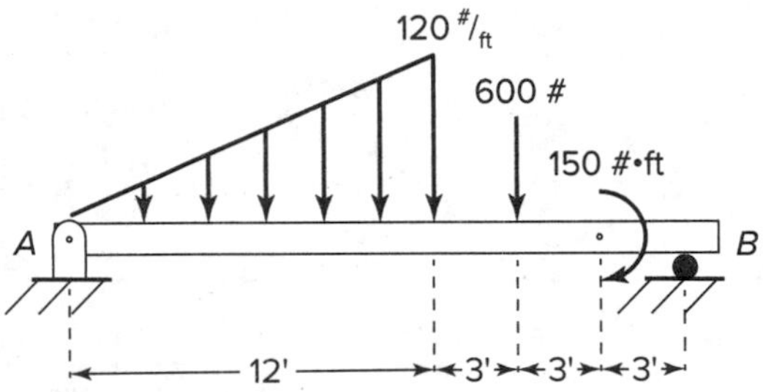

STEP 1: Find global equilibrium. Draw a free body diagram of the beam and solve for the vertical reactions at A_y and B_y (note, the horizontal reaction at A is zero because there are no horizontal forces). Don't forget that the 120 #/ft triangular distributed load can be converted into a point force via finding the "area" of the distributed load and applying it at the centroid for a triangle. In this case the "area" is $(\frac{1}{2})(120)(12) = 720$ # and it is applied at $(\frac{2}{3})(\text{base}) = (\frac{2}{3})(12) = 8'$.

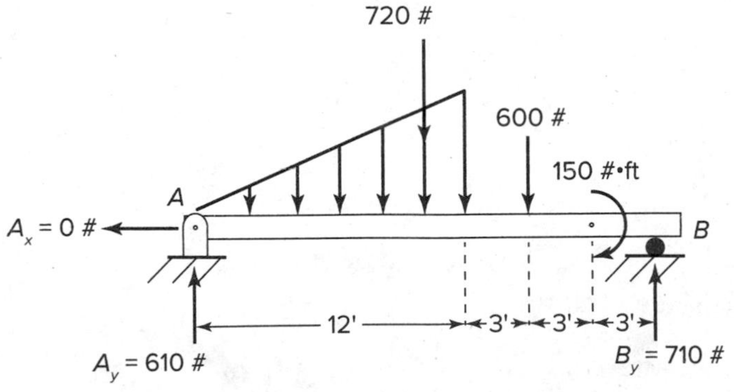

$$\Sigma M_A = 0 = -720(8) - 600(15) - 150 + B_y(21)$$
$$B_y = 710\ \#$$

$$\Sigma F_x = 0 = A_x$$
$$A_x = 0$$

$$\Sigma F_y = 0 = B_y + A_y - 720 - 600 = 710 + A_y - 720 - 600$$
$$A_y = 610\ \#$$

STEP 2: Section (cut) the beam at the point of interest and draw the FBD of either side of the beam.

Here are the diagrams for both sides for reference. Note the use of the positive sign convention for the internal loads.

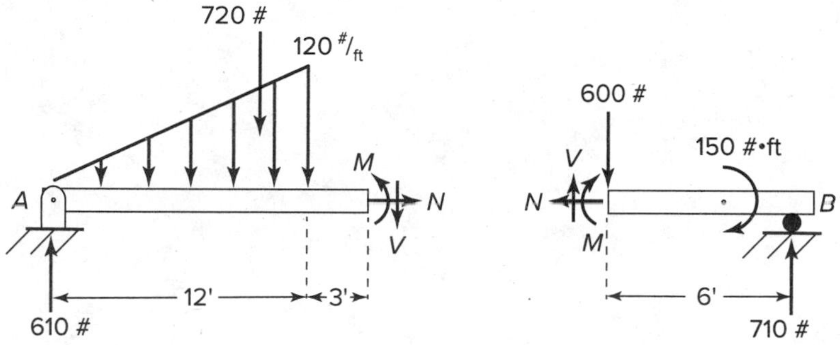

STEP 3: Use either free body diagram (either one should give you the exact same answers) and solve for *M*, *N*, and *V* (the internal forces).

Using the left free body diagram:

$$\Sigma F_x = 0 = N \therefore N = 0$$
$$\Sigma F_y = 0 = 610 - V - 720 \therefore V = -110\ \#$$

Note: The positive sign convention is just an assumption of the direction. The negative sign means it is opposite to what was assumed so *V* is actually pointing up rather than down.

Finally,

$$\Sigma M_{Cut} = 0 = M + 720(15 - 8) - 610(15)$$
$$M = 4{,}110\ \#\cdot\text{ft}$$

INTERNAL FORCES IN BEAMS

For the beam below, find the internal shear force and bending moment at a point 7 ft from *A*. Press pause on video lesson 57 once you get to the workout problem. Only press play if you get stuck.

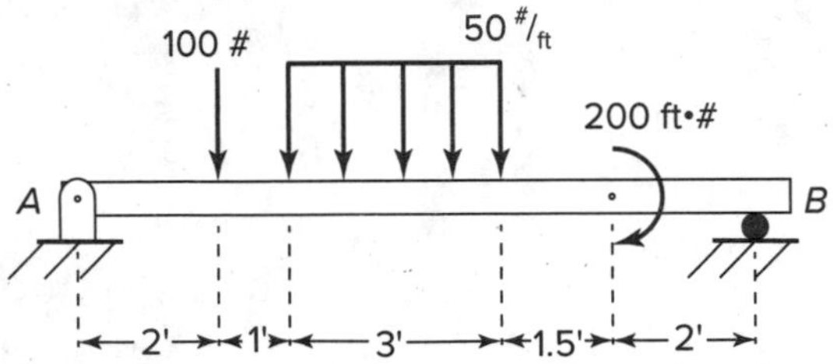

WATCH VIDEO

STATICS LESSON 58
Internal Forces *M, N, V* on a Frame Problem

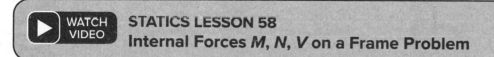

Internal force frame problem:

- Break the body into parts using the recipe for frame problems in order to find global equilibrium. If this is still fuzzy, you may need to review Level 6, Method of Solving Frame Problems Recipe.
- For determining *M*, *N*, and *V*, section (cut) the beam at the point of interest, draw the FBD, and solve using $\Sigma F_x = 0$, $\Sigma F_y = 0$, and $\Sigma M_z = 0$.

Find the internal forces at point *F*. The radii of the frictionless pulleys are 0.6 ft.

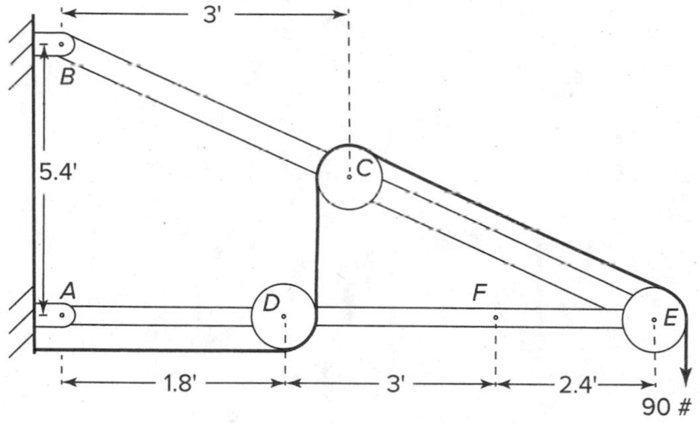

✓ TEST YOURSELF 7.1

SOLUTION TO TEST YOURSELF: Internal Forces in Beams

In the problem below, compute the moment, shear, and axial force in the beam at a position 3' from the support pin *A*. The radius of the wheel at point *B* is 12" and it weighs 50 lbs.

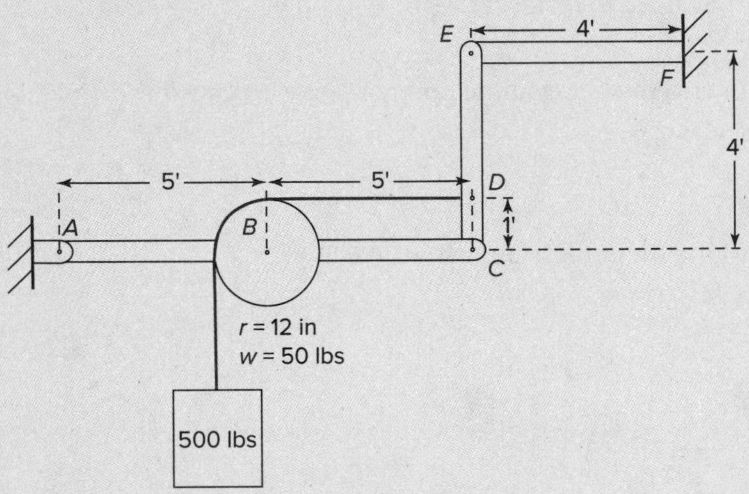

ANSWERS

$M = 825$ ft·#

$N = 125$ #

$V = 275$ #

WATCH VIDEO **STATICS LESSON 59**
Sketch the Shear and Moment Diagram

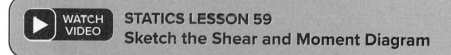

Knowing the worst-case shear force (V) and bending moment (M) is important for determining if a beam will fail. Thus, we graph the entire shear force (V) and bending moment (M) as it isn't always apparent where the worst case is located. Note: typically, we do not sketch N along the length of the beam. You'll see why later when you study stress and displacements in beams.

This section will discuss two methods for drawing shear/moment diagrams:

1. The graphic method
2. The equation method

The Graphic Method

- Uses areas of common shapes
- Utilize table of geographic properties of common shapes found in the back of the book

Before graphing the shear moment diagrams, it is important to understand the relationship of one graph relative to the next.

- Shear diagram (V) is the integral of the load curve (the original given loaded beam).
- Moment diagram is the integral of the shear curve.
- Draw these graphs one under the other since they are related to each other.

The Order of the Lines

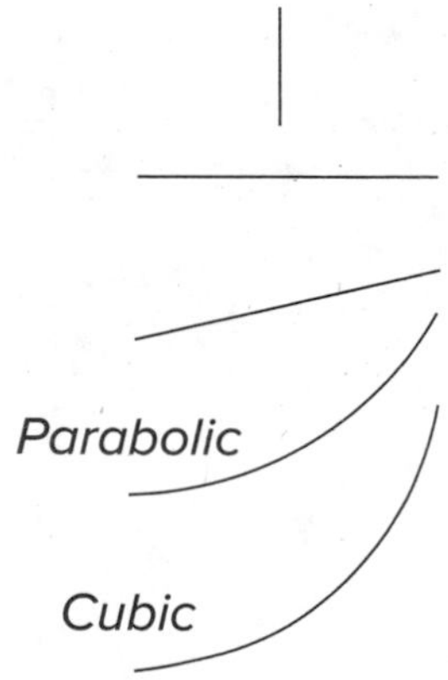

- The top line is a concentrated load.
- The next graph down would be horizontal, y = constant.
- If we integrate, we will have an x (linear with constant slope).
- If we integrate again, we will have an x^2 (parabolic).
- If we integrate again, we will have an x^3 (cubic).

Note: If you understand this concept, you should always know what to expect on the next graph as you go from load (L) to shear (V) to the moment (M) graph.

Shear and Moment Diagram—Graphic Method Recipe

STEP 1: Find global equilibrium for the beam. Also be on the lookout for pinned beams made up of multiple members, which require techniques in the frame problem level to solve global equilibrium.

Note: It is OK to convert distributed loads to a concentrated load for global equilibrium. However, it must be treated as a distributed load later when sketching the V and M diagrams.

STEP 2: Construct and align the axes of the shear and moment diagram so that they align with the diagram of the load curve.

STEP 3: Mark where discontinuities occur on the diagram. Discontinuities are places on the load curve where the function describing the load changes. These are called "everywhere something interesting is happening."

Note: Concentrated moments don't affect the shear force diagram, only the moment diagram.

STEP 4: Graph the shear force diagram. Imagine yourself, like me below, walking across the beam from left to right. As you walk across the beam you have a "load backpack" that is accumulating the loads that you encounter as you walk along. "Van Halen" forces (concentrated loads) make you *jump* (it's a song from the '80s!) up or down on the graph.

Shear and Moment Diagram—Graphic Method Recipe (continued)

STEP 5: Calculate the "area" of each shape on the V diagram. Shapes are the distance in the x-direction multiplied by the force (i.e., "height in the y-direction"). The units of these "areas" are the units of a moment since it is force times a distance.

- Add a positive or a negative to each area depending on whether the "area" is above or below the axis. These positive and negatives dictate whether the slope on the M diagram is "uphill" or "downhill."
- Put a circle around the areas so as not to confuse them with the V diagram values.

STEP 6: Plot the M diagram from the "areas" derived in the previous step by adding or subtracting the area at each point. The only detail missing is how to deal with concentrated moments. A concentrated moment will make you *jump* on your moment diagram the same way a concentrated force made us *jump* on our V diagram. The following silly rhyme will help you to aways get whether you jump up or down correct!

In the kitchen, the clock is above and the counter is below!

When drawing the diagrams from left to right:

- CCW moments make us jump down on our graphs (below)
- CW moments make us jump up on our diagrams (above)

PITFALL

If you solve for global equilibrium incorrectly, your shear force or moment diagram will not end up at zero. You won't know this until the very last step of the problem causing you to have to restart the entire problem. Thus, it is always wise to double-check your global equilibrium.

PRO TIPS

- Any time you have a pin connection along your beam, the value for moment should always be zero as pins cannot support a moment load!
- A tricky concept for students to graph is the parabolic curve. Is it convex up or concave down or something else? Here's an easy way to always get it right:
  - If the arrows on the load curve are pointing downwards, the slope on the *V* diagram will always be downhill, and uphill if the arrows are pointed upwards.
  - Now for concavity: Look at the shape on the load diagram. Are you accumulating the load rapidly and then slowly, or slowly and then rapidly? If the triangle is *fat* at the beginning and *skinny* at the end, then your graph slope will be fast then slow. Think about snow skiing: Fast slope is a black diamond (steep), and slow slope is the bunny slope (flattening out). This little trick will ensure you always get the concavity of your curve correct.

EXAMPLE: SHEAR/MOMENT DIAGRAM

Draw the shear/moment diagram for the following loaded beam.

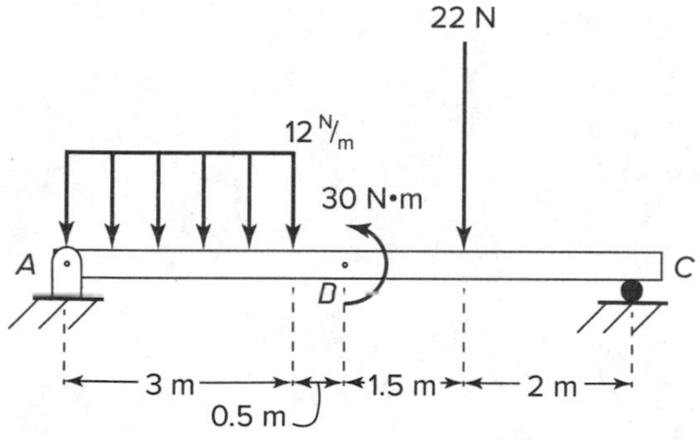

STEP 1: Find global equilibrium for the beam. ***Note:*** By inspection, we assumed both of the reaction forces to be upward.

$$\Sigma M_A = 0 = -12(3)(1.5) + 30 - 22(5) + C_y(7)$$
$$C_y = 19.14\text{ N}$$
$$\Sigma F_y = 0 = C_y + A_y - 12(3) - 22$$
$$= 19.14 + A_y - 36 - 22$$
$$A_y = 38.86\text{ N}$$
$$\Sigma F_x = 0 = A_x$$

STEP 2: Construct your axes as shown below.

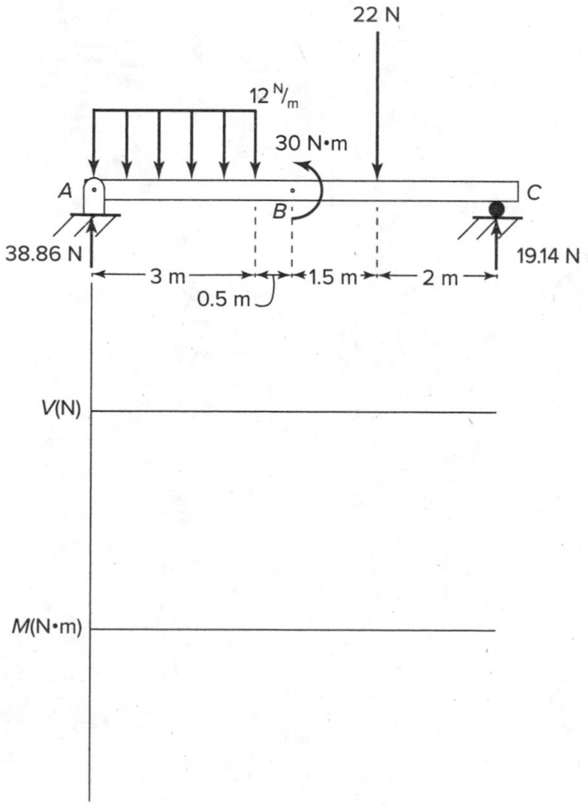

STEP 3: Mark where discontinuities occur on the diagram.

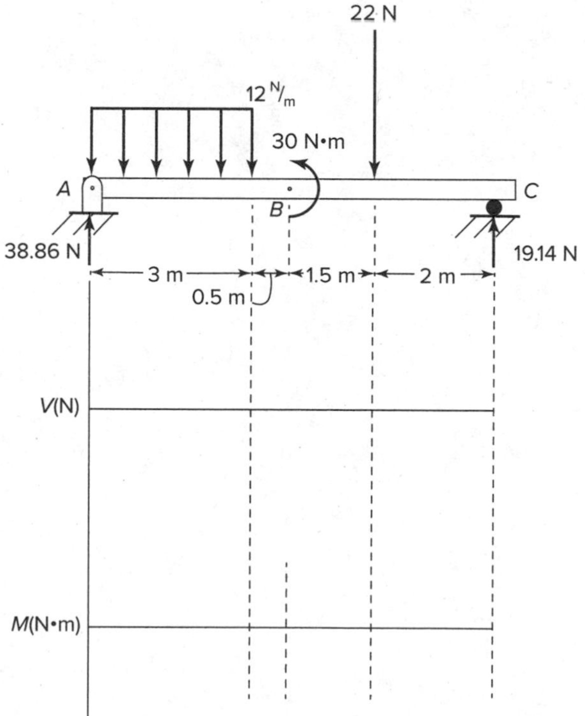

STEP 4: Graph the shear force diagram.

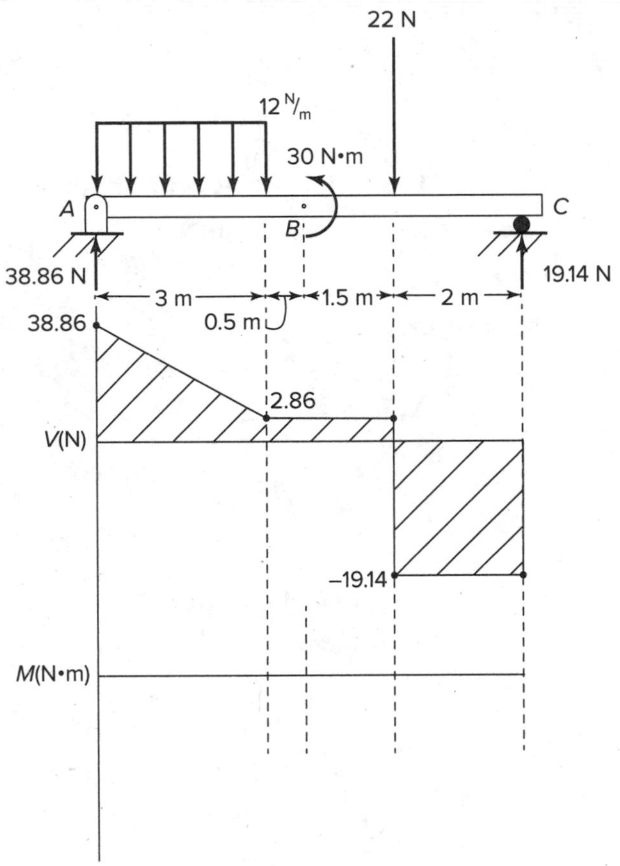

STEP 5: Calculate the "area" of each shape on the *V* diagram.

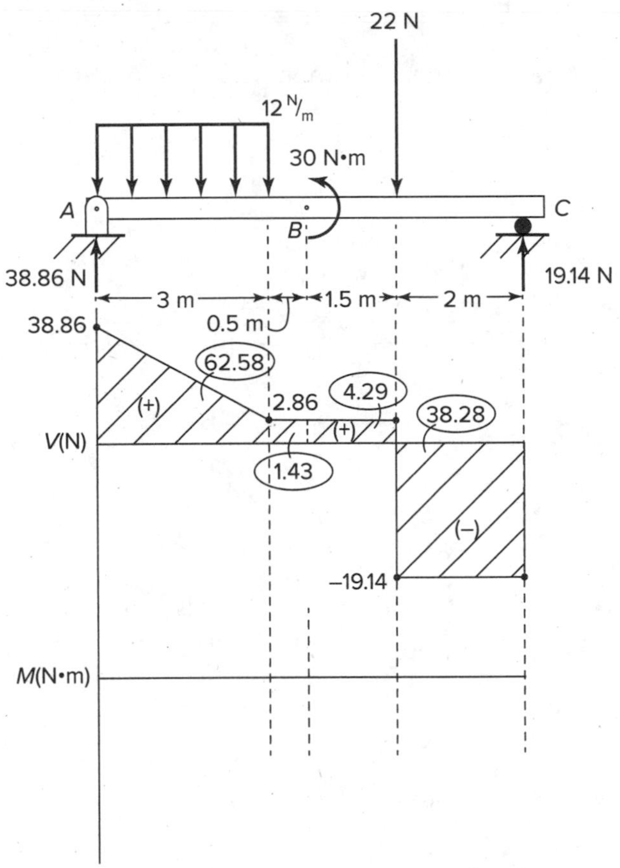

Note: We had to break up the area between the distributed load and point force due to the concentrated moment.

$$\text{Area } 1 = (0.5)(3)(38.86 - 2.86) + (2.86)(3)$$
$$= 62.58$$
$$\text{Area } 2 = (2.86)(0.5) = 1.43$$
$$\text{Area } 3 = (2.86)(1.5) = 4.29$$
$$\text{Area } 4 = (19.14)(2) = 38.28$$

STEP 6: Plot the *M* diagram from the "areas" derived in the previous step.

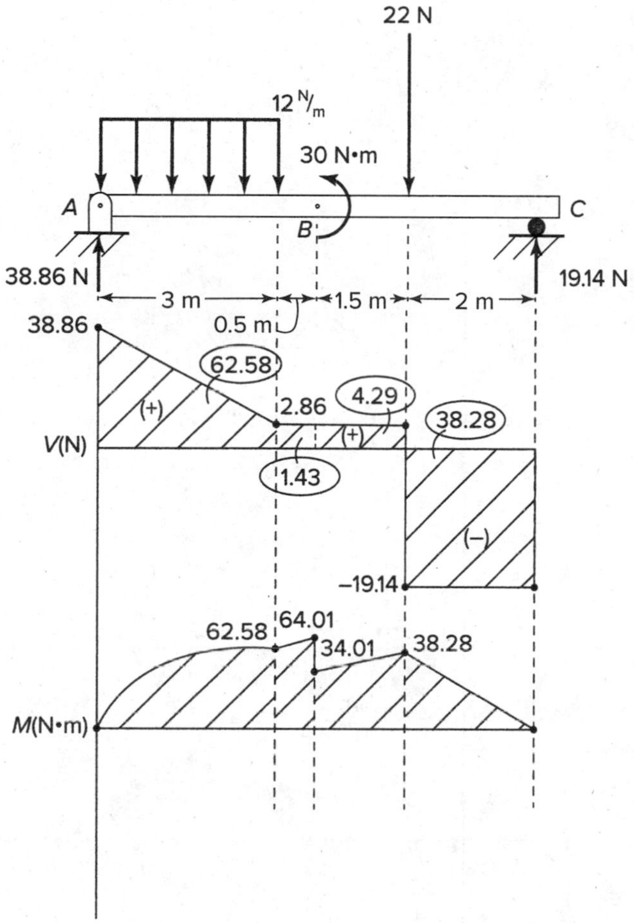

Note: This particular problem had some calculated numbers that were slightly rounded. If your graph is within acceptable rounding error to zero, you can just assume it returns to zero as it *always* should. If your moment diagram doesn't return to zero . . . you need to go all the way back to step 1 and check your global equilibrium . . . *arghhhhh.*

From these diagrams, maximum moment or shear force can be determined easily.

✓ TEST YOURSELF 7.2

SOLUTION TO TEST YOURSELF:
Internal Forces in Beams

Draw the correct concavity for the moment diagram given the shear diagram. Assume the moment starts at zero.

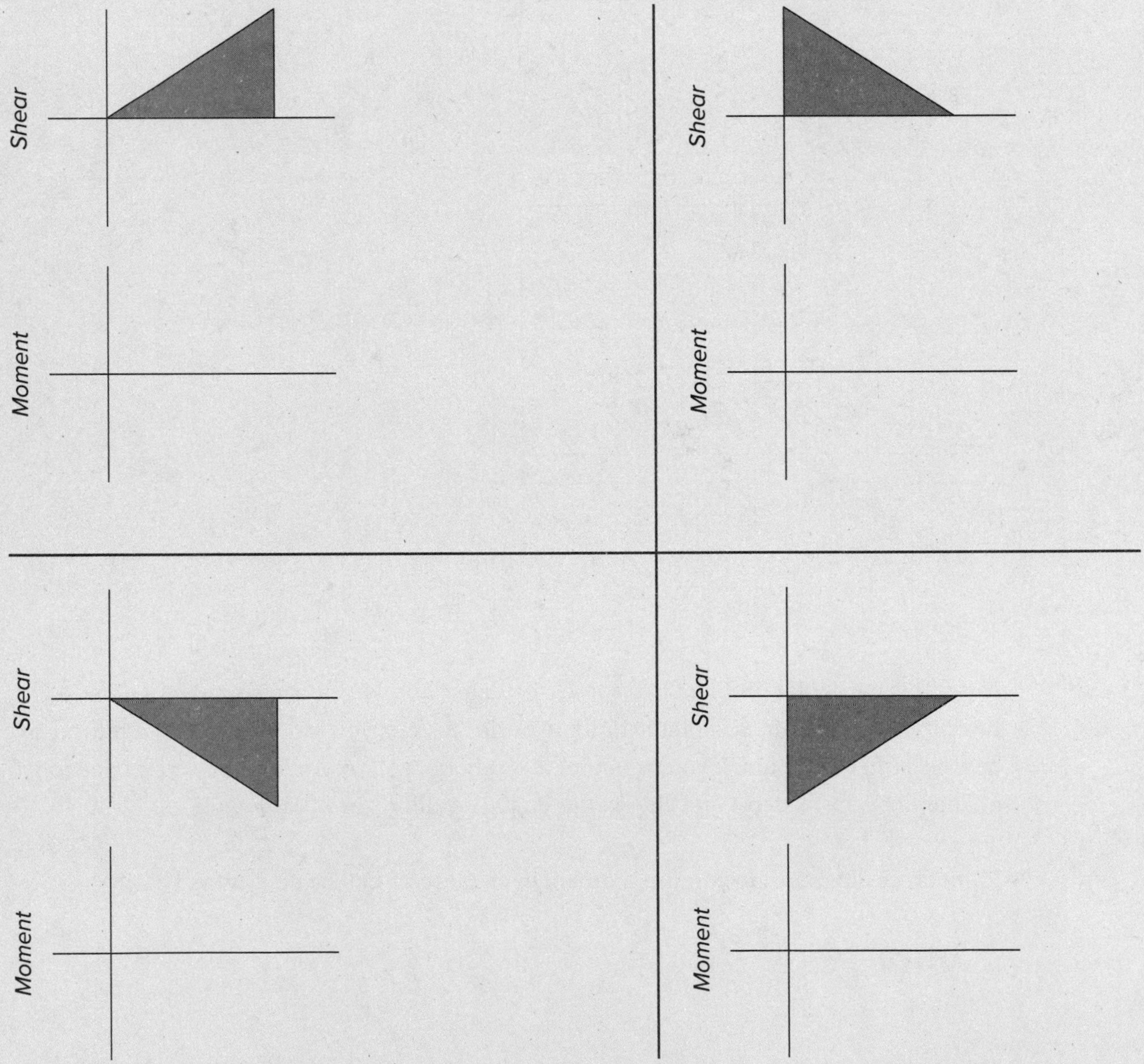

ANSWER
See Test Yourself 7.2 solutions at the back of the book.

Complete the shear/moment diagram for the given loaded beam.

Press pause on video lesson 59 once you get to the workout problem. Only press play if you get stuck.

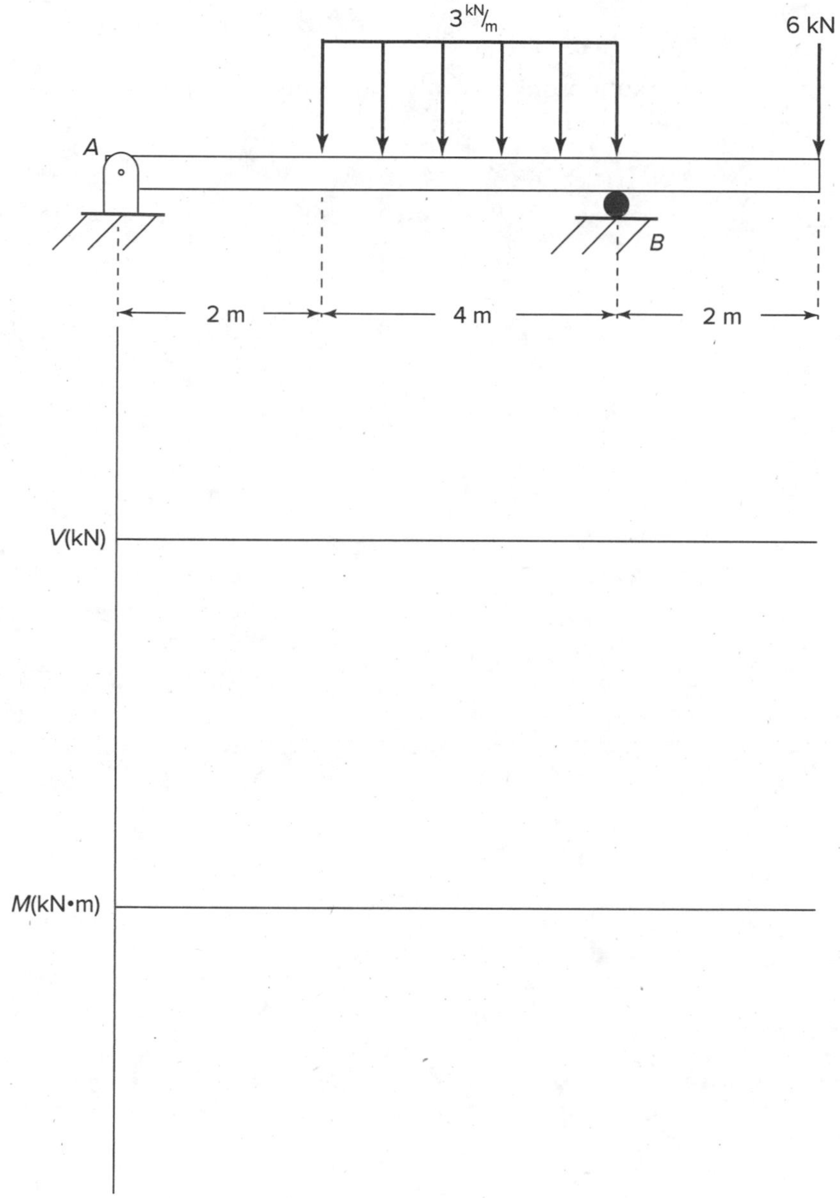

WATCH VIDEO

STATICS LESSON 60
Shear and Moment Diagram with Moments

Complete the shear/moment diagram for the given loaded beam.

Press pause on the video once you get to the workout problem. Only press play if you get stuck.

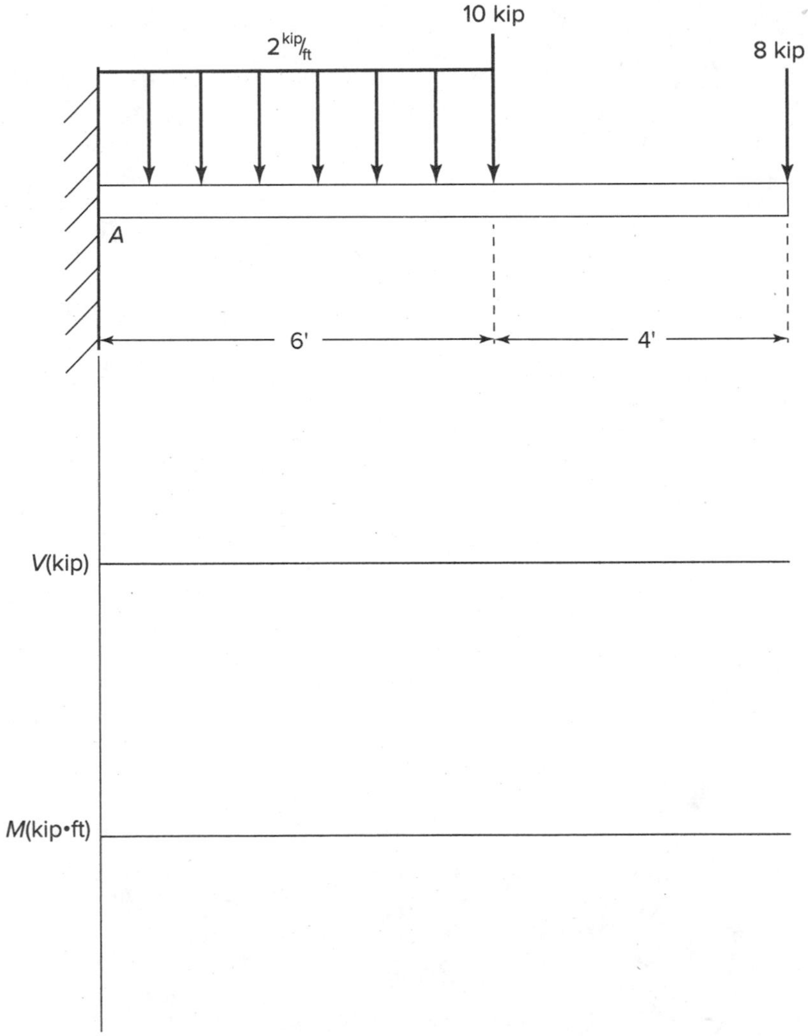

RECALL: There is a differential relationship between the load, shear, and moment diagrams. Shear is related to the integral of the load, and moment is related to the integral of the shear. For sketching *V* and *M* in most cases, compute these integrals by remembering that the integral is nothing more than the area under the curve!

✓ TEST YOURSELF 7.3

SOLUTION TO TEST YOURSELF: Internal Forces in Beams

For the beam shown, sketch the shear and moment diagram, labeling key points on each.

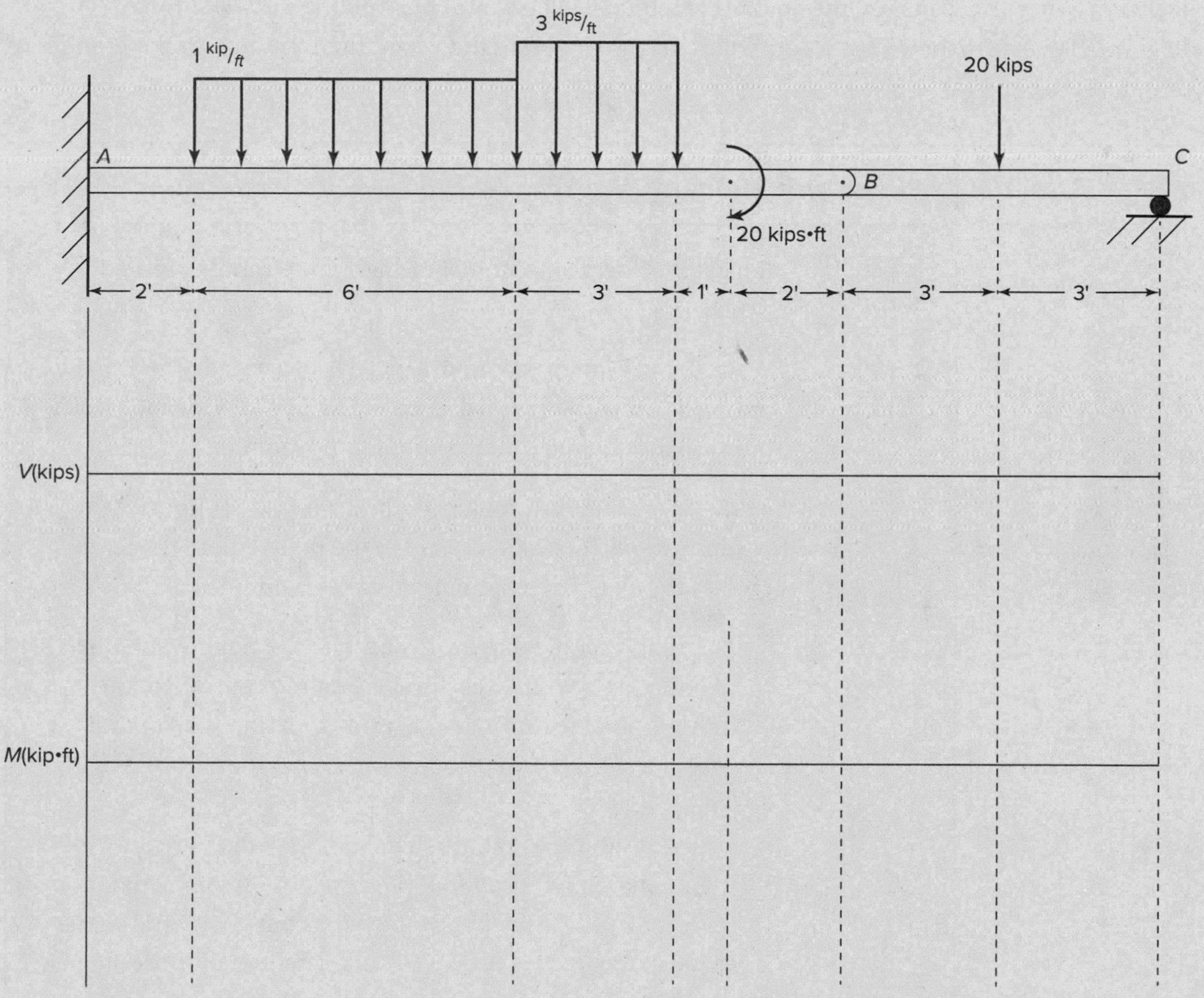

ANSWER

See Test Yourself 7.3 solutions at the back of the book.

WATCH VIDEO **STATICS LESSON 61**
Shear and Moment Diagram by the Equation Method

The Equation Method

The graphic method works on all problem types except a very select few. The easiest way to identify if you are going to have to use the equation method will be while constructing the shear force (V) diagram. If your V diagram has a parabolic curve that crosses the x-axis, then the equation method is your ticket to success!

Shear and Moment Diagram—Equation Method Recipe

STEPS 1-4: Repeat the Shear and Moment Diagram—Graphic Method Recipe.

STEP 5: When the V diagram goes from a positive to a negative and crosses the x-axis, create a vertical dashed line at this new location as it is the local max or min (humpy doo) for the moment.

Note: The reason the graphic method won't work here is that we don't know the distance over to the point where the curve crosses the axis. Two common mistakes made:

- Students assume that the distance to the point where the parabolic curve crosses the axis is at the centroid of the triangle load on the load diagram. It's a trap (in Admiral Akbar's voice!). That distance is never at the centroid of that triangle.
- Students assume that the shape on the V diagram is an ex-parabolic shape (spandrel) (i.e., the outside of the parabola part) or a parabolic shape which is in our geometric shape table in the back of the book. This is neither of those shapes, and the area formulas will not get you back to zero.

STEP 6: Section the original load curve at a distance of x and draw the free body diagram of the left side of the beam (don't forget to include the internal forces M, N, and V). Note you can do the right side of the beam but you have to be really careful with your math!

Shear and Moment Diagram—Equation Method Recipe (continued)

STEP 7: Write the equation for the sloped line for the given load curve in order to find the height (y) of the curve at distance x.

STEP 8: Covert the distributed load to a concentrated load as a function of x. Write an equation for $\Sigma F_y = 0$ and $\Sigma M_z = 0$. Solve these equations for shear force (V) and bending moment (M).

Note: Since the moment diagram is the integral of the shear diagram, we expect to see that relationship in these two equations.

STEP 9: Since the local maximum/minimum of the moment occurs when $V = 0$, using the V equation set it equal to zero and solve for distance x. Once the value of x has been determined, substitute it into the moment equation and solve for the value of moment. Also find the moment at the endpoint of this section.

Note: You probably already know what the endpoint value should be. You can check that you have the correct equation by seeing if you get this answer.

STEP 10: Complete the moment diagram using the derived values for M. Remember, the moment diagram must also always end up back at zero.

EXAMPLE: SHEAR AND MOMENT DIAGRAM

Draw the shear/moment diagram for the following loaded beam.

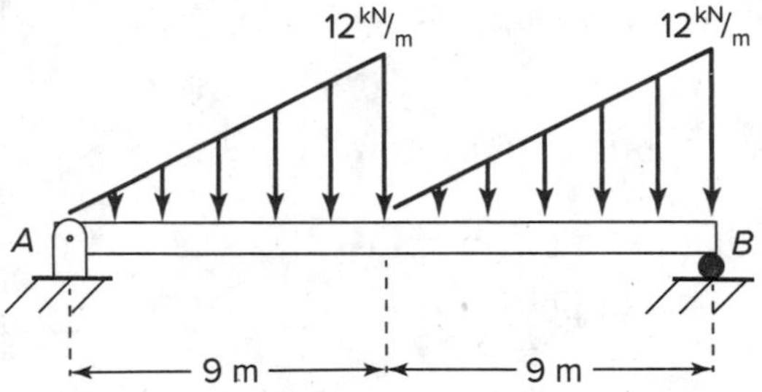

STEP 1: Find global equilibrium for the beam. Don't forget that the 12 kN/m triangular distributed load can be converted into a point force via finding the "area" of the distributed load and applying it at the centroid for a triangle. In this case the "area" is $(\frac{1}{2})(12)(9) = 54$ kN and it is applied at $(\frac{2}{3})(\text{base}) = (\frac{2}{3})(9) = 6$ m. Thus, the first triangular load occurs 6 m meter from the left, and the second triangular load it is $9 + 6 = 15$ m from the left.

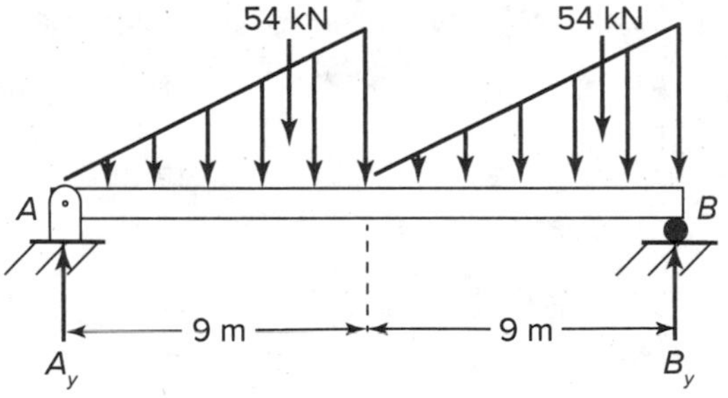

$$\Sigma M_A = 0 = -54(6) - 54(15) + B_y(18)$$
$$B_y = 63 \text{ kN}$$

$$\Sigma F_y = 0 = A_y + B_y - 54 - 54 = A_y + 63 - 54 - 54$$
$$A_y = 45 \text{ kN}$$

STEPS 2&3: Construct your diagrams as shown and add the discontinuities.

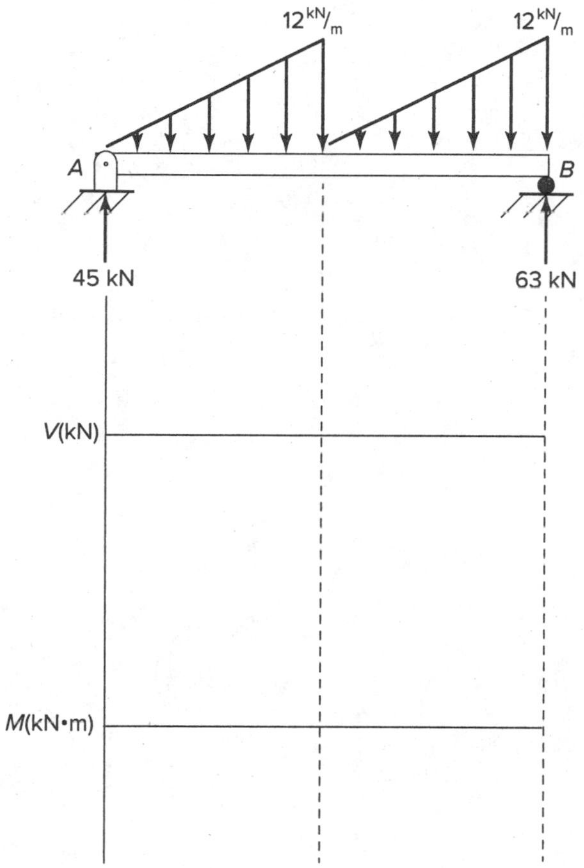

STEP 4: Construct the shear (V) diagram using the same techniques learned in the graphic method recipe.

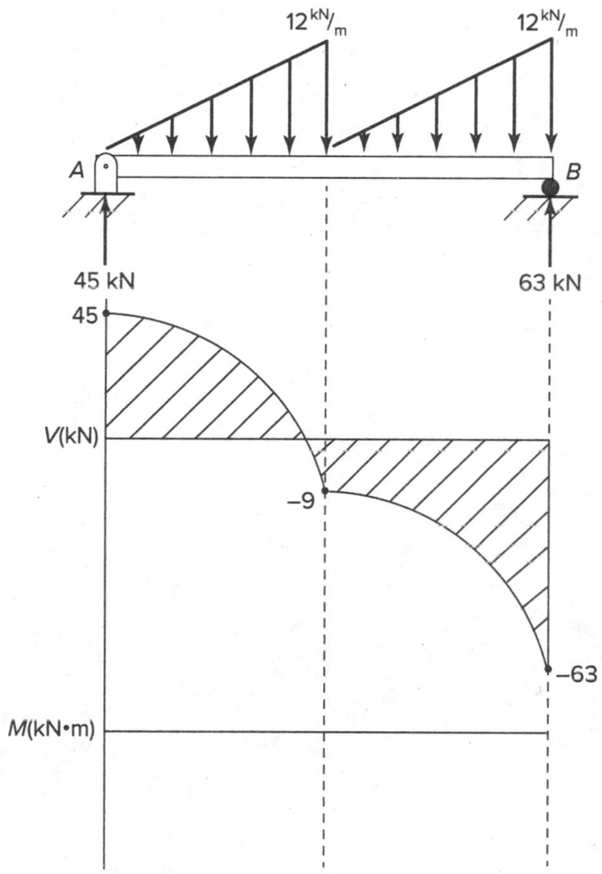

STEP 5: Create a new point of interest at that local max for the moment.

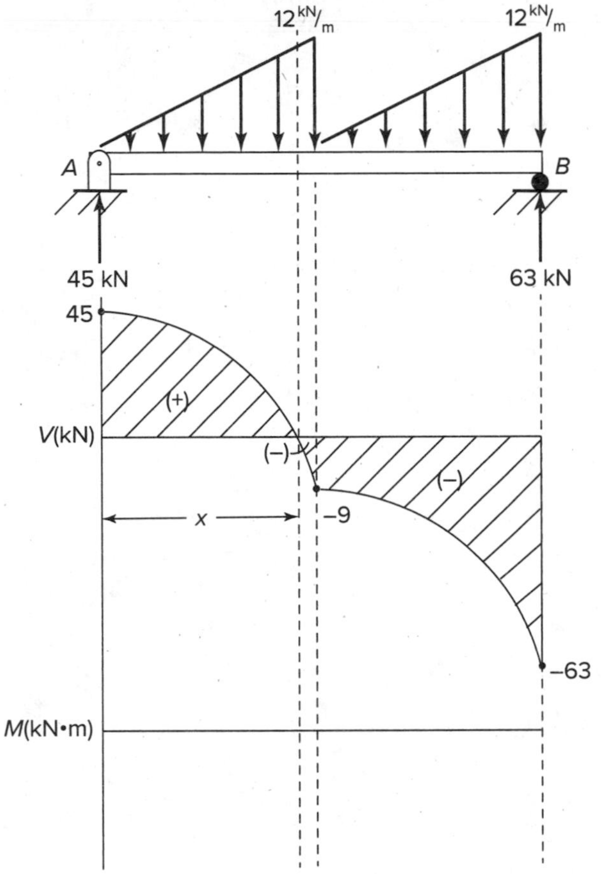

STEP 6: Section the original load curve at a distance of x and draw the free body diagram of the left side of the beam.

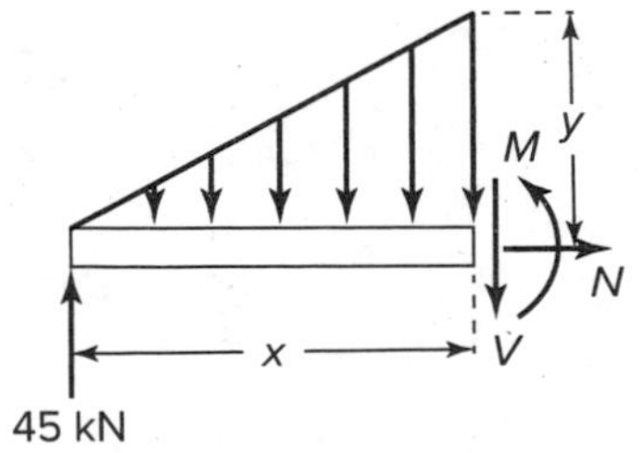

STEP 7: Write the equation for the sloped line in order to find the height of the curve at distance x.

$y = mx + b$

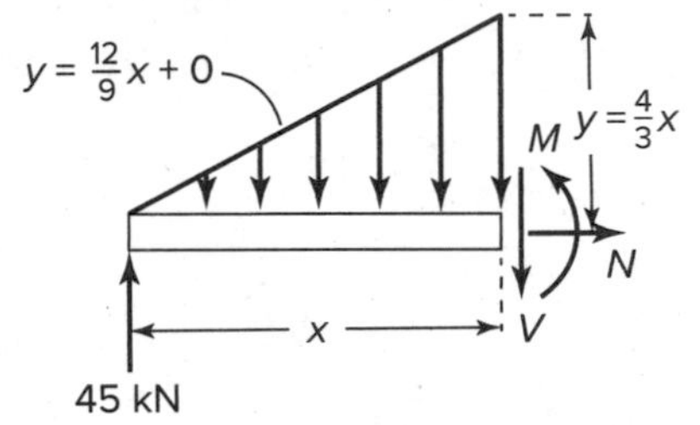

STEP 8: Convert the distributed load to a concentrated load and write the sum of the forces in the y direction and moment equation.

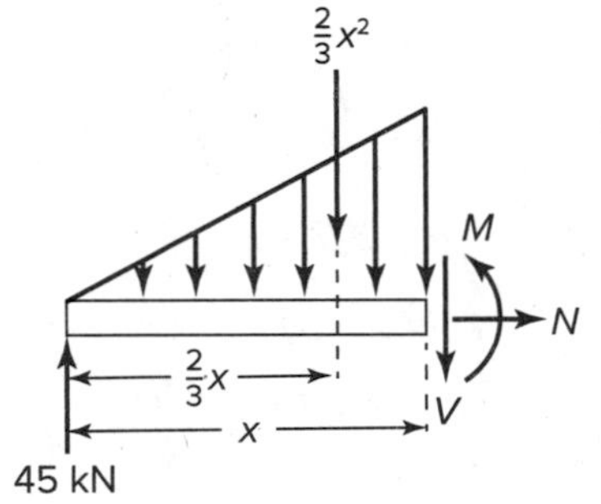

For converting the distributed load to a concentrated load multiply one-half base (x) by the height $(\frac{4}{3})(x)$.

$$\text{Concentrated load} = \frac{1}{2}\left(\frac{4}{3}x\right)(x) = \frac{2}{3}x^2$$

$$\Sigma F_y = 0 = 45 - V - \frac{2}{3}x^2$$

$$V = 45 - \frac{2}{3}x^2$$

$$\Sigma M_x = 0 = \frac{2}{3}x^2\left(x - \frac{2x}{3}\right) - 45(x) + M$$

$$= \frac{2}{3}x^2\left(\frac{x}{3}\right) - 45(x) + M$$

$$M = 45x - \frac{2}{9}x^3$$

STEP 9: Determine the local maximum or minimum of the moment as well as the moment at a distance of 9 m (midpoint on the beam).

Local Maximum or Minimum

Find the position x when $V = 0$ when using the equations found in step 8.

$$0 = 45 - \frac{2}{3}x^2$$

$$45 = \frac{2}{3}x^2$$

$$x^2 = 67.5$$

$$x = 8.22 \text{ m}$$

Substitute x into the moment equation:

$$M = 45(8.22) - \frac{2}{9}(8.22)^3$$

$$M = 246.47 \text{ kN·m}$$

Moment at the Endpoint

At 9 m, what is the value of M?

$$M = 45(9) - \frac{2}{9}(9)^3$$

$$M = 243 \text{ kN·m}$$

STEP 10: Complete the moment diagram using the derived values for M.

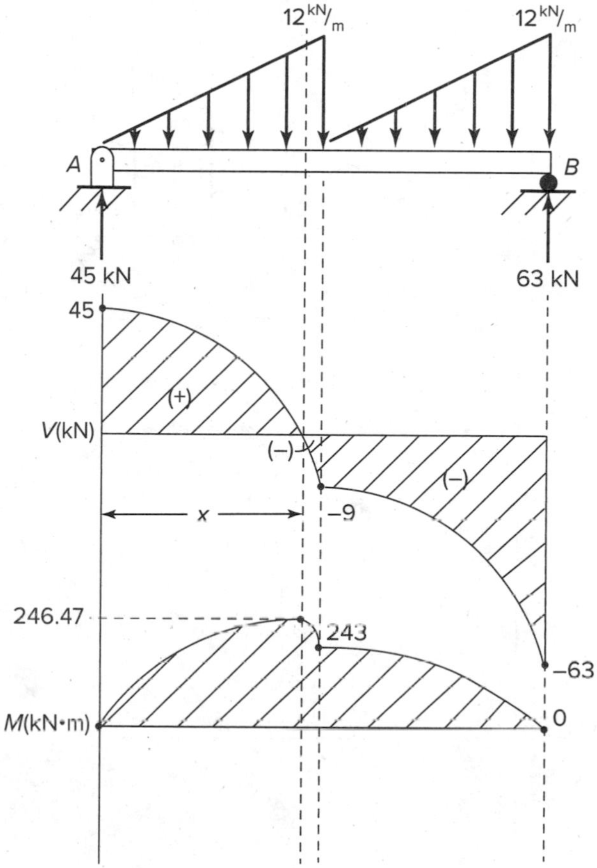

You should now be able to master any shear moment diagram problem, both using the graphic method as well as the equation method.

Complete the shear/moment diagram for the given loaded beam.

Press pause on video lesson 61 once you get to the workout problem. Only press play if you get stuck.

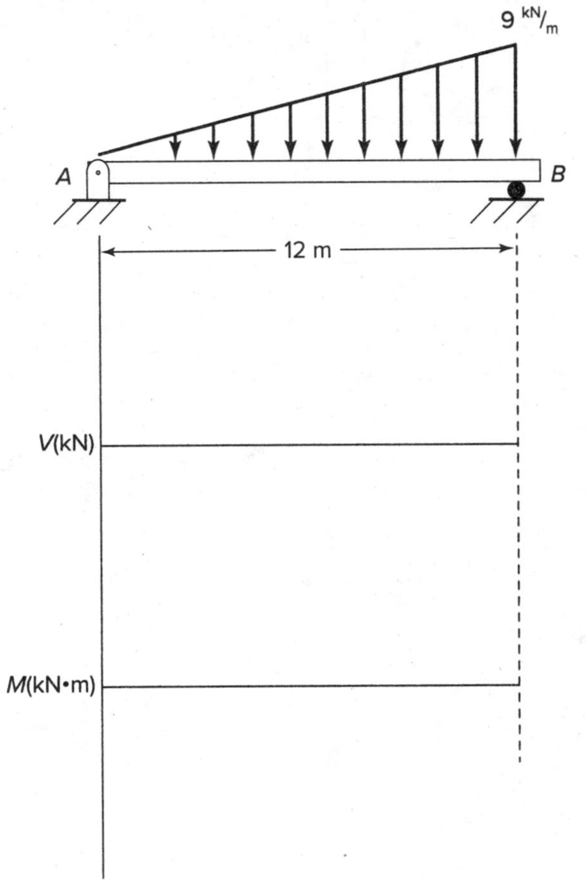

✓ TEST YOURSELF 7.4

SOLUTION TO TEST YOURSELF: Internal Forces in Beams

For the simply supported beam shown, sketch the shear and moment diagram. Label key points on the shear and moment diagrams.

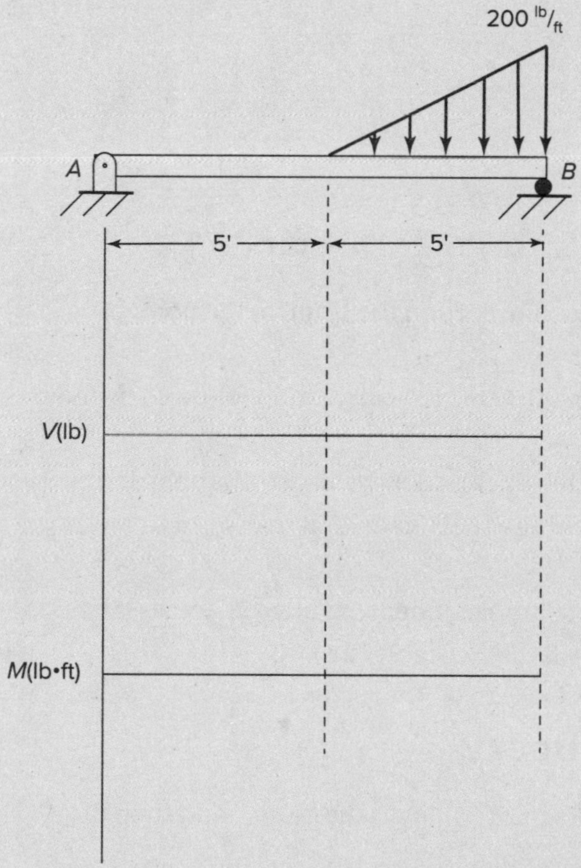

ANSWER

See Test Yourself 7.4 solutions at the back of the book.

KEY TAKEAWAYS

Internal forces are created via external loading being applied to a system. These internal loads are either axial (N), shear (V), and/or bending moment (M).

Main Takeaways

- How to calculate M, N, and V at a point of interest
- How to construct shear/moment diagrams
  - Graphic method
  - Equation method

How to Calculate *M, N, V* at a Point of Interest Summary

- You can find these after you find global equilibrium and section the beam at the point of interest.
- It is important to utilize the sign convention as it is the expected convention followed by most textbooks.
- Challenge questions can include frame problems in which you have to take apart the frame to solve for global equilibrium and then section to find internal forces. You may need to review the previous level to polish up your frame skills.
- For any problem in this level, once you have your beam sectioned and internal forces labeled, you solve like a typical statics problem using $\Sigma F_x = 0$, $\Sigma F_y = 0$, and $\Sigma M_z = 0$.

How to Construct Shear-Moment Summary

- The shear diagram (V) is the integral of the load diagram (which is the given), and then the moment diagram (M) is the integral of the shear diagram (V).
- **Graphic method.** This is definitely the preferred method as it is much faster and works on nearly all problems. The only time this method doesn't work is when you have triangular distributed load on the shear diagram and it crosses the x-axis as a parabolic curve. It's the M diagram that you can't really do as you won't know the local maximum or minimum! You should have success following the given recipe.
- **Equation method.** This technique employs writing a function for a load on the load diagram and integrating it to write a function for V and for M. This allows us to mathematically solve for any important points on the beam including mins and maxes. For the equation method, you can always just use the techniques to construct the V diagram determined via the graphical technique. The equation method only has to be utilized when the V diagram is parabolic and crosses the x-axis.

KEY TAKEAWAYS

RECIPES

- Solving Internal Forces at a Given Point Recipe
- Shear and Moment Diagram—Graphic Method Recipe
- Shear and Moment Diagram—Equation Method Recipe

PRO TIPS

Internal Forces

- When cutting a beam, it generates three components: ***M***, ***N***, and ***V***. Otherwise, it can be treated as a regular equilibrium problem. You can remember with my handy (silly) jingle . . .

  ♫♫♩♩♩ Every time you cut the beam, you must have an ***M***, ***N***, ***V***!

- When cutting a beam and using the positive sign convention, feel free to use either free body diagram (right or left side) as you will get the same answer. *Note:* All answers to internal force questions should have signs in accordance with the positive sign convention.

Shear/Moment Diagrams

- Any time you have a pin connection along your beam, the value for moment should always be zero as pins cannot support a moment load!
- A tricky concept for students to graph is the parabolic curve. Is it convex up or concave down or something else? Here's an easy way to always get it right.
  - If the arrows on the load curve are pointing downwards, the slope on the ***V*** diagram will always be downhill, and uphill if the arrows are pointed upwards.

(continued on next page)

PRO TIPS

Shear/Moment Diagrams *(continued)*

- Now for concavity: Look at the shape on the load diagram. Are you accumulating the load rapidly and then slowly, or slowly and then rapidly? If the triangle is *fat* at the beginning and *skinny* at the end, then your graph slope will be fast then slow. Think about snow skiing: Fast slope is a black diamond (steep), and slow slope is the bunny slope (flattening out). This little trick will ensure you always get the concavity of your curve correct.

PITFALLS

Internal Forces

- It is easy to get the positive convention for the shear force, *V*, in the wrong direction. When using the left-side free body diagram, a trick for remembering the sign convention is "Don't get *left* at the store or you'll feel *down*." Silly, right? But I bet you'll remember it!

Shear/Moment Diagrams

- If you solve for global equilibrium incorrectly, your shear force or moment diagram will not end up at zero. You won't know this until the very last step of the problem causing you to have to restart the entire problem. Thus, it is always wise to double-check your global equilibrium.

Level 8

Friction

WATCH VIDEO

STATICS LESSON 62
Friction Is Fun, Box on an Incline

This level is solved the exact same way as previous techniques, there is just one additional force (due to friction).

Friction is a force that resists sliding or rolling of a solid object. Friction will be referred to as a friction force, but it is actually a reaction force that resists motion. Friction force by itself cannot move a body! Friction is lazy. It only uses what it needs to prevent motion, but it isn't necessarily at its maximum.

- μ_s: Static coefficient of friction (not moving). Note you can also have this type with constant velocity motion, but this is very, very rare.
- μ_k: Dynamic coefficient of friction (moving).
  - Static coefficient of friction is always going to be higher than dynamic coefficient of friction as it is difficult to start moving a body (μ_s) but then becomes easier to slide once in motion (μ_k).
- N: The normal force, i.e., contact force. "What would a bathroom scale read?"

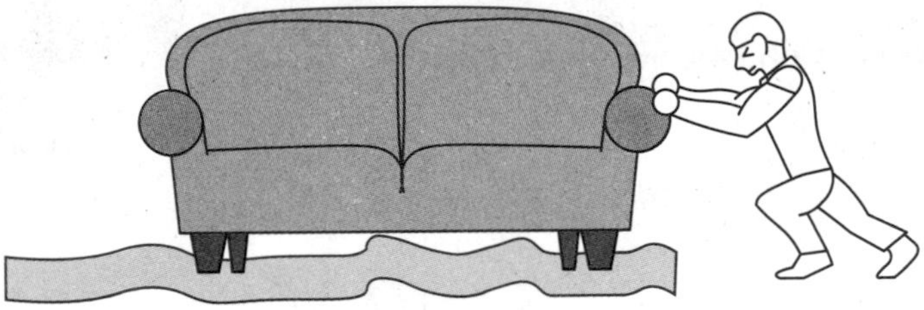

- Friction is $F = \mu_s N$ (notice how the equation spells *fun*): used only for impending motion (here in Texas we call it "fixin' to move").
  - If you push against a wall, the max you can push is the coefficient of friction between your feet and the carpet multiplied by your weight. If you push more than this, you will slip and hit the floor.
  - If you are not at the brink of impending motion, then the friction is between 0 and $\mu_s N$ (max). You will need more information to solve for the friction variable.
- Friction is assumed to be 0 if the problem statement states *smooth* surface or *frictionless*.

FRICTION

PRO TIPS

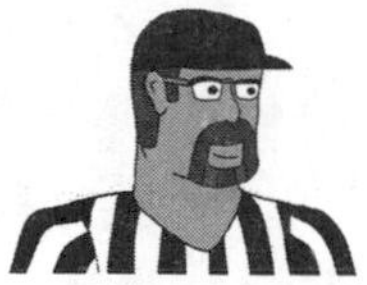

- If there is anything you can do to be successful in this level, it is to practice FBDs. Practice, practice, practice!
- Sometimes it might be easier to tilt your *x-y* coordinate axis so that it aligns with the unknowns you are trying to solve, especially any time you have two unknowns that are perpendicular to each other.

PITFALL

You have to guess the direction of the friction correctly, unlike when solving other problem types. If you guess it incorrectly, you won't simply get a negative sign. You will get a wrong answer.

The typical, quintessential, simple friction problems include a box on plane and ladder leaning on a wall.

CHALLENGE QUESTION

You have two sports cars of the same size and weight. One has big fat racing tires, and one has skinny bicycle tires made of the exact same rubber compound. If you set the brakes on both on an incline, which one would slide first?

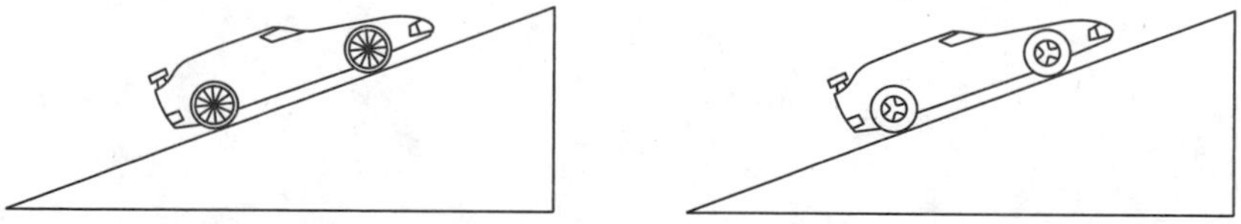

ANSWER

While the actual details of friction present some practical challenges, for the case described, the tire width does not matter. They would slide at the exact same time.

Solving Simple Friction Recipe

STEP 1: Read the question and visualize the motion. Identify what moves and which direction. Think about all the different cases for slipping (i.e., what happens if the block is light or really heavy, slipping left versus right, uphill versus downhill). If you don't get the visual picture correct, you are going to assume your frictions backwards.

STEP 2: Construct your FBD for the case in question (i.e., min or max case). This will set the direction of motion for the object and for friction.

STEP 3: Use your equations of equilibrium to solve for unknowns ($\Sigma F_x = 0$, $\Sigma F_y = 0$, and/or $\Sigma M_z = 0$).

STEP 4: Repeat for the other case, which means if you solved for min repeat to find max.

EXAMPLE: SIMPLE FRICTION

A ladder leans against a frictionless wall. The static coefficient of friction between the ladder and the floor is $\mu_s = 0.35$. Find the minimum angle (θ) before the ladder slips and falls. The uniform ladder weighs 30 # and there is a 160 # person on the ladder at the position shown.

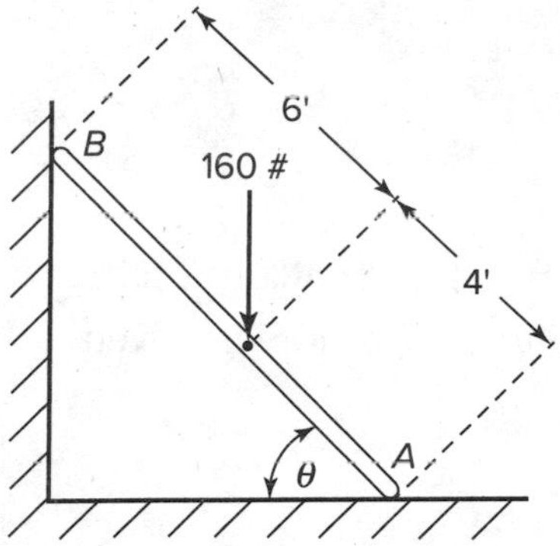

STEP 1: Visualize the motion. In this case, the top of the ladder will slide down the wall and the bottom of the ladder will move to the right.

STEP 2: Construct your FBD. Note the "free body" is the ladder.

There is a friction force for every normal force except at the wall (B); as the problem statements says, it is frictionless.

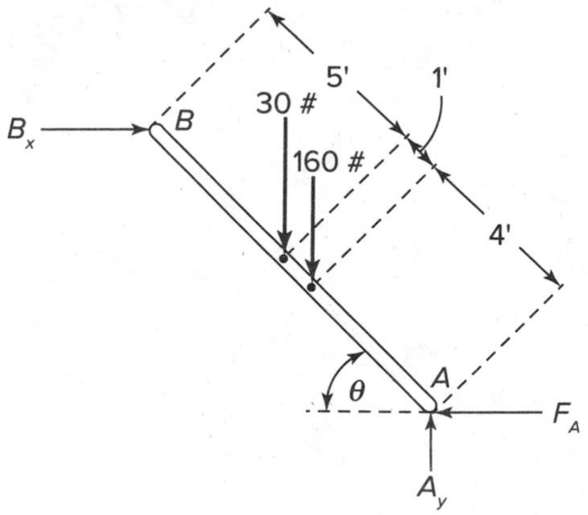

STEP 3: Use your equations of equilibrium to solve for unknowns ($\Sigma F_x = 0$, $\Sigma F_y = 0$, and/or $\Sigma M_z = 0$).

$\Sigma F_x = 0 = B_x - F_A => B_x = F_A$

Note: Since the ladder is on the verge of slipping, we use equation $F = \mu_s N$. This equation then becomes $F_A = \mu_s A_y = 0.35\ A_y$. Since $B_x = F_A$, then $B_x = 0.35(A_y)$.

$\Sigma F_y = 0 = A_y - 190\ \# => A_y = 190\ \#$

substituting A_y (normal force) into the friction equation would yield:

$B_x = 0.35(190) = 66.5\ \#$

The final equation is the moment about point *A*. ***Note:*** The same answer will be obtained if the moment is summed about point *B* (the equation will simply be a bit more involved).

$\Sigma M_A = 0 = (160)cos(\theta)(4) + (30)cos(\theta)(5) - (B_x)sin(\theta)(10)$

$(10)(B_x)sin(\theta) = (640)cos(\theta) + (150)cos(\theta)$

$$\frac{sin(\theta)}{cos(\theta)} = \frac{79}{B_x}$$

$$tan(\theta) = \frac{79}{66.5}$$

$\theta = 49.91°$

Therefore, any angle less than 49.91° would result in the ladder slipping and falling.

STEP 4: No other cases for this problem.

FRICTION

Box 1 weighs 100 #. If the static coefficient of friction is 0.35, find the range of the weight of box 2 (W_2) to maintain equilibrium.

Press pause on video lesson 62 once you get to the workout problem. Only press play if you get stuck.

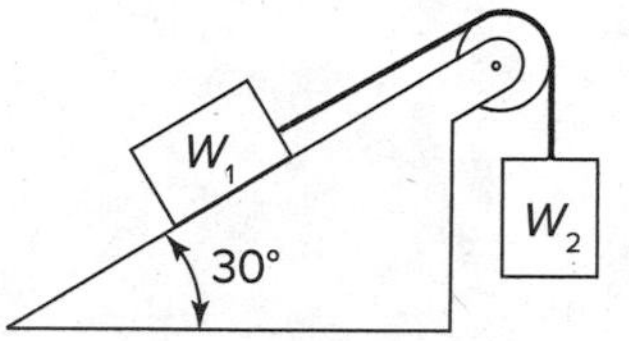

✓ TEST YOURSELF 8.1

SOLUTION TO TEST YOURSELF: Friction

What force (P) is required to push the block uphill?

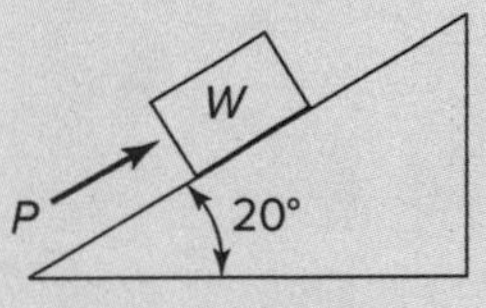

$W = 100$ lbs
$\mu_s = 0.40$

ANSWER
$P = 71.79$ lbs

WATCH VIDEO

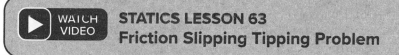

STATICS LESSON 63
Friction Slipping Tipping Problem

You can identify friction slipping versus tipping problems when the book asks "find the force where the motion occurs." You need to ask yourself, "What is *motion*?" These motions include slipping or tipping for all bodies in the problem (single items, combined items, etc.).

When tipping occurs, you need to identify the point of rotation (i.e., where is it rotating about?). This is the location where you will take your moment.

PITFALL

You need to figure out what scenarios of motion can happen and pick the smallest value. The smallest value is what is going to occur first. You will never get to the bigger values as the little one happens first.

Slipping or Tipping Recipe

STEP 1: Read the question and visualize the motion. Identify what moves and its direction. Think about all the different cases for both slipping and tipping. If you don't get the visual picture correct, you are going to assume your frictions backwards. Write out each of the possible motion conditions.

STEP 2: Construct your FBD for each case.

STEP 3: Use your equations of equilibrium to solve for unknowns ($\Sigma F_x = 0$, $\Sigma F_y = 0$, and/or $\Sigma M_z = 0$).

STEP 4: Repeat for the other cases.

STEP 5: The smallest force that causes motion is the case that will occur (i.e., the answer).

EXAMPLE: SLIPPING OR TIPPING

Find the force P that causes motion to occur for the given stack of boxes. Boxes A, B, and C weigh 40 lbs, 35 lbs, and 60 lbs, respectively. The boxes are uniformly loaded. The coefficient of static friction between box A and the floor is $\mu_s = 0.45$, between A and B is $\mu_s = 0.35$, and between B and C is $\mu_s = 0.4$.

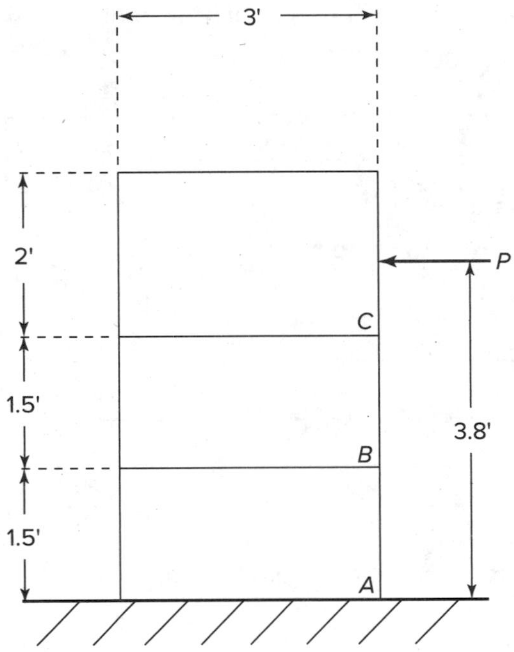

STEP 1: Identify all conditions of motion:

1. Whole stack of boxes slides
2. Boxes C and B remain together and slide relative to box A
3. Only box C slides
4. Whole stack tips
5. Box C and B remain together and tip
6. Box C tips

STEPS 2, 3, AND 4: Construct your FBD for each case and use your equations of equilibrium to solve for unknowns. Repeat for the other cases. ***Note:*** Sliding can always be found using $\Sigma F_x = 0$ and $\Sigma F_y = 0$ only. Tipping will require the use of a moment equation ($\Sigma M_z = 0$).

FRICTION

#1. WHOLE STOCK OF BOXES SLIDES

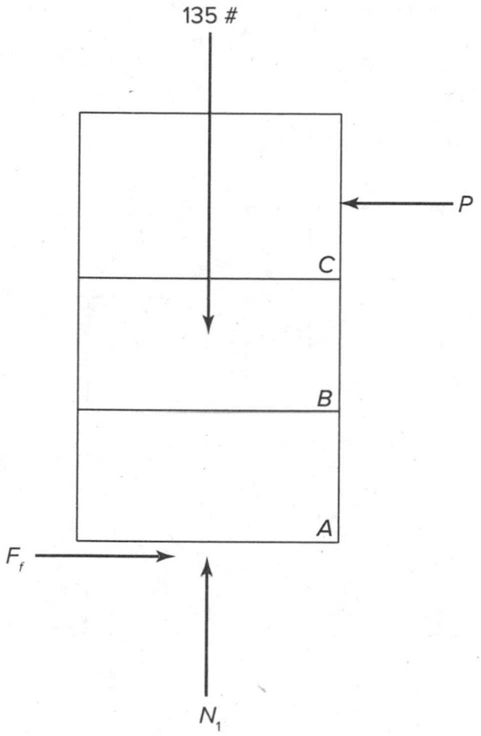

$\Sigma F_x = 0 = F_f - P$

$P = \mu_s N_1 = 0.45 N_1$

$\Sigma F_y = 0 = N_1 - 135$

$N_1 = 135$ #

Substituting N_1 into the first equation:

$P = 0.45(135) = 60.75$ #

#2. BOXES C AND B REMAIN TOGETHER AND SLIDE RELATIVE TO BOX A

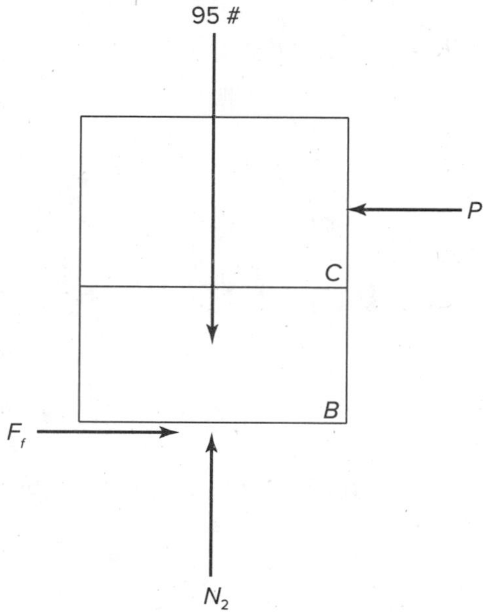

$\Sigma F_x = 0 = F_f - P$

$P = \mu_s N_2 = 0.35 N_2$

$\Sigma F_y = 0 = N_2 - 95$

$N_2 = 95$ #

Substituting N_2 into the first equation:

$P = 0.35(95) = 33.25$ #

#3. ONLY BOX C SLIDES

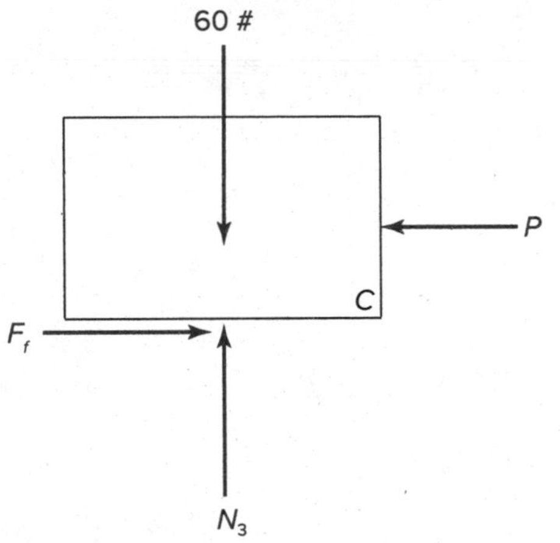

$$\Sigma F_x = 0 = F_f - P$$
$$P = \mu_s N_3 = 0.4N_3$$
$$\Sigma F_y = 0 = N_3 - 60$$
$$N_3 = 60\ \#$$

Substituting N_3 into the first equation:

$\mathbf{P = 0.4(60) = 24\ \#}$

#4. WHOLE STACK TIPS

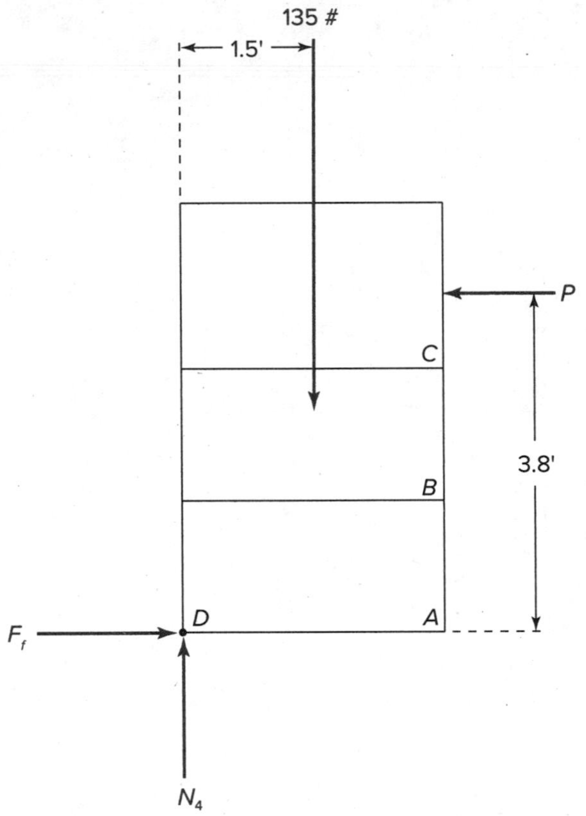

$$\Sigma M_D = 0 = P(3.8) - 135(1.5)$$

$$\mathbf{P = \frac{135(1.5)}{3.8} = 53.3\ \#}$$

Note: Always take the moment about the point at which the body will rotate to find the tipping force.

#5. BOX C AND B REMAIN TOGETHER AND TIP

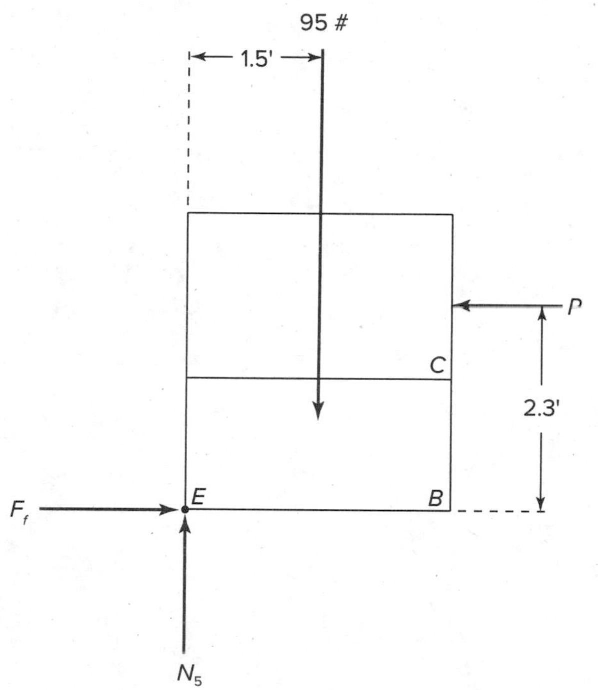

$$\Sigma M_E = 0 = P(2.3) - 95(1.5)$$

$$\mathbf{P = \frac{95(1.5)}{2.3} = 62\ \#}$$

#6. BOX C TIPS

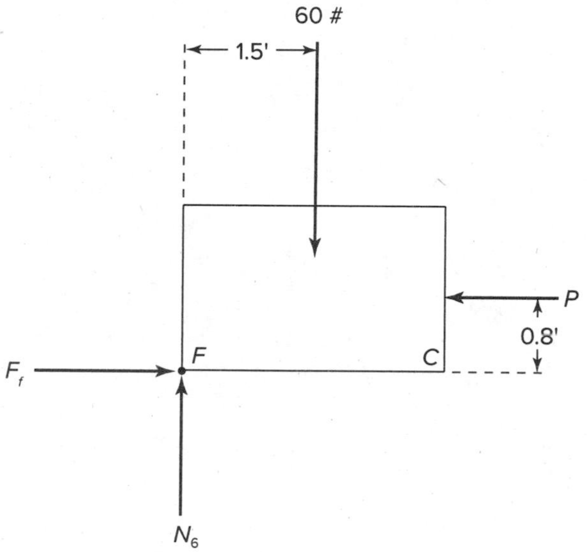

$$\Sigma M_F = 0 = P(0.8) - 60(1.5)$$

$$\mathbf{P = \frac{60(1.5)}{0.8} = 112.5\ \#}$$

STEP 5: Compare the results and select the force with the smallest value. The smallest represents the motion that will occur before all others.

$P_1 = 60.75\ \#$	$P_4 = 53.3\ \#$
$P_2 = 33.25\ \#$	$P_5 = 62\ \#$
$P_3 = 24\ \#$	$P_6 = 112.5\ \#$

For our result, it is clear to see that scenario 3, only box *C* slides, will be the first to move. It is very important to test all scenarios to ensure you have selected the correct scenario.

PRO TIP

These questions are professors' favorites as they are thought questions. Problems that ask for forces that cause "motion" are interesting because the motion scenarios are not defined but must be identified by the student.

If each box weighs 150 #, find the force *P* where motion occurs.

Assume the following:

$\mu_s = 0.65$ box to box

$\mu_s = 0.35$ box to floor

Press pause on video lesson 63 once you get to the workout problem. Only press play if you get stuck.

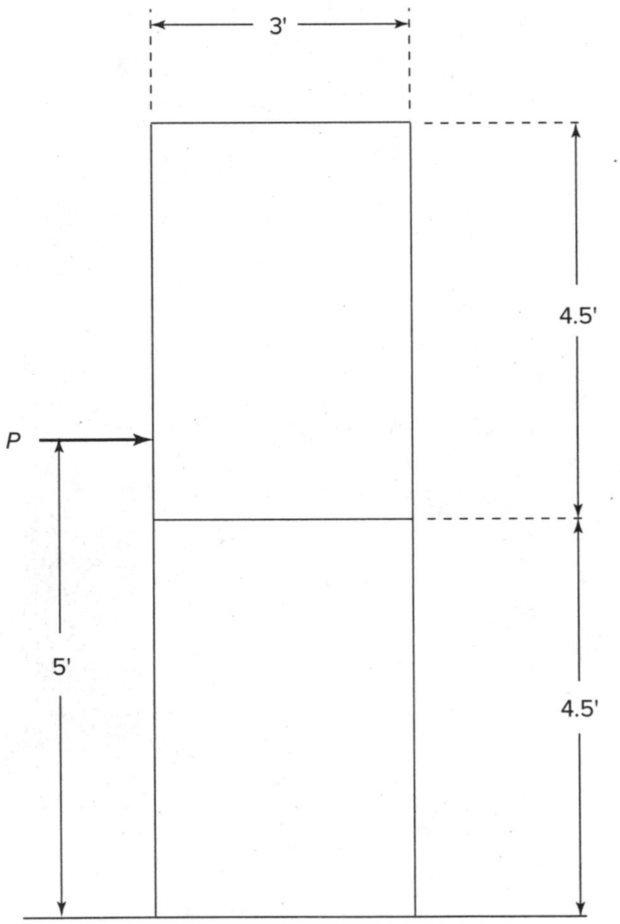

✓ TEST YOURSELF 8.2

SOLUTION TO TEST YOURSELF: Friction

What will happen first? Will the block tip or slip?

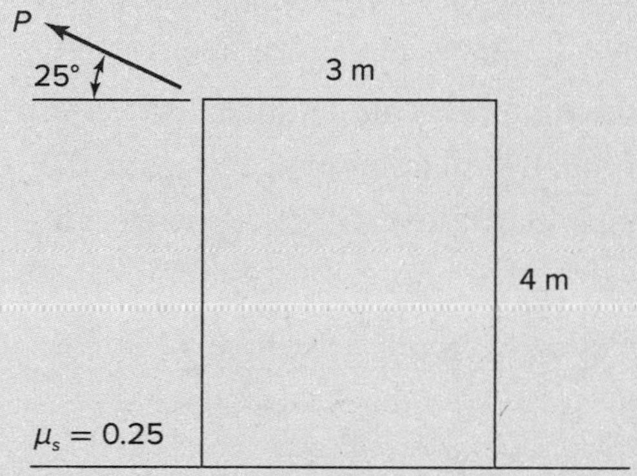

ANSWER

The box slips without tipping.

WATCH VIDEO

STATICS LESSON 64
Friction On Wedges, Example Problems

- Wedge problems are when both objects slide relative to each other.
- Make sure to get good at drawing FBDs as writing the equations of equilibrium should then be easy.
- Figuring out the direction of friction when interacting with stationary objects (i.e., wall) is easy. (If the block slides to the left, then the friction is going to the right.) It is more difficult when you have non-stationary objects (two objects sliding on each other). Remember when drawing the FBD, make yourself the free body. Think about another block sliding against you: what would you feel? The thing that you "feel" is the friction.

Wedge Friction Problem Recipe

STEP 1: Read the question and visualize the motion (which blocks slide and in what direction).

STEP 2: Construct your FBD for each case.

STEP 3: Identify which of the friction forces are impending motion such that $F = \mu_s N$.

STEP 4: Use your equations of equilibrium to solve for unknowns ($\Sigma F_x = 0$, $\Sigma F_y = 0$, and/or $\Sigma M_z = 0$).

STEP 5: Take the information learned on the previous FBD and apply it to the next FBD.

PRO TIPS

- **If you aren't given any distances in a problem, then you can't take the moment, so solve using $\Sigma F_x = 0$ and $\Sigma F_y = 0$.**
- **When you think of a normal force, you should ask yourself, "what would a bathroom scale read?"**

PITFALL

Don't forget any time two bodies are in contact, there is a normal force! And any time there is a normal force there is a friction force unless you see the words "smooth surface."

EXAMPLE: WEDGE FRICTION

If block A weighs 125 N and block B weighs 250 N, find the force P required to lift block B. Friction between block A and the floor is $\mu_s = 0.4$, between blocks A and B, $\mu_s = 0.32$, and friction between block B and the wall is $\mu_s = 0.45$.

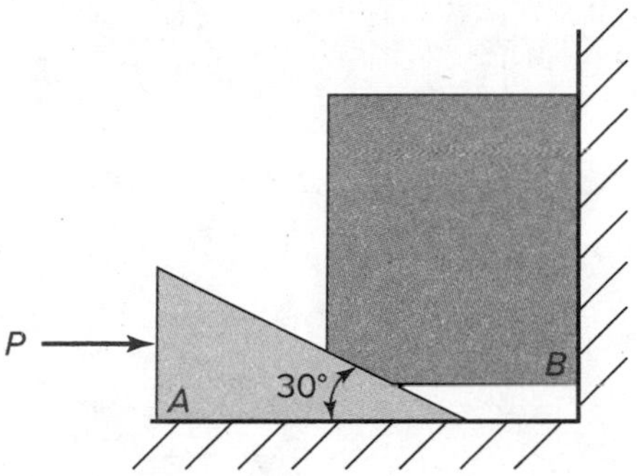

STEP 1: Force P causes block A to move to the right, which in turn pushes block B upward.

STEP 2: Draw FBD of each component in this system.

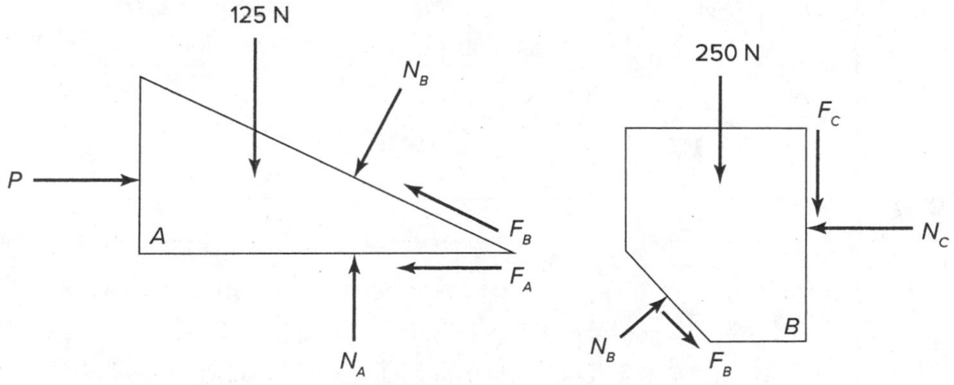

Note: It is not important what you call the normal forces as long as they are different values. Calling every normal force "N" might lead you to believe they are all equal, but they are not!

STEP 3: Ensure that every surface with friction is going to have impending motion. For surface(s) with impending motion, substitute $\mu_s N$ for the F_f. If there is no impending motion, the friction will simply be another unknown force, F. For this particular problem, all surfaces have impending motion and so all friction forces can be replaced with $\mu_s N$. Let's redraw the FBDs.

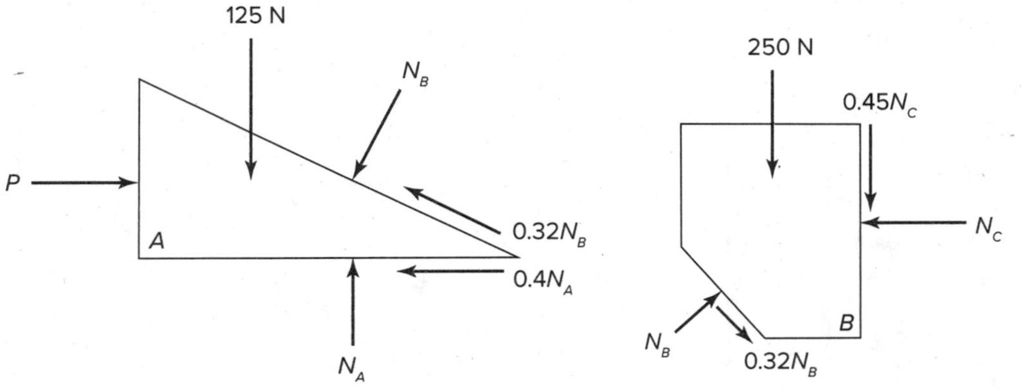

STEP 4: Use your equations of equilibrium to solve for unknowns $\Sigma F_x = 0$, $\Sigma F_y = 0$, and/or $\Sigma M_z = 0$.

Begin analysis with the FBD with the fewest number of unknowns. In this problem, that would be the rectangular wedge as it only has two unknowns, N_B and N_C. Write equations of equilibrium for that FBD.

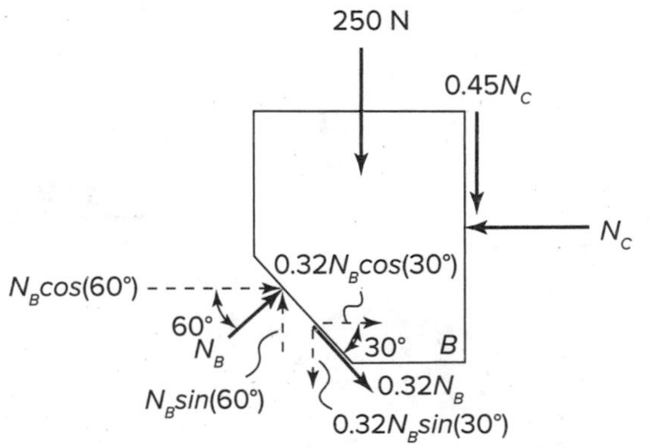

$$\Sigma F_x = 0 = N_B cos(60°) + (0.32)N_B cos(30°) - N_C$$
$$0.777N_B - N_C = 0$$
$$\Sigma F_y = 0 = N_B sin(60°) - (0.32)N_B sin(30°) - 0.45N_C - 250$$
$$0.706N_B - 0.45N_C = 250$$
$$N_B = 701.6 \text{ N}$$
$$N_C = 545.1 \text{ N}$$

FRICTION

STEP 5: Take this information and apply to the next free body diagram. Write equations of equilibrium for the next free body diagram and solve.

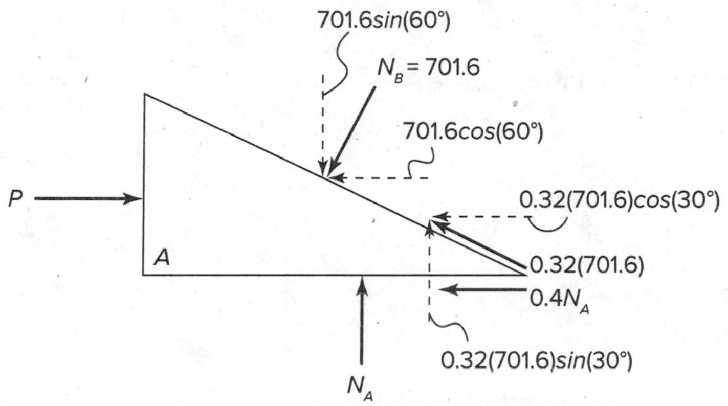

$\Sigma F_x = 0 = P - 701.6cos(60°) - 0.32(701.6)cos(30°) - 0.4N_A$

$0 = P - 350.8 - 194.4 - 0.4N_A$

$P - (0.4)N_A = 545.2$

$\Sigma F_y = 0 = N_A - 701.6sin(60°) + 0.32(701.6)sin(30°)$

$0 = N_A - 607.6 + 112.3$

$N_A = 495.3 \text{ N}$

Substituting N_A into the previous equation:

$P - (0.4)(495.3) = 545.2$

$P = 545.2 - 198.12$

$P = 347.08 \text{ N}$

✓ TEST YOURSELF 8.3

SOLUTION TO TEST YOURSELF: Friction

For the block problems below, assume that the motion is in the direction of the applied force (P) and all surfaces have friction; draw the free body diagrams for each block.

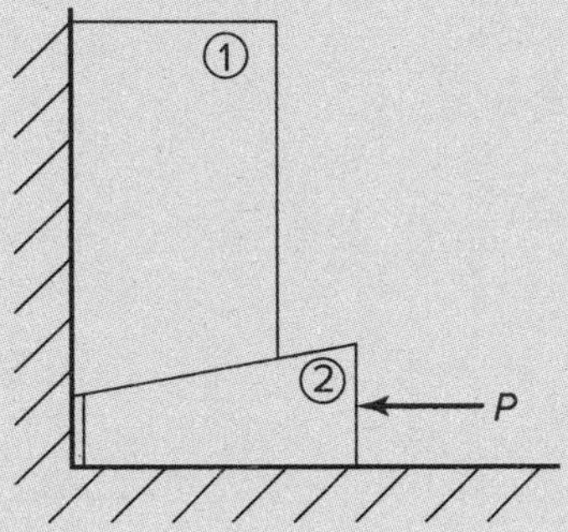

1

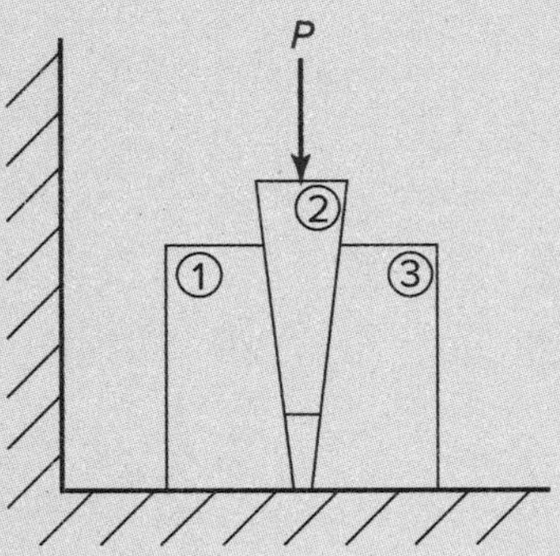

2

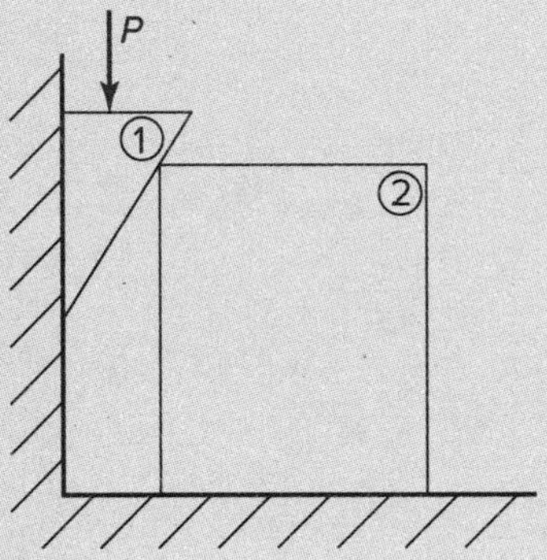

3

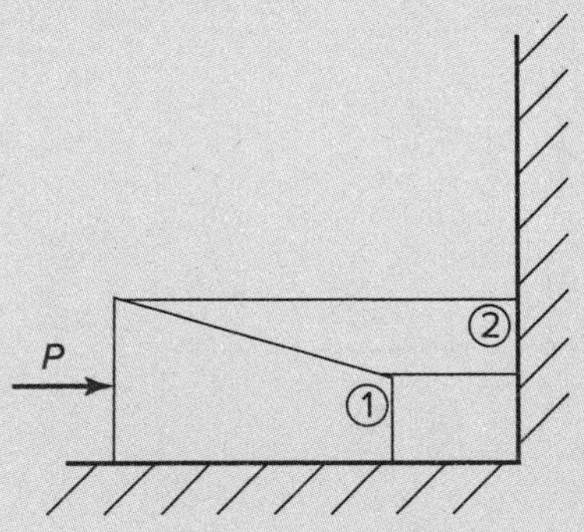

4

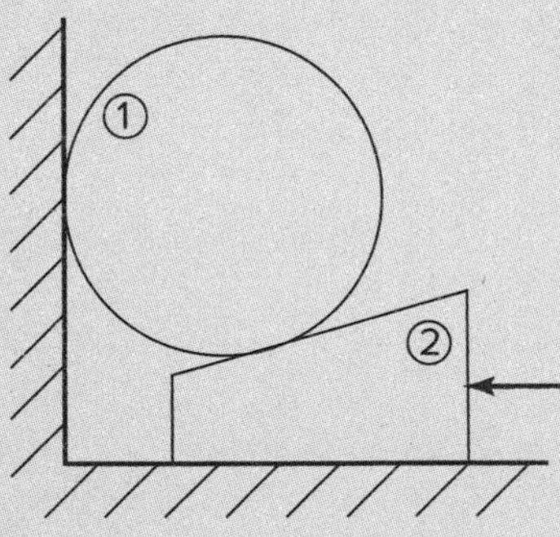

5

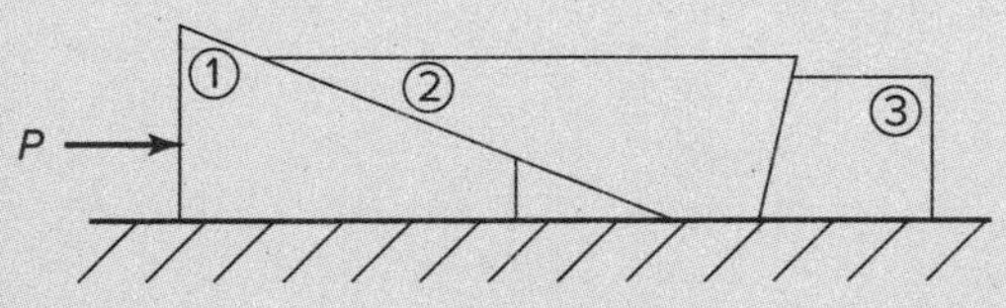

6

ANSWERS

See Test Yourself 8.3 solutions at the back of the book.

Find the smallest horizontal force P that will cause motion in block B. Block $A = 300$ N, block $B = 120$ N, and block $C = 600$ N.

Assume the following:

$\mu_s = 0.4$ between blocks and the floor

$\mu_s = 0.2$ between all blocks.

Press pause on video lesson 64 once you get to the workout problem. Only press play if you get stuck.

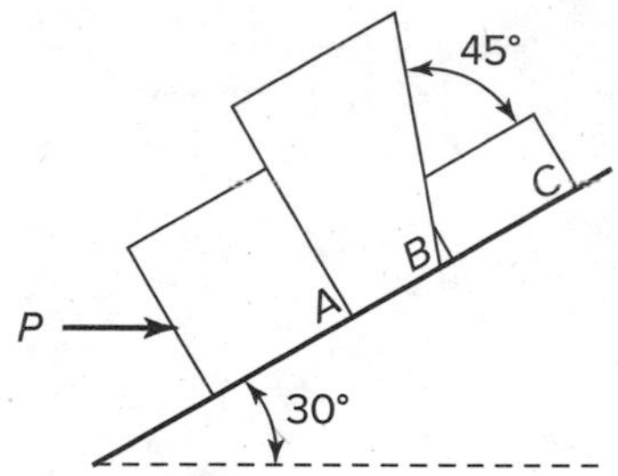

WATCH VIDEO

STATICS LESSON 65
Challenging Friction Wedge Problem with Roller

Find the smallest weight of the wedge that will cause motion.

Assume the following:

$\mu_s = 0.5$ at points A and C

$\mu_s = 0.6$ at point B

Press pause on the video once you get to the workout problem. Only press play if you get stuck.

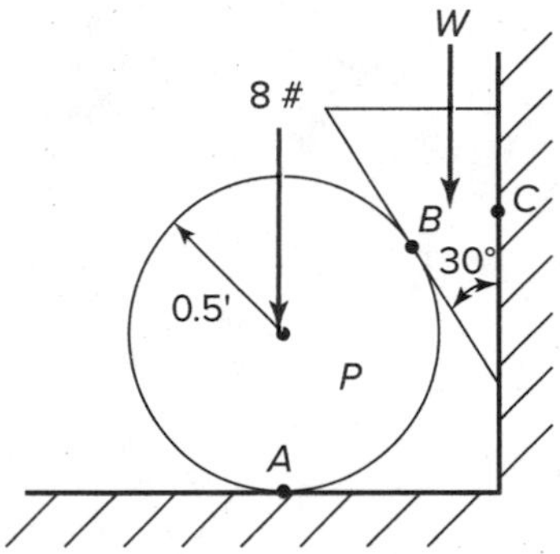

✓ TEST YOURSELF 8.4

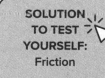

SOLUTION TO TEST YOURSELF: Friction

What force (P) is required to move block B upward?

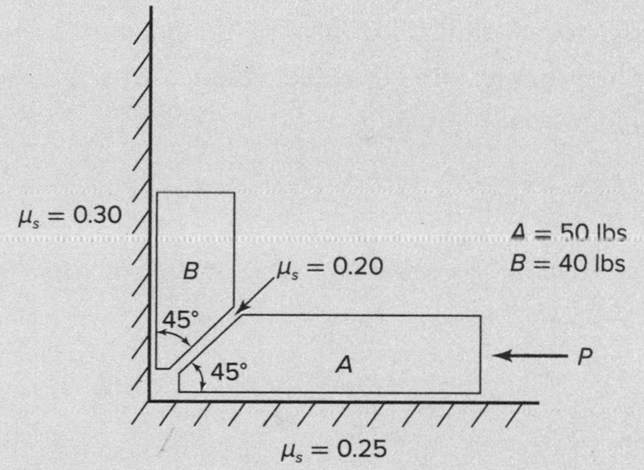

ANSWER

$P = 139.8$ lbs

WATCH VIDEO

STATICS LESSON 66
Belt Friction Problem

- Typically, we assume pulleys are perfectly frictionless, but for belts there is a friction between the belt and what it is rubbing against (i.e., roller, rock, tree, etc.).
- You can avoid getting this problem wrong by using your intuition. The force to lift the object plus the friction of the roller is much more force then simply lifting the object alone. Thus, it is easy to identify which tension on what side of the roller has more force.

$$T_1 = T_2 e^{\mu_s \beta}$$

- For belt friction:
  - T_2: Force that opposes the motion
  - T_1: Force in the direction of motion
  - β: Angle of contact in radians
  - μ_s: The static coefficient of friction

Belt Friction Recipe

STEP 1: Draw FBD of the system.

STEP 2: Determine the contact angle between the surface and the belt, rope, chain, cable, etc. (make sure this angle is in radians!).

STEP 3: Using the belt friction equation, find the unknown:

$T_1 = T_2 e^{\mu_s \beta}$

T_2 will always be the smaller tension
T_1 will always be the largest tension of the two

Note: This assumption is true only in a tension situation, when something is being raised or pulled.

PITFALL

A common mistake on belt friction is to incorrectly assume what is T_1 and T_2, i.e., flip-flopping them. Use your intuition to determine from the problem statement, which one of the tensions will be the larger or smaller. T_2 will always be the smaller tension; T_1 will always be the larger tension of the two.

EXAMPLE: BELT FRICTION

Determine the force (P) required to lift the box if the rope is passing around a barrel and the static coefficient between the barrel and the rope is 0.45. The weight of the box is 100 #.

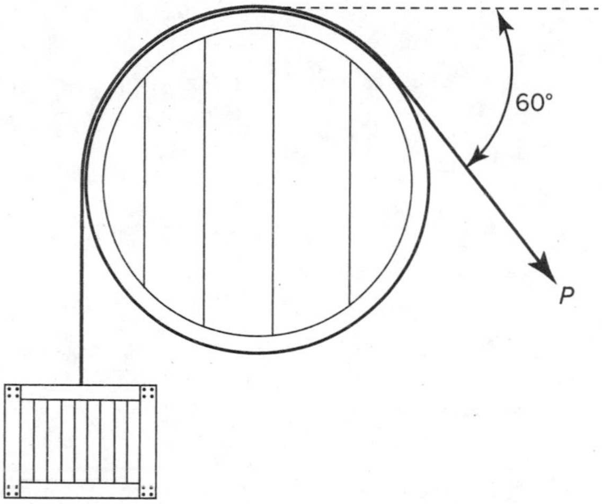

STEP 1: Draw FBD of the system.

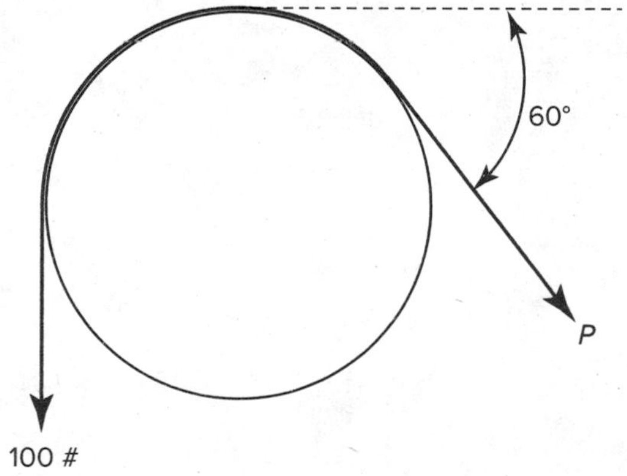

STEP 2: Determine the angle of the wrap of the rope around the pulley (make sure this angle is in radians!).

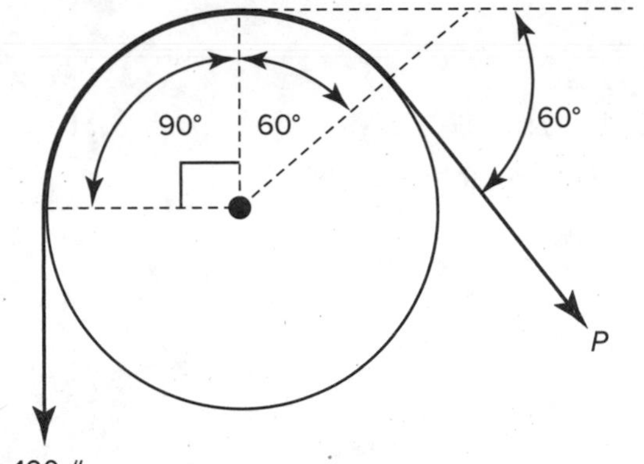

$150° = 2.62$ radians $= \beta$

STEP 3: Using the belt friction equation, find the force P.

$T_1 = T_2 e^{\mu_s \beta}$

T_2 will always be the smaller tension
T_1 will be the larger tension of the two

Here, the pulley force (P) will be the larger tension since it has to lift the box plus overcome the belt friction.

$T_1 = P = 100(e)^{(0.45)(2.62)}$
$P = 325.1$ lbs

Find the force required to begin to raise Cliff with the rope and then find the force required such that Cliff remains in equilibrium (i.e., he doesn't fall and is just chillin').

Assume the following:

$\mu_s = 0.2$ for edge of the rock

Press pause on video lesson 66 once you get to the workout problem. Only press play if you get stuck.

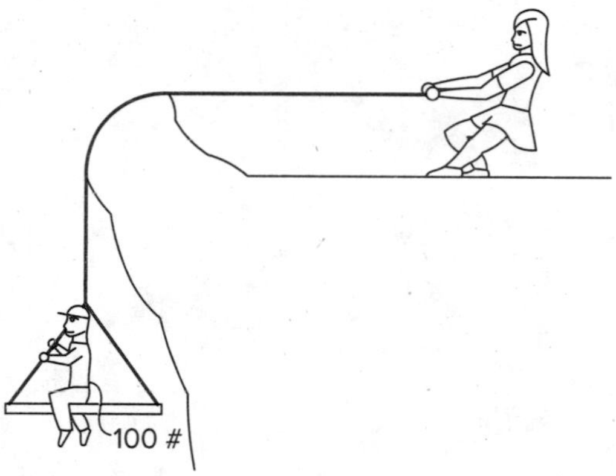

TEST YOURSELF 8.5

SOLUTION TO TEST YOURSELF: Friction

What is the minimum μ_s to hold the system in equilibrium?

Note: In this problem, we are not assuming that the pulley is frictionless.

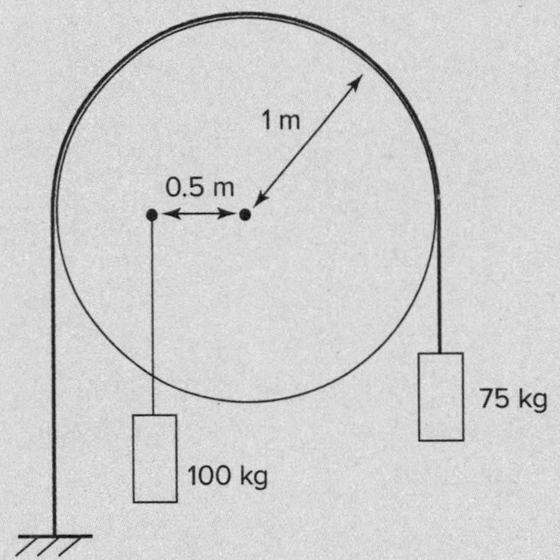

ANSWER

$\mu_s = 0.35$

KEY TAKEAWAYS

Friction problems are solved like any other statics problems except that we introduced one new element "the friction force" which is technically not a force but rather a resistance to motion.

Main takeaways from Level 8 include how to calculate:

- Simple friction
- Slipping friction versus tipping
- Wedges
- Belt friction

Simple Friction Summary

- We consider simple friction problems as ones that generally have a single component, and those problems can be solved with one set of equations: $\Sigma F_x = 0$, $\Sigma F_y = 0$, or $\Sigma M_z = 0$.
- You have to understand which direction the friction goes, otherwise if you assume the direction wrong you will get an incorrect answer.
- Don't be fooled if you are given both μ_s and μ_k. Read the problem carefully because if everything is stationary, then μ_K is the distractor. The only time you will use μ_k is if you have motion with constant velocity.

Slipping Friction Versus Tipping Summary

- The key to these problems is identifying all of the possible scenarios for motion such as all slipping scenarios and all tipping scenarios.
- Once all the different types of motion are identified, it is very important that you analyze each one of these scenarios and compare the force required for motion for all scenarios. The smallest force identified is the scenario that happens first (i.e., the answer).
- To analyze tipping problems, always identify the point at which your FBD will rotate about. This point is where you will take your moment about to find the unknown tipping force.

KEY TAKEAWAYS

Wedges Summary

- You need to practice drawing FBD problems over and over. Definitely utilize the Test Yourself problems in this level. If you get good at drawing the FBD, writing the equations of equilibrium should be easy.
- Figuring out the direction of friction when interacting with stationary objects (i.e., wall) is the easier of the two types. (If the block slides to the left, then the friction is going to the right.) It is more difficult when you have non-stationary objects (two objects sliding on each other). You have to remember when drawing the FBD, make yourself the free body. Think about another block sliding against you: what would you feel? The thing that you "feel" is the friction.

Belt Friction Summary

- Typically, we assume pulleys are perfectly frictionless, but for belts there is a friction between the belt and what it is rubbing against (i.e., roller, rock, tree, etc.).
- You can avoid getting this problem wrong by using your intuition. The force to lift the object plus the friction of the roller is much more force then simply lifting the object alone. Thus, it is easy to identify which tension on what side of the roller has more force.
- For belt friction:
  - T_2: Force that opposes the motion
  - T_1: Force in the direction of motion
  - β: Angle of contact in radians

KEY TAKEAWAYS

RECIPES

- **Solving Simple Friction Recipe**
- **Slipping or Tipping Recipe**
- **Wedge Friction Problem Recipe**
- **Belt Friction Recipe**

PRO TIPS

Simple Friction

- **If there is anything you can do to be successful in this level, it is to practice FBDs. Practice, practice, practice!**
- **Sometimes it might be easier to tilt your *x-y* coordinate axis so that it aligns with the unknowns you are trying to solve, especially any time you have two unknowns that are perpendicular to each other.**

Slipping or Tipping

- **These questions are professors, favorites as they are thought questions. Problems that ask for forces that cause "motion" are interesting because the motion scenarios are not defined but must be identified by the student.**

Wedge Friction

- **If you aren't given any distances in a problem, then you can't take the moment, so solve using $\Sigma F_x = 0$ and $\Sigma F_y = 0$.**
- **When you think of a normal force, you should ask yourself, "what would a bathroom scale read?"**

KEY TAKEAWAYS

PITFALLS

Simple Friction

- You have to guess the direction of the friction correctly, unlike when solving other problem types. If you guess it incorrectly, you won't simply get a negative sign. You will get a wrong answer.

Slipping or Tipping

- You need to figure out what scenarios of motion can happen and pick the smallest value. The smallest value is what is going to occur first. You will never get to the bigger values as the little one happens first.

Wedge Friction

- Don't forget any time two bodies are in contact, there is a normal force! And any time there is a normal force there is a friction force unless you see the words "smooth surface."

Belt Friction

- A common mistake on belt friction is to incorrectly assume what is T_1 and T_2, i.e., flip-flopping them. Use your intuition to determine from the problem statement, which one of the tensions will be the larger or smaller. T_2 will always be the smaller tension; T_1 will always be the larger tension of the two.

Level 9

Second Moment of Area (Moment of Inertia)

INTRO TO MOI

COMPOSITE SHAPES METHOD

CALCULUS METHOD

MISSION ACCOMPLISHED

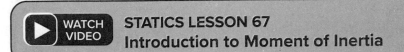

WATCH VIDEO **STATICS LESSON 67**
Introduction to Moment of Inertia

Moment of Inertia (I)

- Also called "second moment of area"
- Geometric property of the cross-section of a beam
- Describes the flexibility of beams (bendiness of the beam)
- Used to calculate bending stress in future Mechanics of Materials (Solids) course

Neutral Axis

Imagine a beam is bending. The top fibers of the beam shown below would be in compression, and the bottom fibers are in tension. At some point in the beam, it is in neither tension nor compression (i.e., it is neutral), and that location is the neutral axis (also called the *centroidal axis*). Note this is true for straight beams only not curved beams.

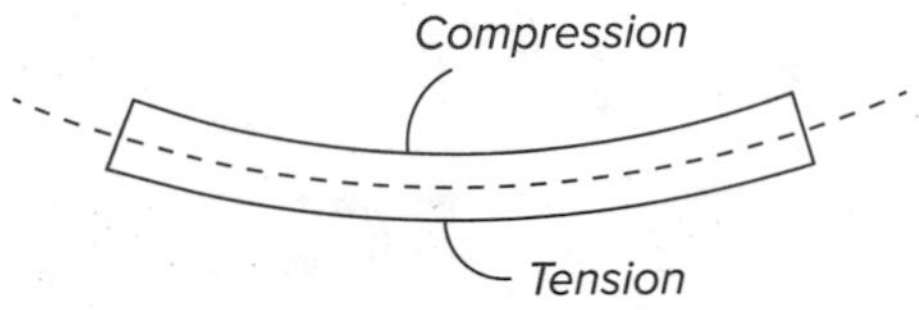

The more area you have farther away from the neutral axis the "less bendy" that cross-section becomes. The diving board is so flexible because the neutral axis is located so close to the top and bottom of the beam. For the I-beam, there is a significant amount of material far away from the neutral axis making it much stiffer.

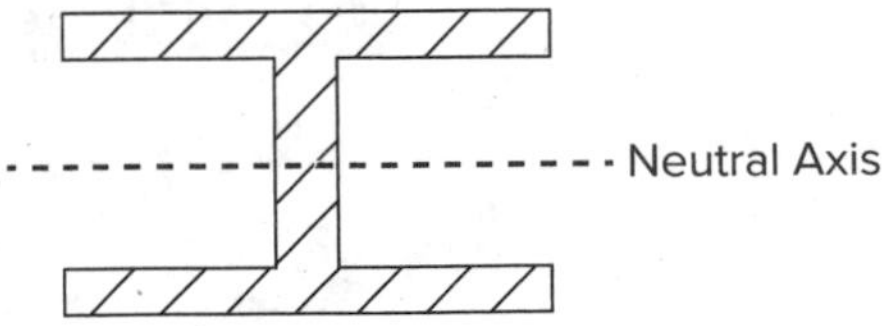

I-beam
Cross-section that is rigid

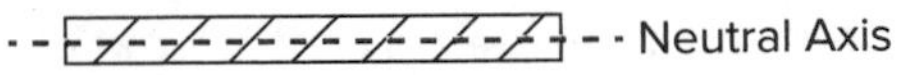

Diving board
Cross-section that is flexible

All shapes in the real world always bend around their neutral axis. If your book provides an arbitrary x and y axis, this is for practice only and not what would really happen.

CHALLENGE QUESTION

Why is it called an I-beam and not an H-beam?

Second Moment of Area of Common Geometric Shapes

b = base; h = height, and r = radius

	Shape	MOI about *x*-axis	MOI about *y*-axis
Rectangle	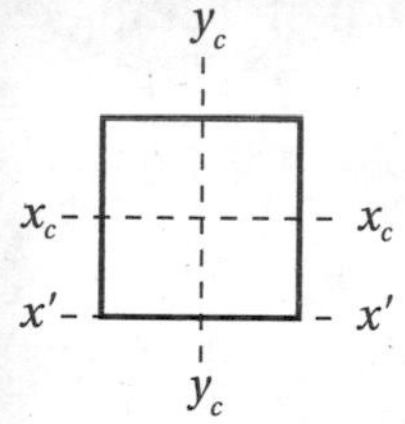	$I_{x_c} = \left(\frac{1}{12}\right)bh^3$ $I_{x'} = \left(\frac{bh^3}{3}\right)$	$I_{y_c} = \left(\frac{1}{12}\right)hb^3$
Triangle	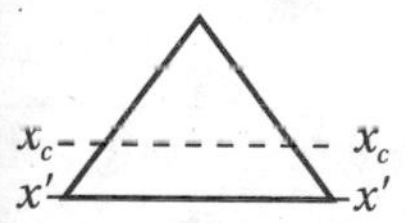	$I_{x_c} = \left(\frac{1}{36}\right)bh^3$ $I_{x'} = \left(\frac{bh^3}{12}\right)$	
Circle	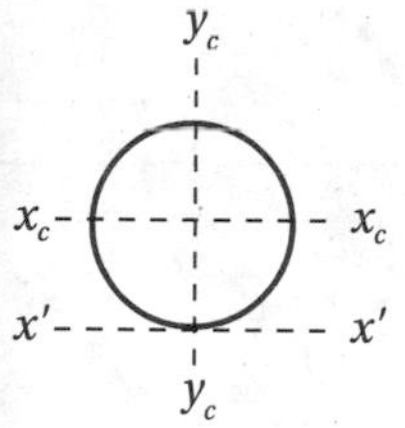	$I_{x_c} = \left(\frac{1}{4}\right)\pi r^4$ $I_{x'} = \left(\frac{5\pi r^4}{4}\right)$	$I_{y_c} = \left(\frac{1}{4}\right)\pi r^4$
Semi-circle	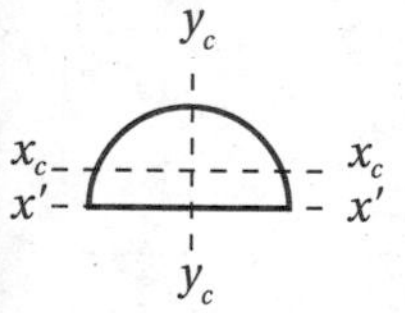	$I_{x_c} = \left(\frac{\pi r^4}{8}\right) - \left(\frac{8r^4}{9\pi}\right)$ $I_{x'} = \left(\frac{1}{8}\right)\pi r^4$	$I_{y_c} = \left(\frac{1}{8}\right)\pi r^4$
Quarter-circle		$I_{x_c} = \left(\frac{\pi r^4}{16}\right) - \left(\frac{4r^4}{9\pi}\right)$ $I_{x'} = \left(\frac{\pi r^4}{16}\right)$	

Note: The only time that you can use I_{x_c} or I_{y_c} is if the piece-parts' centroid lies on the neutral axis of the whole-part; otherwise you will have to utilize the parallel axis theorem (see next video).

ANSWER

Because it is always oriented so it has the maximum area moment of inertia (strong bending axis).

STAR CONCEPT

MOMENT OF INERTIA NOTATION

- I_{x_c} and I_{y_c} refer to the MOI of an individual piece about its centroid
- I_x and I_y refer to the MOI of the entire composite shape about its centroid, made up of multiple pieces
- $I_{x'}$ and $I_{y'}$ refer to the MOI for the axis shown (i.e., not at the centroid)

- For most shapes you can use the tables provided. If the shape isn't composed of rectangles, squares, triangles, and circles, use calculus.

$$I_{x'} = \int_0^y y^2 dA \qquad I_{y'} = \int_0^x x^2 dA$$

x = strip width
y = strip height
dA = differential area

- The unit for the second area moment of inertia may seem strange (length[4]). So what does it represent? I is a geometric property of a cross-section that describes a beam's ability to resist bending. It is not a space/time dimension.

PRO TIPS

- **Most of the time bending occurs around the x_c-axis because of the loading direction, which is typically due to gravity (exception is wind loading).**
- **When doing I_{y_c}, rather than use different equations just rotate your head by 90°.**

PITFALL

If the beam is not symmetric, then you will first need to calculate the centroid of the cross-section *before* calculating the moment of inertia.

WATCH VIDEO

STATICS LESSON 68
Parallel Axis Theorem, Moment of Inertia

Parallel Axis Theorem for composite shapes: $I_x = \Sigma(I_{x_c} + Ad^2)_i$

I_x: The second area moment of inertia for the whole-part

I_{x_c}: The piece-parts' second area moment of inertia about its centroid (see common shapes table)

A: The piece-parts' area

d: Distance from the centroid of the piece-part to the neutral axis of the whole-part

i: The number of shapes you have

Σ: Add these up for each simple shape

Moment of Inertia Composite Shapes Recipe

STEP 1: Break the composite shape into as many simple shapes as necessary to calculate the centroid (neutral axis), refer to the table, Properties of Geometric Shapes and Areas. If you are calculating moment of inertia bending around the x-axis, calculate $\bar{y}$; if you are calculating moment of inertia bending around the y-axis, calculate $\bar{x}$.

$$\bar{x} = \frac{\Sigma x_i A_i}{\Sigma A_i}, \quad \bar{y} = \frac{\Sigma y_i A_i}{\Sigma A_i}$$

Recall: Here i represents a particular piece of the entire object you are analyzing. The x and y are the distance to the center of that particular piece as measured from the datum. Thus, you are essentially adding up all the pieces. If a piece is a void, don't forget to subtract it! See Level 5 for an example problem.

Note: If you have a symmetric beam, try to pick composite shapes so that you don't need to use the parallel axis theorem. In other words, try to pick piece-parts whose centroid lies on the neutral axis of the whole-part.

STEP 2: Using the same composite shapes, refer to the table, Second Moment of Area of Common Geometric Shapes to calculate the area moment of inertia using the equation $I_x = \Sigma(I_{x_c} + Ad^2)_i$. See above for variable definitions.

Note: Check your math as this is a very easy place to make a calculator error.

EXAMPLE: MOI W/O PARALLEL AXIS THEOREM

Find I_x about the centroidal axis for the whole part.

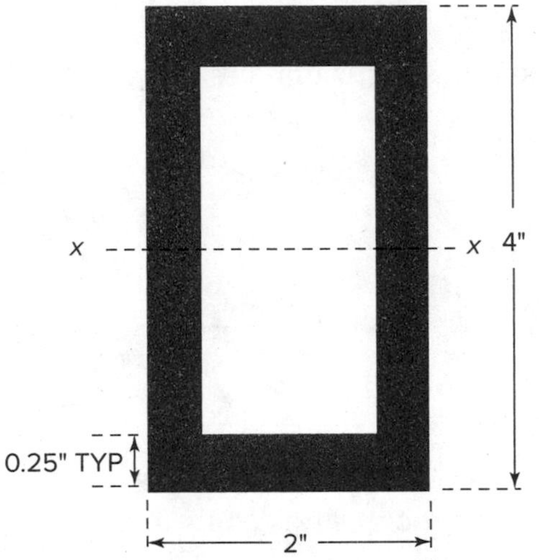

Note: TYP means typical. For this problem, it means that the wall thickness is 0.25" everywhere.

STEP 1: By observation, we have already determined the centroid is 2" from the bottom.

STEP 2: Break the composite shape into as many simple shapes as necessary to calculate the area moment of inertia.

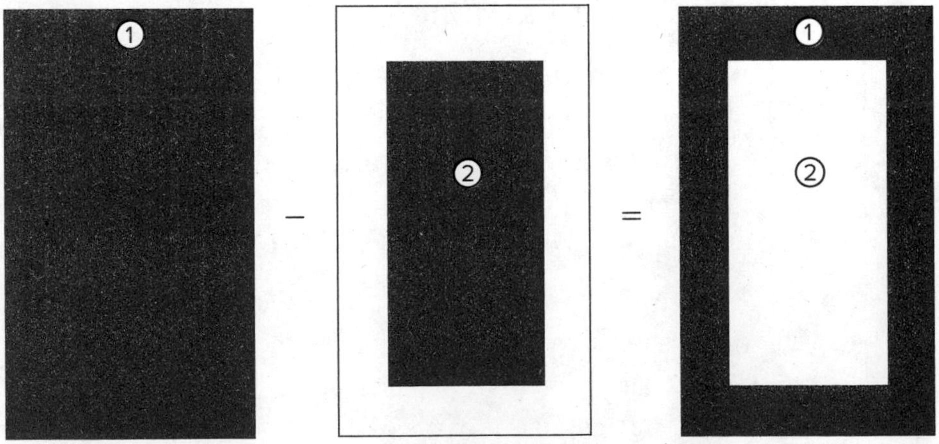

Calculate the area moment of inertia using the equation, $I_x = \Sigma(I_{x_c} + Ad^2)_i$. Note in this case, the distance for this equation is 0 for both the outer rectangle and inner rectangles since each of these piece-part centroid lies on the neutral axis of whole-part..

$$I_x = I_{x_c \text{ outer rect}} - I_{x_c \text{ inner rect}} = \frac{1}{12}(2)(4)^3 - \frac{1}{12}(1.5)(3.5)^3 = 5.31 \text{ in}^4$$

Find I_x about the centroidal axis without using the parallel axis theorem.

Press pause on video lesson 68 once you get to the workout problem. Only press play if you get stuck.

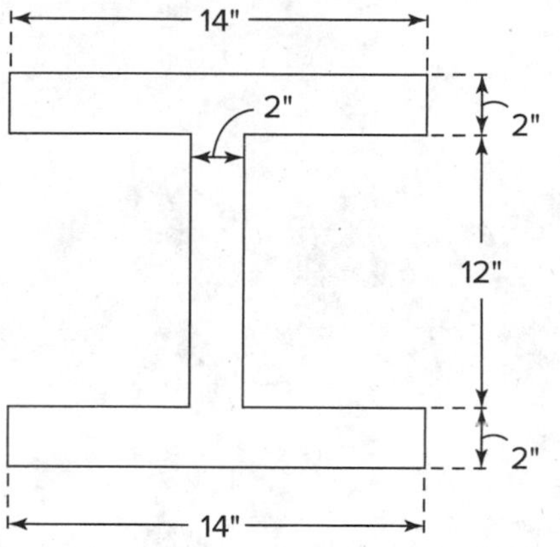

✓ TEST YOURSELF 9.1

SOLUTION TO TEST YOURSELF: Second Moment of Area

Consider the beam with the cross-section shown below. Compute the area moment of inertia about the x-axis through the centroid of the cross-section.

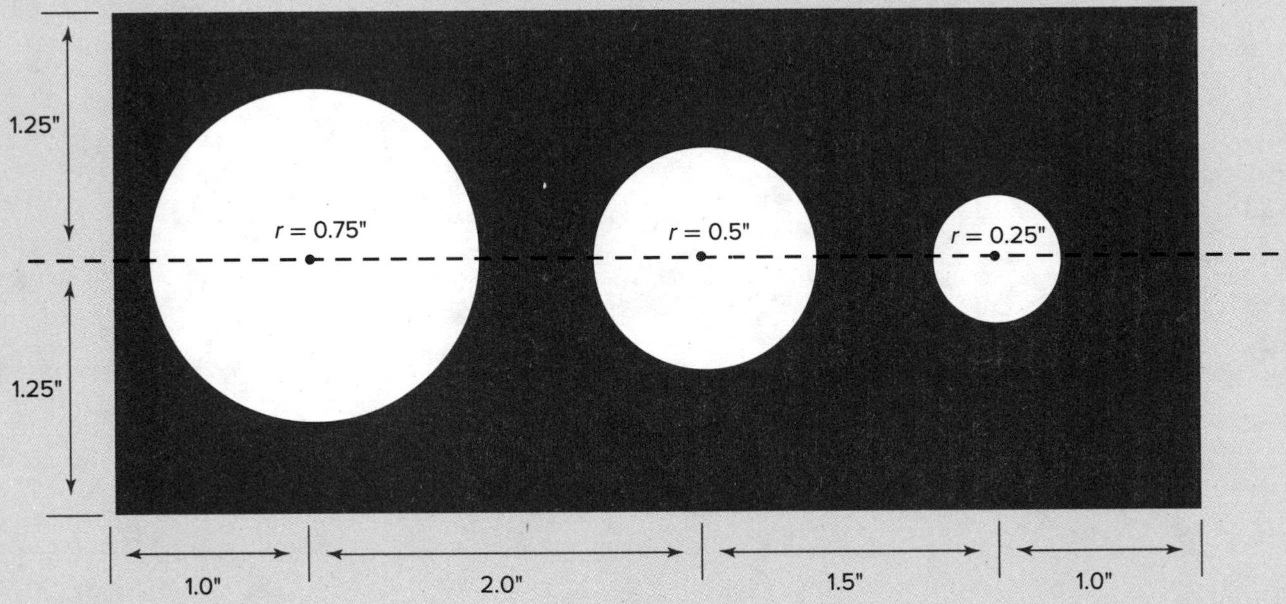

ANSWER

$I_x = 6.86\ \text{in}^4$

EXAMPLE: MOI W/ PARALLEL AXIS THEOREM

Find I_x about the centroidal axis. Note that we worked this problem earlier without using the parallel axis theorem, and you will see here we get the same answer if we use the parallel axis theorem.

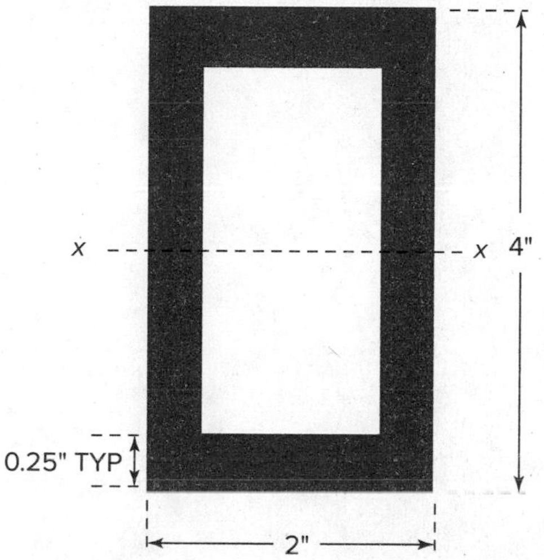

STEP 1: By observation, we have already determined the centroid is 2" from the bottom.

STEP 2: Calculate the area moment of inertia using the equation, $I_x = \Sigma(I_{x_c} + Ad^2)_i$.

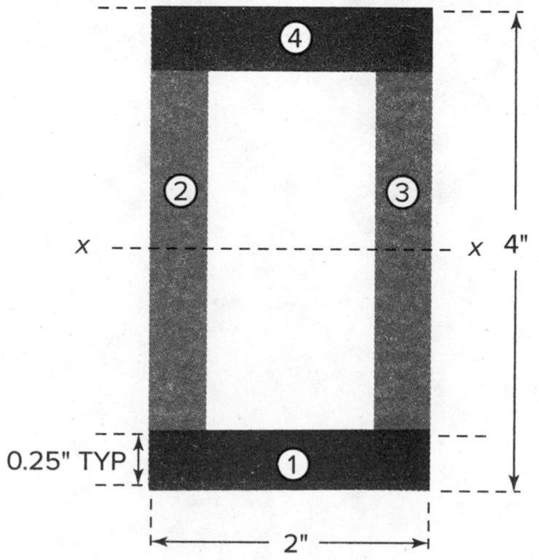

$$I_x = I_{x_1} + I_{x_2} + I_{x_3} + I_{x_4}$$

$$I_x = \left[\frac{1}{12}(2)(0.25)^3 + (0.5)(1.875)^2\right] + \left[\frac{1}{12}(0.25)(3.5)^3 + (0.875)(0)^2\right]$$

$$+ \left[\frac{1}{12}(0.25)(3.5)^3 + (0.875)(0)^2\right] + \left[\frac{1}{12}(2)(0.25)^3 + (0.5)(1.875)^2\right] = 5.31 \text{ in}^4$$

Find I_x about the centroidal axis using the parallel axis theorem.

Press pause on video lesson 68 once you get to the workout problem. Only press play if you get stuck.

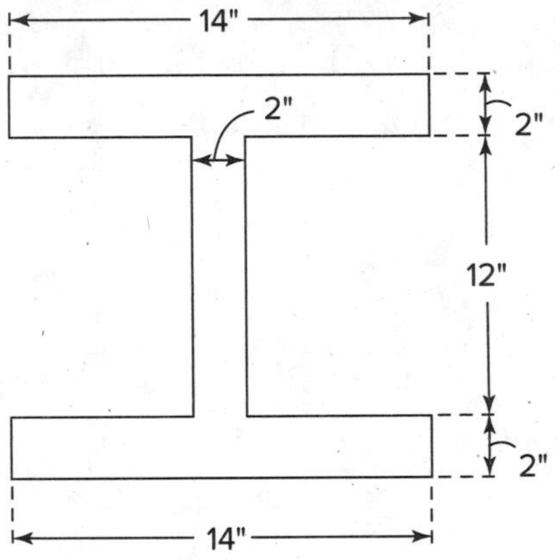

✓ TEST YOURSELF 9.2

SOLUTION TO TEST YOURSELF: Second Moment of Area

Consider the beam with the cross-section shown below. Compute the moment of inertia about the x-axis through the centroid of the cross-section.

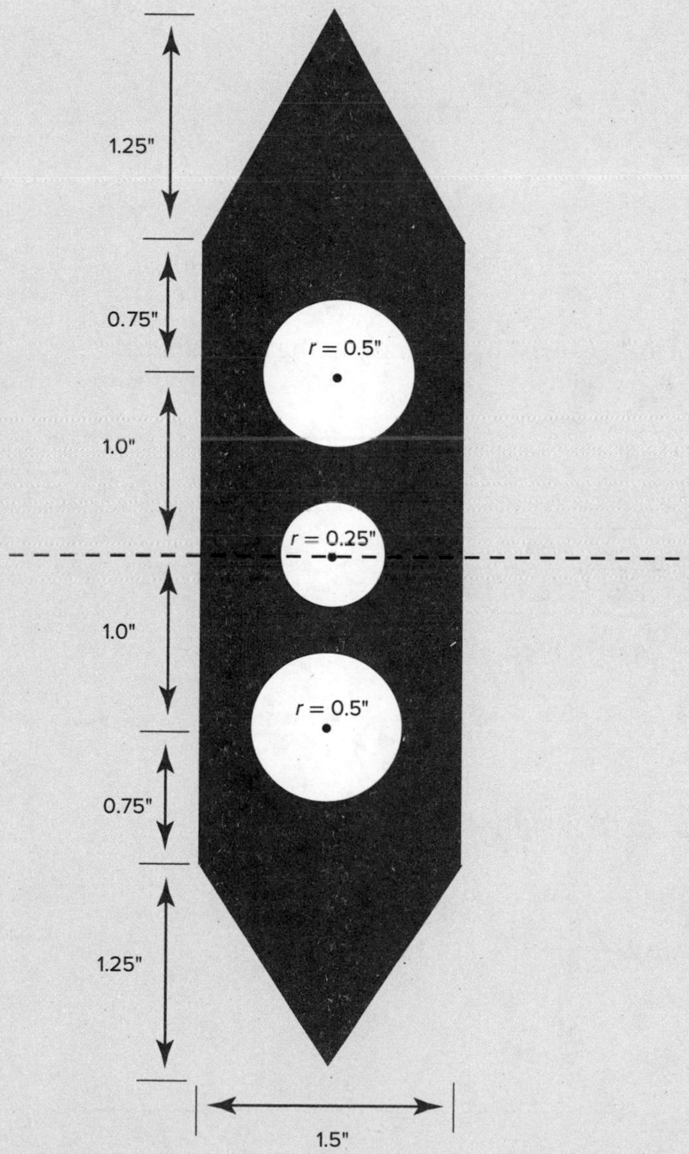

ANSWER

$I_x = 12.65\ \text{in}^4$

EXAMPLE: COMPOSITE SHAPE W/ CENTROID

Find I_x for the given shape.

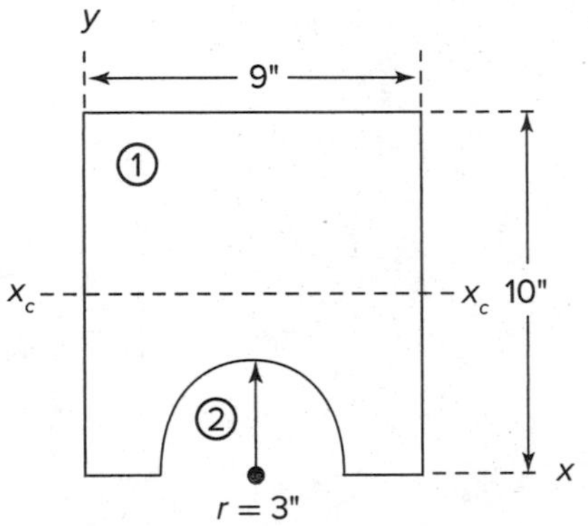

STEP 1: Determine the location of the neutral axis (x_c-axis) by finding $\bar{y}$ of the given shape.

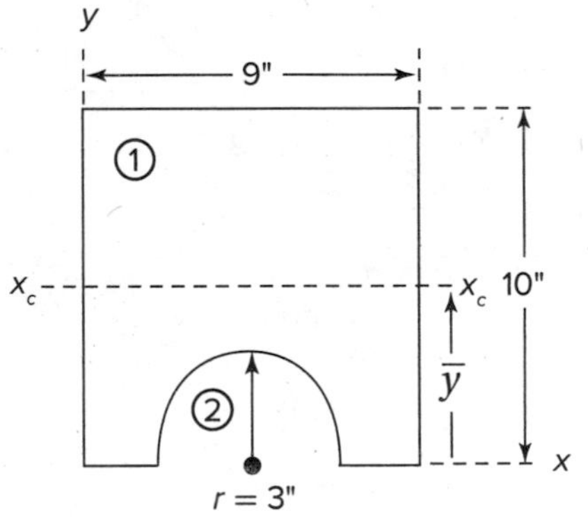

	y	A	yA
1	5	90	450
2	1.27	-14.14	-18.0

$\Sigma A = 75.86$, $\Sigma yA = 432$

$$\bar{y} = \frac{432}{75.86} = 5.70 \text{ in}$$

STEP 2: Find I_x using parallel axis theorem: $I_x = \Sigma(I_{x_c} + Ad^2)_i$.

Calculate the area MOI.

$$I_x = \left[\frac{1}{12}(9)(10)^3 + (90)(0.70)^2\right] - \left[\frac{\pi}{8}(3)^4 - \frac{(8 \times 3^4)}{(9 \times \pi)} + (14.14)(4.43)^2\right]$$

$$I_x = 794.1 - 286.4 = 507.7 \text{ in}^4$$

PRO TIP

- **When in doubt, always use the parallel axis theorem. If the piece-parts' centroid is on the neutral axis, the value of $d = 0$ and the Ad^2 portion of the equation goes away.**
- **When calculating d, all you need to do is simply pick the centroid of the whole part and subtract the centroid of the piece-part.**

WATCH VIDEO

STATICS LESSON 69
Moment of Inertia, Composite Shape Method

Find I_y about the neutral axis of the shaded area. ***Note:*** This is not asking you to find the I about the y-axis shown (that would be $I_{y'}$).

Press pause on the video once you get to the workout problem. Only press play if you get stuck.

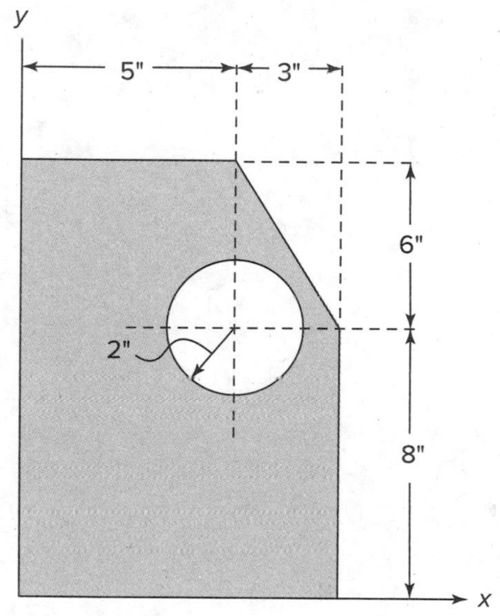

✓ TEST YOURSELF 9.3

SOLUTION TO TEST YOURSELF: Second Moment of Area

Consider the beam with the cross-section shown below. This cross-section was created by welding together two rectangular cross-sections 15 in x 2 in and 20 in x 5 in and a semi-circular cross-section (radius = 8 in). Compute the moment of inertia about the x-axis through the centroid of the cross-section.

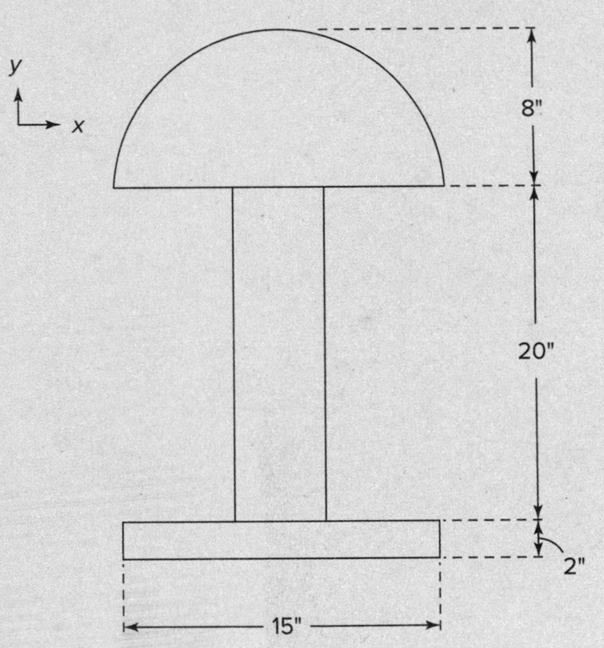

ANSWER

$I_x = 20{,}984$ in^4

WATCH VIDEO

STATICS LESSON 70
Moment of Inertia, Calculus Method

Use the following equations to calculate the area moment of inertia about the given axis.

$I_{x'} = \int y^2 \, dA$
$I_{y'} = \int x^2 \, dA$

x: Strip width
y: Strip height
dA: Differential area

Note: This looks a lot like the centroid by calculus equation except the x and y are squared.

Moment of Inertia Calculus Method Recipe

STEP 1: Recall the definition of area moment of inertia:
$I_{x'} = \int y^2 \, dA$
$I_{y'} = \int x^2 \, dA$
Draw one differential area "strip" to help set up the equations. For $I_{x'}$ use horizontal strips, and for $I_{y'}$ use vertical strips.

STEP 2: Write a function for dA exactly as we did when calculating centroids by calculus.

STEP 3: Simply plug in dA into the moment of inertia equations using limits from our stack of strips.

STEP 4: Solve.

EXAMPLE: MOI USING CALCULUS METHOD

Find the moment of inertia about the x-axis.

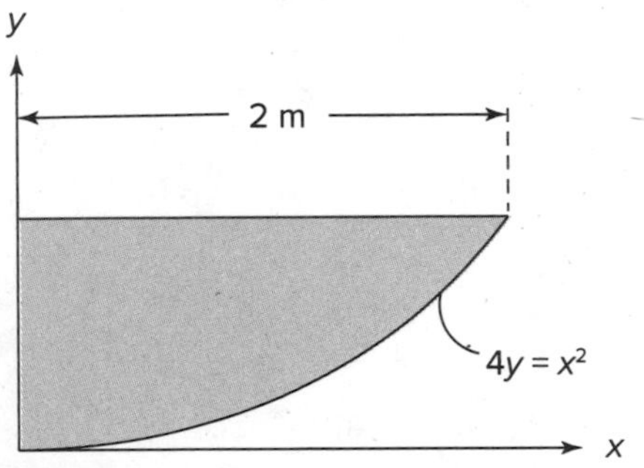

STEP 1: Recall the definition of area moment of inertia:

$I_{x'} = \int y^2\, dA$

$I_{y'} = \int x^2\, dA$

Draw one differential area "strip" to help set up the equations. For $I_{x'}$ use horizontal strips, and for $I_{y'}$ use vertical strips.

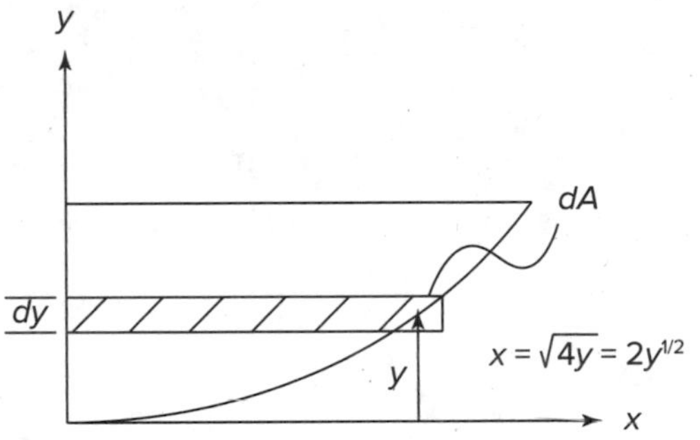

STEP 2: Write a function for dA exactly as we did when we calculated centroids by calculus:

$dA = 2y^{1/2}dy$

STEP 3: Simply plug in dA into the area moment of inertia equations using limits from our stack of strips. In this case, limits are from 0 to 1 (to find the upper limit, use $4y = x^2$ and substitute 2 m for x value—and solve for y).

$$I_{x'} = \int_0^1 (y^2)(2y^{1/2})dy = \int_0^1 (2y^{5/2})dy$$

STEP 4: Solve.

$$I_{x'} = \left(\frac{4}{7}\right)(y^{7/2})\Big|_0^1 = \left(\frac{4}{7}\right) = 0.571 \text{ m}^4$$

Find I_y, about the y-axis.

Press pause on video lesson 70 once you get to the workout problem. Only press play if you get stuck.

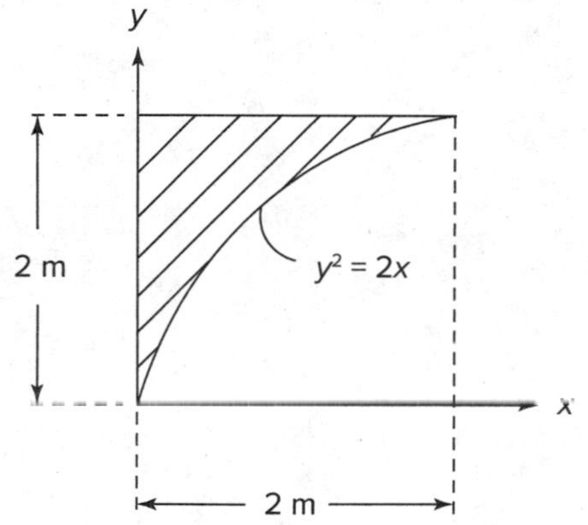

✓ TEST YOURSELF 9.4

SOLUTION TO TEST YOURSELF: Second Moment of Area

Consider the beam with the cross-section shown below. Compute the moment of inertia about the x-axis.

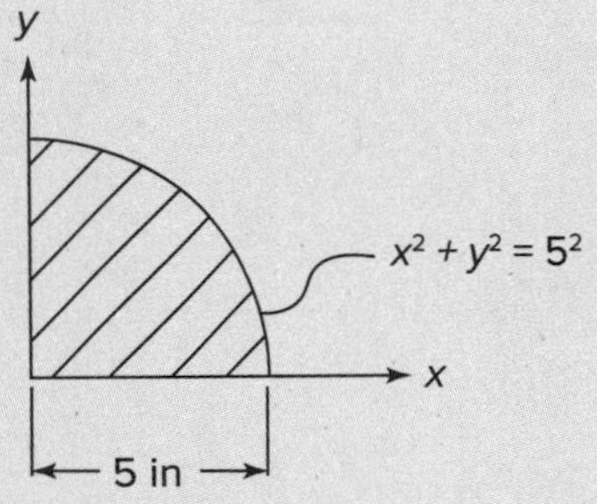

ANSWER

$I_x = 122.7 \text{ in}^4$

WATCH VIDEO

STATICS LESSON 71
Statics Conclusion and Farewell! What's Next?

Below is a summary everything that we covered in this statics unbook. As you go over the list, see if you recall how to do each of these things.

- Adding vectors
  - 2D
  - 3D
- Equilibrium of a particle
  - 2D
  - 3D
- Rigid bodies
  - 2D: $M = Fd$
  - 3D: $\vec{M} = \vec{r} \times \vec{F}$
  - $M_{W.A.} = (\vec{r} \times \vec{F}) \cdot \hat{\lambda}_{W.A.}$ (Wacky axis)
- Equilibrium of rigid bodies
  - 2D: Reaction forces, 3 equations, 3 unknowns
  - 3D: Reaction forces, 6 equations, 6 unknowns
- Trusses, frames, machines
  - Method of Joints
  - Method of Sections
- Centroids
  - Composite shape method
  - Calculus method
  - Pappus Guldinus
- Internal forces
  - M, N, V at a point
  - Shear/moment diagrams
- Friction
  - $F = \mu N$ equation
  - Slipping versus tipping
  - Wedges/screws
  - Belt friction
- Moment of inertia
  - Parallel axis theorem
  - Calculus method

The next course in the mechanics sequence that you will take is called either Strengths, Solids, or Mechanics of Materials. In that class, you will utilize the forces solved in this unbook and the material above to calculate stresses and strains in structures, which will allow you to predict failure.

KEY TAKEAWAYS

The main takeaway from Level 9 is how to calculate the second moment of area (i.e., moment of inertia).

Second Moment of Area or Moment of Inertia (MOI)

- Describes the *bendiness* of the beam
- Calculated using composite shapes (refer to the table, Second Moment of Area of Common Geometric Shapes) or calculus method

$$I_{x'} = \int_0^y y^2 dA \qquad I_{y'} = \int_0^x x^2 dA$$

Neutral Axis

- Location in the beam where it is neither in tension nor compression
- Occurs at the centroidal axis for straight beams
- When a beam is not symmetric, you will have to find the neutral axis first before you can find the MOI using the composite shapes or calculus method

$$\bar{x} = \frac{\Sigma x_i A_i}{\Sigma A_i}, \; \bar{y} = \frac{\Sigma y_i A_i}{\Sigma A_i}$$

Parallel Axis Theorem

- Applies when a composite piece-parts' centroid you have chosen does not fall on the neutral axis of the entire part
  $I_{xx} = I_{xc} + Ad^2$ or for composite shapes: $I_{xx} = \Sigma(I_{xc} + Ad^2)_i$
- When comparing beams, the one with the most area farthest away from the neutral axis will have the largest I

KEY TAKEAWAYS

RECIPES

- **Moment of Inertia Composite Shapes Recipe**
- **Moment of Inertia Calculus Method Recipe**

PRO TIPS

Intro to MOI

- **Most of the time bending occurs around the x_c-axis because of the loading direction, which is typically due to gravity (exception is wind loading).**
- **When doing I_{y_c}, rather than use different equations just rotate your head by 90°.**

Composite Shapes Method

- **If you are struggling with whether or not to use the parallel axis theorem, then use it every time! If the piece-parts' centroid is on the neutral axis, the value of $d = 0$ and the Ad^2 portion of the equation goes away.**

PITFALL

Intro to MOI

- **If the beam is not symmetric, then you will first need to calculate the centroid of the cross-section *before* calculating the moment of inertia.**

Test Yourself Solutions

Test Yourself 1.1

Since point A is the origin $\vec{r}_A = \vec{0}$

$$\vec{r}_{AB} = \vec{r}_B - \vec{r}_A = (1\hat{i} + 3\hat{j} - 2\hat{k}) - \vec{0}$$

$$\boxed{\vec{r}_{AB} = 1\hat{i} + 3\hat{j} - 2\hat{k}}$$

$$\vec{r}_{AC} = \vec{r}_C - \vec{r}_A = \boxed{-2\hat{i} + 2\hat{j} + 1\hat{k}}$$

$$\vec{r}_{AD} = \vec{r}_D - \vec{r}_A = \boxed{-2.5\hat{i} - 2.5\hat{j} - 1\hat{k}}$$

$$\vec{r}_{AE} = \vec{r}_E - \vec{r}_A = \boxed{2.5\hat{i} - 2.5\hat{j} + 4\hat{k}}$$

$$\vec{r}_{BC} = (-2\hat{i} + 2\hat{j} + \hat{k}) - (\hat{i} + 3\hat{j} - 2\hat{k})$$
$$= -3\hat{i} - \hat{j} + 3\hat{k}$$

$$\vec{r}_{BE} = (2.5\hat{i} - 2.5\hat{j} + 4\hat{k}) - (\hat{i} + 3\hat{j} - 2\hat{k})$$
$$= 1.5\hat{i} - 5.5\hat{j} + 6\hat{k}$$

$$\vec{r}_{DE} = (2.5\hat{i} - 2.5\hat{j} + 4\hat{k}) - (-2.5\hat{i} - 2.5\hat{j} - \hat{k})$$
$$= 5\hat{i} + 5\hat{k}$$

Test Yourself 1.2

1.

$$\vec{F} = 300\left[\sin 30\hat{i} + \cos 30\hat{j}\right] \text{lb}$$
$$= (150\hat{i} + 259.8\hat{j}) \text{ lb}$$

2.

$$\vec{F} = 150\left[\cos 40\hat{i} - \sin 40\hat{j}\right] \text{N}$$
$$= (114.9\hat{i} - 96.4\hat{j}) \text{N}$$

3.

$$\vec{F} = 55\left[-\cos 25\hat{i} + \sin 25\hat{j}\right] \text{Kips}$$
$$= (-49.8\hat{i} + 23.2\hat{j}) \text{ Kips}$$

4.

$$\vec{F} = 353\left[-\sin 42\hat{i} + \cos 42\hat{j}\right] \text{lb}_f$$
$$= (-236.2\hat{i} + 262.3\hat{j}) \text{ lb}_f$$

5.

$$\vec{F} = 8\left[-\sin 25\hat{i} - \cos 25\hat{j}\right] \text{kN}$$
$$= (-3.38\hat{i} - 7.25\hat{j}) \text{ kN}$$

6.

$$\vec{F} = 756\left[\cos 30\hat{i} + \sin 30\hat{j}\right] \text{N}$$
$$= (654.7\hat{i} + 378.0\hat{j}) \text{ N}$$

Test Yourself 1.3

$$\hat{\lambda}_{AB} = \frac{1\hat{i} + 3\hat{j} - 2\hat{k}}{\sqrt{1^2+3^2+2^2}} = 0.267\hat{i} + 0.802\hat{j} - 0.534\hat{k}$$

$$\hat{\lambda}_{AC} = \frac{-2\hat{i} + 2\hat{j} + 1\hat{k}}{\sqrt{2^2+2^2+1^2}} = -0.667\hat{i} + 0.667\hat{j} + 0.333\hat{k}$$

$$\hat{\lambda}_{AD} = \frac{-2.5\hat{i} - 2.5\hat{j} - 1\hat{k}}{\sqrt{2.5^2+2.5^2+1^2}} = -0.680\hat{i} - 0.680\hat{j} - 0.272\hat{k}$$

$$\hat{\lambda}_{AE} = \frac{2.5\hat{i} - 2.5\hat{j} + 4\hat{k}}{\sqrt{2.5^2+2.5^2+4^2}} = 0.468\hat{i} - 0.468\hat{j} + 0.749\hat{k}$$

Test Yourself 1.4

$\vec{F}_1 \Rightarrow \phi = -20°$

$\theta_z = 60°$

$|F_1| = 200$

$$\vec{F}_1 = 200\left[\cos(-20)\sin 60\,\hat{i} + \sin(-20)\sin 60\,\hat{j} + \cos 60\,\hat{k}\right]$$

$$\boxed{\vec{F}_1 = 162.8\hat{i} - 59.24\hat{j} + 100\hat{k}}$$

$\vec{F}_2 \Rightarrow |F_2| = 350$

$\theta_x = 120$

$\theta_z = 135$

$\theta_y = ?$

$$\cos^2 120 + \cos^2 135 + \cos^2\theta_y = 1$$

$$\cos^2\theta_y = 0.25$$

$$\theta_y = 60°$$

$$F_x = 350\cos 120 = -175$$

$$F_y = 350\cos 60 = 175$$

$$F_z = 350\cos 135 = -247.5$$

$$\boxed{\vec{F}_2 = -175\hat{i} + 175\hat{j} - 247.5\hat{k}}$$

$$\vec{F}_3 = 120\hat{j}$$

$$\vec{F}_4 = 225\hat{\lambda}$$

$$\hat{\lambda} = \frac{-1, -3, 5}{\sqrt{1^2+3^2+5^2}} = -0.169\hat{i} - 0.507\hat{j} + 0.845\hat{k}$$

$$\boxed{\vec{F}_4 = -38.03\hat{i} - 114.08\hat{j} + 190.13\hat{k}}$$

$$\therefore \boxed{\vec{F}_R = -50.2\hat{i} + 121.7\hat{j} + 42.6\hat{k}}$$

Test Yourself 2.1

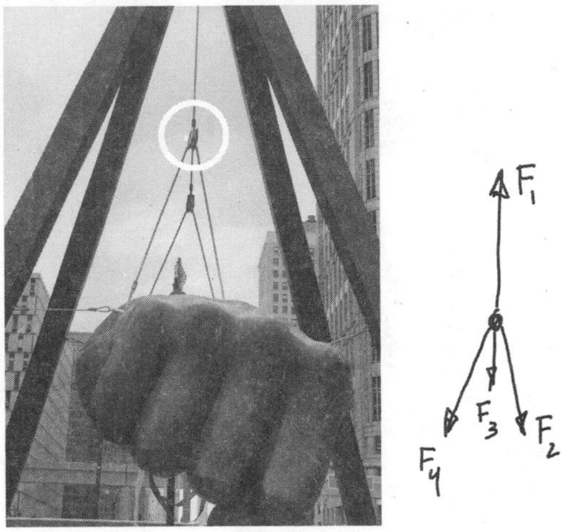

1.

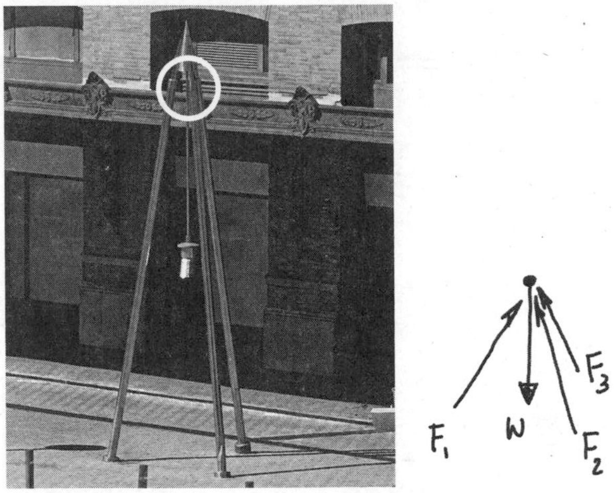

2.

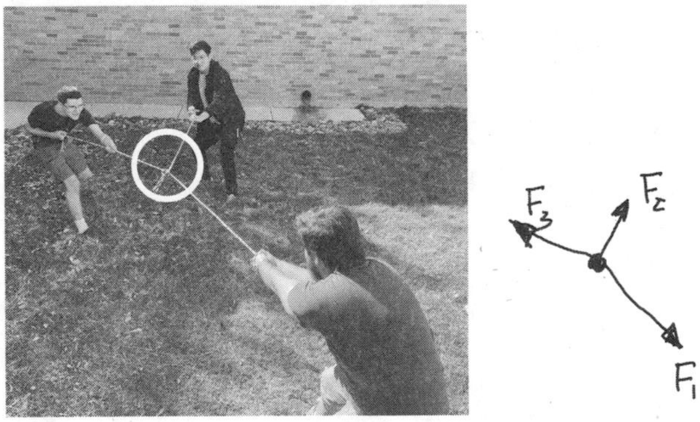

3.

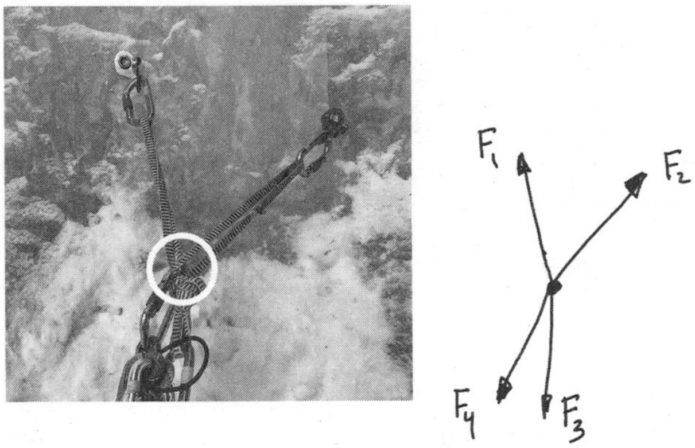

4.

Test Yourself 2.2

1. FBD A

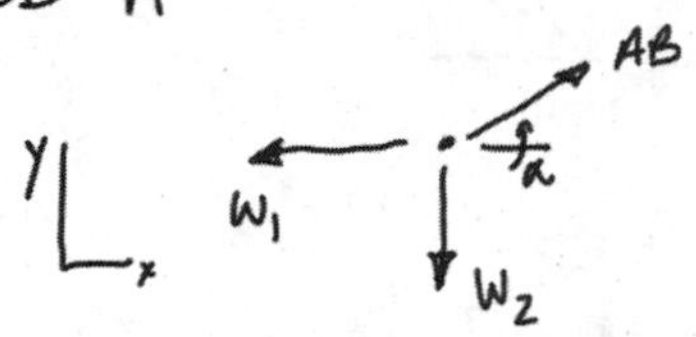

$\Sigma F_x = 0$

$-W_1 + AB\cos\alpha = 0$

$AB\cos\alpha = AB_x = W_1$

$\Sigma F_y = 0$

$-W_2 + AB\sin\alpha = 0$

$AB\sin\alpha = AB_y = W_2$

$\therefore AB^2 = AB_x^2 + AB_y^2 = W_1^2 + W_2^2$

$AB = \sqrt{100^2 + 175^2} = 201.6$ lb

$\tan\alpha = \frac{W_2}{W_1} = \frac{175}{100} = 1.75$

$\alpha = 60.26°$

FBD B

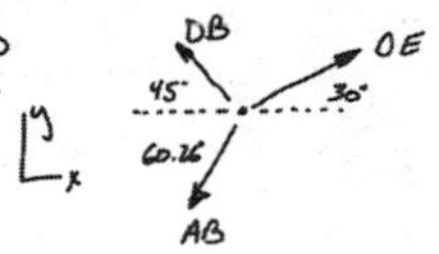

$\Sigma F_x = 0$

$-AB\cos 60.26 - DB\cos 45 + DE\cos 30 = 0$ —①

$\Sigma F_y = 0$

$DB\sin 45 + DE\sin 30 - AB\sin 60.26 = 0$ —②

Solve ① & ② for DB & DE

DB = 105.2 lb

DE = 201.3 lb

2.

FBD A

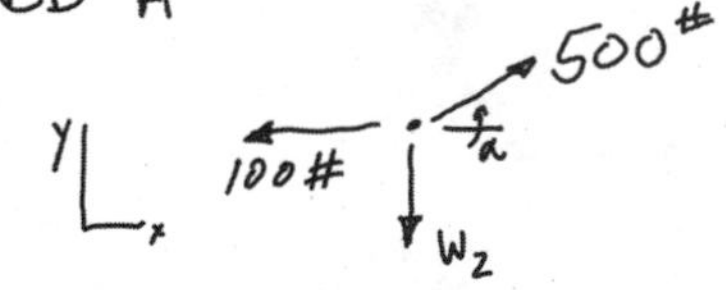

$\Sigma F_x = 0$

$-100 + 500\cos\alpha = 0$

$\cos\alpha = \frac{100}{500} = 0.20$

$\therefore \alpha = 78.5°$

$\Sigma F_y = 0$

$-W_2 + 500\sin(78.5) = 0$

$W_2 = 500\sin(78.5)$

$\therefore W_2 = 490\#$

Test Yourself 2.3

1.

FBD A

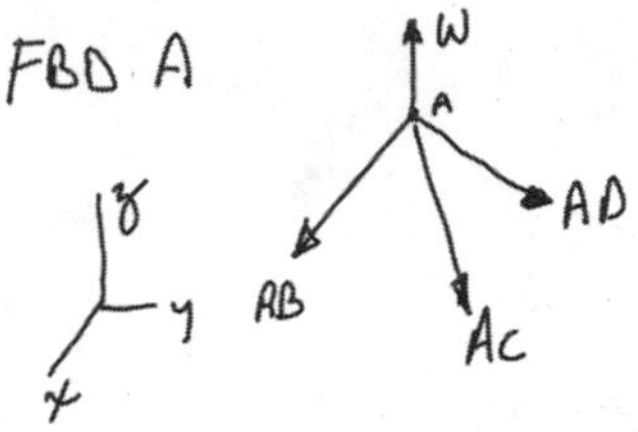

$$\vec{AB} = AB\,\frac{1.5\hat{i} - 1.5\hat{j} - 4\hat{k}}{\sqrt{1.5^2 + 1.5^2 + 4^2}} = \frac{1.5\hat{i} - 1.5\hat{j} - 4\hat{k}}{4.528}\,AB$$

$$\vec{AC} = AC\,\frac{1.5\hat{j} - 4\hat{k}}{\sqrt{1.5^2 + 4^2}} = \frac{1.5\hat{j} - 4\hat{k}}{4.272}\,AC$$

$$\vec{AD} = AD\,\frac{-1.5\hat{i} - 4\hat{k}}{\sqrt{1.5^2 + 4^2}} = \frac{-1.5\hat{i} - 4\hat{k}}{4.272}\,AD$$

$$\sum F_x = 0 \quad 0 = \frac{1.5}{4.528}AB - \frac{1.5}{4.272}AD \quad -①$$

$$\sum F_y = 0 \quad 0 = \frac{-1.5}{4.528}AB + \frac{1.5}{4.272}AC \quad -②$$

$$\sum F_z = 0 \quad 0 = W - \frac{4}{4.528}AB - \frac{4}{4.272}AC - \frac{4}{4.272}AD \quad -③$$

Solve:

$AB = 3.77$ kN

$AC = 3.56$ kN

$AD = 3.56$ kN

2. If $\overline{AB}_{MAX} = 5$ kN, what is W_{MAX}?

Re-write equations with $AB = 5$ kN

and W as the unknown:

$$\sum F_x = 0 \quad 0 = \frac{1.5}{4.528}(5) - \frac{1.5}{4.272}AD \quad -④$$

$$\sum F_y = 0 \quad 0 = \frac{-1.5}{4.528}(5) + \frac{1.5}{4.272}AC \quad -⑤$$

$$\sum F_z = 0 \quad 0 = W - \frac{4}{4.528}(5) - \frac{4}{4.272}AC - \frac{4}{4.272}AD \quad -⑥$$

$W = 13.25$ kN $\quad AC = AD = 4.72$ kN

$AB = 5.00$ kN

Test Yourself 3.1

1.
$$\begin{vmatrix} & \hat{i} & \hat{j} & \hat{k} \\ \vec{a} & 3 & -2 & 6 \\ \vec{b} & 1 & 8 & -7 \end{vmatrix}$$

$$= [(-2)(-7)-(6)(8)]\hat{i} - [(3)(-7)-(6)(1)]\hat{j} + [(3)(8)-(-2)(1)]\hat{k}$$

$$\vec{a}\times\vec{b} = -34\hat{i} + 27\hat{j} + 26\hat{k}$$

2.
$$\begin{vmatrix} & \hat{i} & \hat{j} & \hat{k} \\ \vec{a} & 8 & -6 & 7 \\ \vec{b} & -10 & 5 & -3 \end{vmatrix}$$

$$= [(-6)(-3)-(7)(5)]\hat{i} - [(8)(-3)-(7)(-10)]\hat{j} + [(8)(5)-(-6)(-10)]\hat{k}$$

$$\vec{a}\times\vec{b} = -17\hat{i} - 46\hat{j} - 20\hat{k}$$

3.
$$\begin{vmatrix} & i & \hat{j} & \hat{k} \\ a & 12 & 6 & -8 \\ b & -8 & -7 & -6 \end{vmatrix}$$

$$= [(6)(-6)-(-8)(-7)]\hat{i} - [(12)(-6)-(-8)(-8)]\hat{j} + [(12)(-7)-(6)(-8)]\hat{k}$$

$$\vec{a}\times\vec{b} = -92\hat{i} + 136\hat{j} - 36\hat{k}$$

4.
$$\begin{vmatrix} & \hat{i} & \hat{j} & \hat{k} \\ \vec{a} & -9 & 4 & -1 \\ \vec{b} & 8 & -3 & -2 \end{vmatrix}$$

$$= [(4)(-2)-(-1)(-3)]\hat{i} - [(-9)(-2)-(-1)(8)]\hat{j} + [(-9)(-3)-(4)(8)]\hat{k}$$

$$\vec{a}\times\vec{b} = -11\hat{i} - 26\hat{j} - 5\hat{k}$$

Test Yourself 3.2

(1) $-(200\cos 30)(3\sin 60) + (200\sin 30)(8 + 3\cos 60) = M_A$

$-450 + 950$ $\Big\} = \underline{\underline{500}}\,\hat{k}$ ft lb

(2) $(8 + 3\cos 60)\hat{i} + (3\sin 60)\hat{j} \times (200\cos 30\hat{i} + 200\sin 30\hat{j})$

$(8 + 3\cos 60)(200\sin 30)\hat{k} - (3\sin 60)(200\cos 30)\hat{k}$

$950\,k$ $-450\,k$ $\Big\} = \underline{\underline{500}}\,\hat{k}$ ft·lb

Test Yourself 3.3

① Find $\vec{F}$

$$\vec{r}_F = -3\hat{i} + 5\hat{j} + 3\hat{k}$$

$$\hat{r}_F = \frac{-3\hat{i} + 5\hat{j} + 3\hat{k}}{\sqrt{3^2 + 5^2 + 3^2}} = -0.45750\hat{i} + 0.76249\hat{j} + 0.45750\hat{k}$$

② $F = 500\,(\hat{r}) = (-228.75\hat{i} + 381.25\hat{j} + 228.75\hat{k})\,\text{lb}$

③ $\vec{r}_A = -8\hat{i} + 10\hat{j} + 2\hat{k}$ (ft)

④ $\vec{M}_A = \vec{r}_A \times \vec{F}$

$\therefore M_A = 1525\hat{i} + 1373\hat{j} - 762.5\hat{k}$ ft lb

Test Yourself 3.4

1. Point A 9,3,6
Point B 7,5,11
$\theta = 60°$
$|\vec{F}| = 1{,}000$ N

Find the moment of $\vec{F}$ about the line given by the vector $\vec{C}$

① Build $\vec{F}$

$$\hat{AB} = \frac{\vec{B} - \vec{A}}{|\vec{B} - \vec{A}|} = \frac{(7-9)\hat{i} + (5-3)\hat{j} + (11-6)\hat{k}}{\sqrt{2^2 + 2^2 + 5^2}}$$

$$= -0.348\hat{i} + 0.348\hat{j} + 0.870\hat{k}$$

$$\therefore \vec{F} = |\vec{F}| \cdot \hat{AB} = 1000 \cdot \hat{AB}$$

$$\vec{F} = -348\hat{i} + 348\hat{j} + 870\hat{k} \text{ N}$$

② Compute $\vec{r}_{OA} \times \vec{F} = M_o = (9\hat{i} + 3\hat{j} + 6\hat{k}) \times (-348\hat{i} + 348\hat{j} + 870\hat{k})$

$M_o = 522\hat{i} - 9918\hat{j} + 4176\hat{k}$ N·M

③ Find the unit vector of $\vec{C}$

$\hat{C} = \cos 60\,\hat{i} + \cos 30\,\hat{j} + 0\hat{k} = 0.500\hat{i} + 0.866\hat{j}$

④ $M_c = \vec{M}_o \cdot \hat{C} = (522\hat{i} - 9918\hat{j} + 4176\hat{k}) \cdot (0.5\hat{i} + 0.866\hat{j})$

$M_c = -8{,}328$ N·M

2. Cube is 3 ft on each side

$|\vec{F}| = 100$ lbs

What is the moment of $\vec{F}$ about the axis going from B to A?

① $\vec{F} = 100 \left[\dfrac{-3\hat{i} + 3\hat{j} + 3\hat{k}}{\sqrt{3^2 + 3^2 + 3^2}}\right]$

$\vec{F} = -57.74\hat{i} + 57.74\hat{j} + 57.74\hat{k}$ lbs.

② $\widehat{BA} = 3\hat{i} + 3\hat{k} / \sqrt{3^2 + 3^2} = 0.707\hat{i} + 0.707\hat{k}$

③ $M_{BA} = (\vec{r} \times \vec{F}) \cdot \widehat{BA} = 3\hat{k} \times (-57.74\hat{i} + 57.74\hat{j} + 57.74\hat{k}) \cdot \widehat{BA}$

$M_{BA} = -122.5$ ft lbs

Test Yourself 3.5

Compute the moment of the two couples.

① The two couples act on a plane that is 45° looking down the Z-axis

$\hat{n} = \cos 45\,\hat{i} + \cos 45\,\hat{j}$

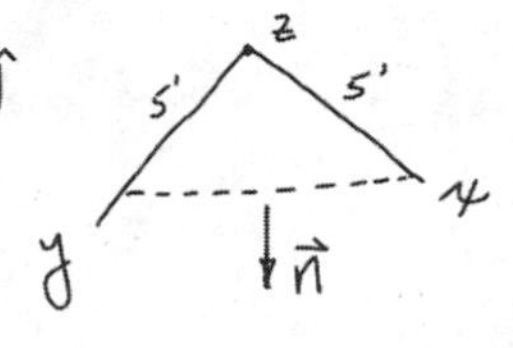

② $(400)(5)(-\hat{n}) +$

$(300)(5\sqrt{2})(\hat{n})$

$[(300)(5\sqrt{2}) - (400)(5)](\cos 45\,\hat{i} + \cos 45\,\hat{j})$

$\vec{M} = 121.3\,(0.707\hat{i} + 0.707\hat{j}) = 85.8\hat{i} + 85.8\hat{j}$

Test Yourself 3.6

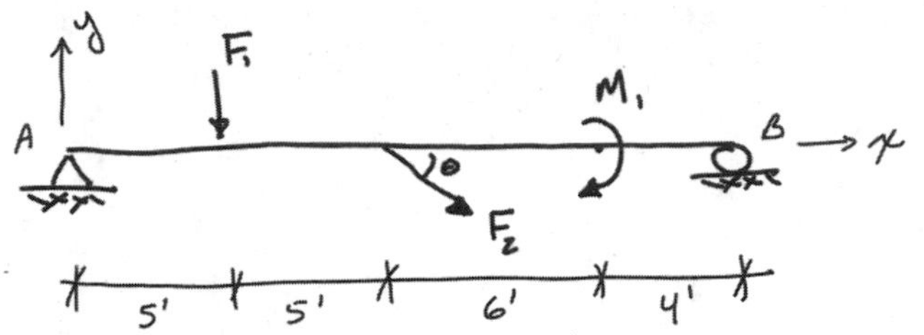

Find the Equivalent load

MAGNITUDES:

$F_1 = 10{,}000$ lb $\qquad \theta = 30°$

$F_2 = 8{,}000$ lb $\qquad M_1 = 15{,}000$ lb·ft

① Find F_R

$F_y = -10{,}000 - 8{,}000 \sin\theta$

$= -14{,}000$ lb

$F_x = 8000 \cos 30 = 6{,}928$ lb

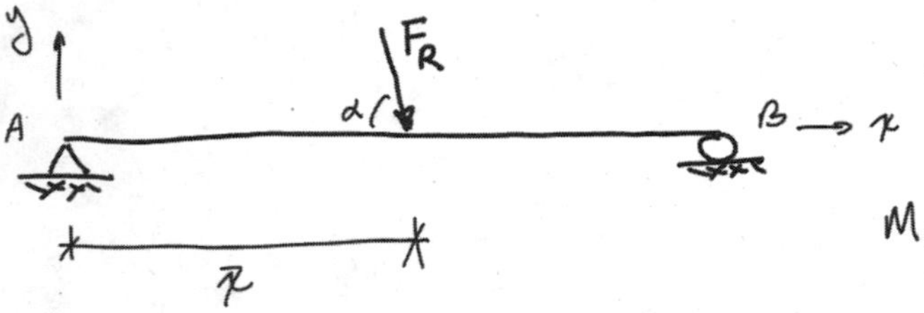

② Take moments of all loads about A in the original system and the Equ. Sys.

$M_A = -(10{,}000)(5) - (8{,}000)\sin 30\,(10) - 15{,}000$

$= -(F_R)_y(\bar{x})$

Note: only y-comp produces a moment at A

Use $-(F_R)_y = -14{,}000$

$-50{,}000 - 40{,}000 - 15{,}000 = -14{,}000\ \bar{x}$

$\bar{x} = 7.5$ ft

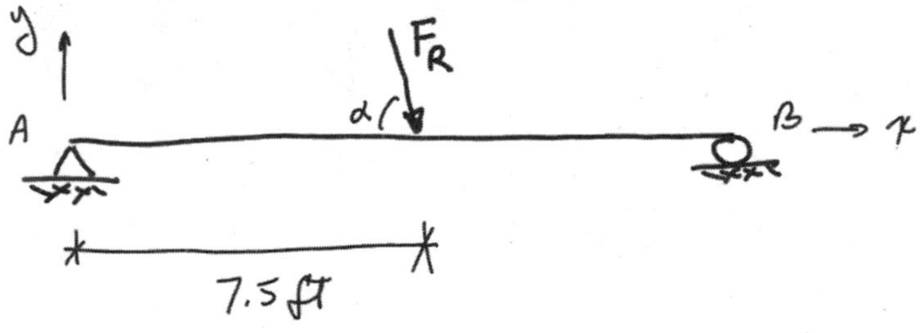

$F_R = -14000\,\hat{j} + 6{,}928\,\hat{i}$

$\tan\alpha = \dfrac{14{,}000}{6928}$

$\alpha = 63.7°$

Test Yourself 4.1

1. Pin and Simple Cable
2. Pin and Point
3. Pin and Single Point
4. Fixed
5. Fixed
6. Pin and Single Point
7. Pin and Simple Cable and Cart is a Roller
8. Pin and Roller
9. Roller

Test Yourself 4.2

1. Member *AB* and the nut in the jaws of the pliers
2. Member *BC*
3. Member *BD*
4. Cable *AC*

Test Yourself 4.3

FBD C

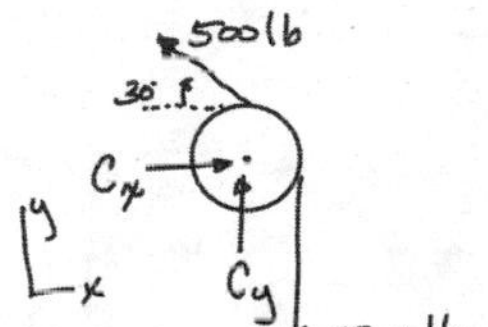

$\Sigma F_x = 0$

$-500\cos 30° + C_x = 0$

$C_x = 433\text{ lb}$

$\Sigma F_y = 0$

$C_y - 500 + 500\sin 30° = 0$

$C_y = 250\text{ lb}$

FBD Beam

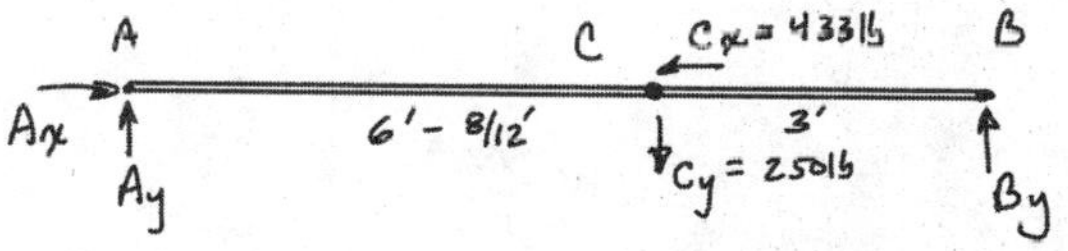

$\Sigma M_A = 0$

$B_y(9) - 250(6 - 8/12) = 0 \quad \therefore \boxed{B_y = 148.15\text{ lb}}$

$\Sigma F_y = 0$

$A_y - 250 + B_y = 0 \qquad B_y = 148.15$

$\therefore \boxed{A_y = 101.85\text{ lb}}$

$\Sigma F_x = 0$

$A_x - C_x = 0 \quad \therefore \boxed{A_x = 433\text{ lb}}$

Test Yourself 4.4

1. FBD

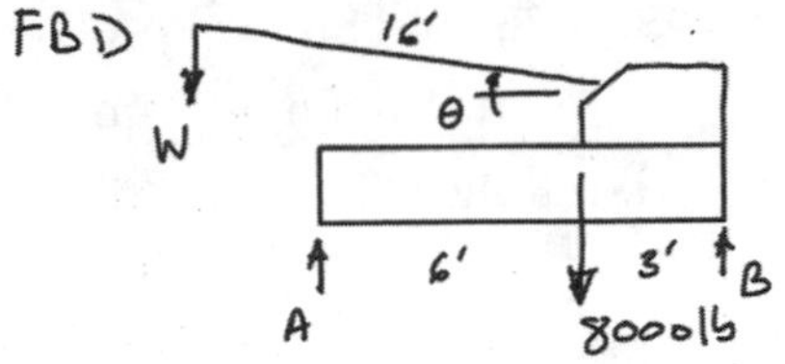

$\theta = 30°$ MAX W AT TIPPING $\therefore B = 0$

$\Sigma M_A = 0 \qquad W\left[16\cos 30 - 6\right] - 8000\,(6) = 0$

$W = \dfrac{48{,}000}{16\cos 30 - 6} = 6{,}110\text{ lb}$

2. USE $W = \dfrac{48{,}000}{16\cos\theta - 6}$ TO FIND $W(\theta)$

θ	W (lb)
0°	4,800
10°	4,920
20°	5,313
30°	6,110
40°	7,671
45°	9,033

NOTE: THE LOAD CAPACITY INCREASES AS θ INCREASES

3. FOR $\theta = 35°$ AND $W = 5{,}000\text{ lb}$ FIND THE REACTIONS

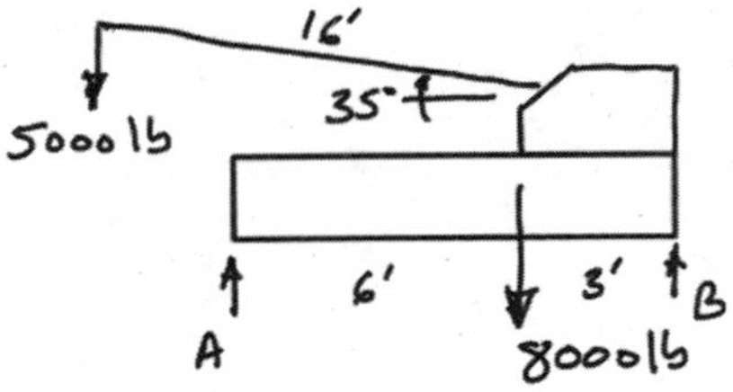

$\Sigma M_A = 0$

$5000\left(16\cos 35 - 6\right) - 8000\,(6) + B\,(9) = 0$

$\therefore B = 1{,}385\text{ lb}$

$\Sigma F_y = 0$

$-5000 - 8000 + 1385 + A = 0$

$\therefore A = 11{,}615\text{ lb}$

Test Yourself 4.5

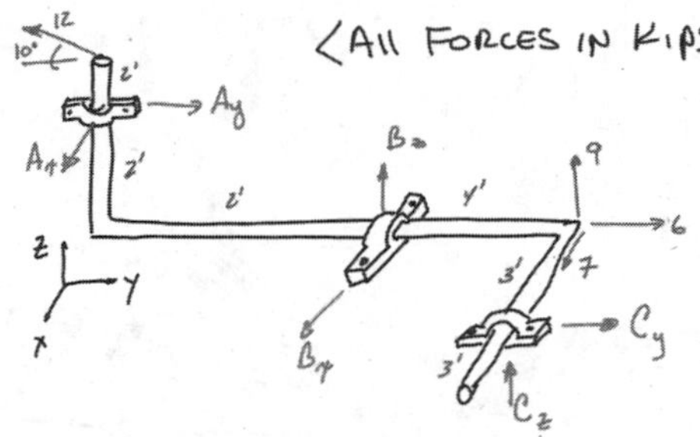

<ALL FORCES IN KIPS>

$\Sigma F_y = 0 \quad A_y + C_y + 6 - 12\cos 10 = 0$

$\Sigma F_x = 0 \quad A_x + B_x + 7 = 0$

$\Sigma F_z = 0 \quad B_z + C_z + 9 + 12\sin 10 = 0$

$\Sigma M_{y\,AXIS} = A_x(2) - C_z(3) = 0$

$\Sigma M_{x\,AXIS} = B_z(2) + 9(6) + C_z(6) + (12\cos 10)4 - A_y(2) = 0$

$\Sigma M_{z\,AXIS} = -B_x(2) + C_y(3) - 7(6) = 0$

SUBSTITUTION. FROM EACH EQUATION, SOLVE FOR A UNIQUE UNKNOWN.

① $C_y = 12\cos 10 - 6 - A_y = 5.818 - A_y$

② $A_x = -7 - B_x$

③ $B_z = -9 - 12\sin 10 - C_z = -11.084 - C_z$

④ $C_z = \frac{2}{3} A_x$

⑤ $A_y = B_z + 27 + 3C_z + 24\cos 10 = B_z + 3C_z + 50.635$

⑥ $B_x = \frac{3}{2} C_y - 21$

START WITH A SIMPLE SUBSTITUTION:

④ → ③

$C_z = \frac{2}{3} A_x$

$B_z = -11.084 - C_z$

$\therefore B_z = -11.084 - \frac{2}{3} A_x$ ⑦

⑦ ; ④ → ⑤

$A_y = B_z + 3C_z + 50.635$ (⑦, ④)

$= \left(-11.084 - \frac{2}{3} A_x\right) + 3\left(\frac{2}{3} A_x\right) + 50.635$

$= -11.084 - \frac{2}{3} A_x + 2A_x + 50.635$

$\therefore A_y = \frac{4}{3} A_x + 39.551$ ⑧

② → ⑧

$A_x = -7 - B_x$

$\therefore A_y = \frac{4}{3}(-7 - B_x) + 39.551$

$= -\frac{28}{3} - \frac{4}{3} B_x + 39.551$

$\therefore A_y = 30.218 - \frac{4}{3} B_x$ ⑨

⑥ → ⑨

$B_x = \frac{3}{2} C_y - 21$

$A_y = 30.218 - \frac{4}{3}\left(\frac{3}{2} C_y - 21\right)$

$\therefore A_y = 58.218 - 2C_y$ ⑩

⑩ → ①

$C_y = 5.818 - A_y$

$= 5.818 - (58.218 - 2C_y)$

$C_y = 5.818 - 58.218 + 2C_y$

$\therefore \boxed{C_y = 52.400}$

FROM ⑥

$B_x = \frac{3}{2} C_y - 21$

$= \frac{3}{2}(52.400) - 21$

$\therefore \boxed{B_x = 57.600}$

FROM ①

$C_y = 5.818 - A_y$

$A_y = 5.818 - C_y$

$= 5.818 - 52.400$

$\therefore \boxed{A_y = -46.582}$

FROM ②

$A_x = -7 - B_x$

$= -7 - 57.600$

$\therefore \boxed{A_x = -64.600}$

FROM ④

$C_z = \frac{2}{3} A_x$

$= \frac{2}{3}(-64.600)$

$\therefore \boxed{C_z = -43.067}$

FROM ③

$B_z = -11.084 - C_z$

$= -11.084 - (-43.067)$

$\therefore \boxed{B_z = 31.983}$

Test Yourself 5.1

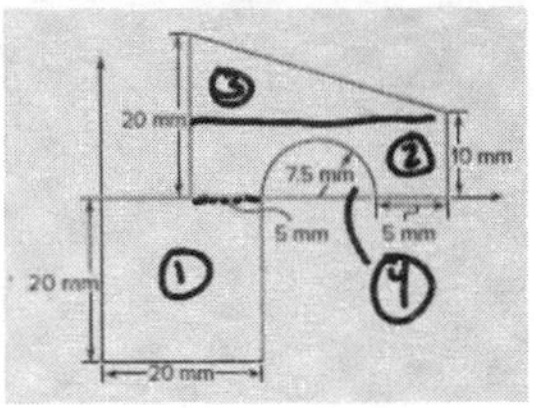

AREA	A_i	x_i	$A_i x_i$
1	400	10	4000
2	250	27.5	6875
3	125	23.33	2916.25
4	-88.36	27.5	-2429.9
	686.64 mm²		11,361.35 mm³

$$\bar{x} = \frac{A_i x_i}{A_i} = \frac{11,361.35}{686.64}$$

$$\boxed{\bar{x} = 16.55 \text{ mm}}$$

AREA	A_i	y_i	$A_i y_i$
1	400	-10	-4000
2	250	5	1250
3	125	13.33	1666.25
4	-88.36	3.183	-281.25
	686.64 mm²		-1365 mm³

$$\bar{y} = \frac{A_i y_i}{A_i} = -\frac{1365}{686.64}$$

$$\boxed{\bar{y} = -1.99 \text{ mm}}$$

Test Yourself 5.2

① $\frac{y}{x} = \tan(90-35) = 1.428$

$\therefore y_1 = 1.428x$

$y_2 = 2x^2$

② SET $y_1 = y_2$ to find A

$1.428x = 2x^2$

$\therefore x = 0.7141$

$y = 1.428(0.7141) = 1.020$

$A\ (0.7141, 1.020)$

③ 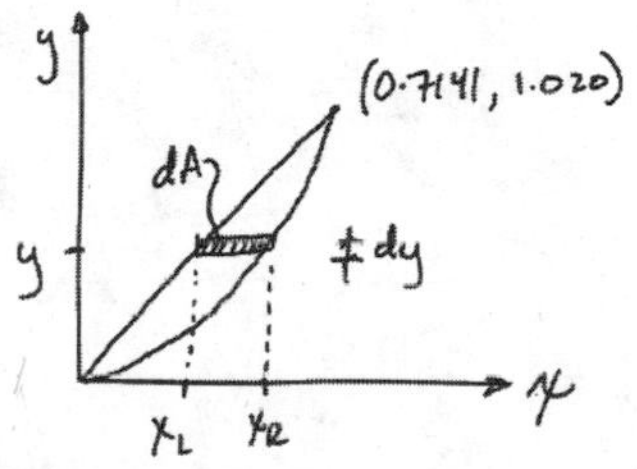

$dA = (x_R - x_L)\,dy$

$y = 2x_R^2 \quad \therefore x_R = \frac{\sqrt{y}}{\sqrt{2}}$

$y = 1.428x_L \quad \therefore x_L = \frac{y}{1.428}$

$dA = \left[\frac{\sqrt{y}}{\sqrt{2}} - \frac{y}{1.428}\right]dy$

$A = \int_0^{1.020} \left[\frac{\sqrt{y}}{\sqrt{2}} - \frac{y}{1.428}\right]dy$

$\therefore A = 0.121331$

④ $A\bar{y} = \int_0^{1.020} \left[\frac{\sqrt{y}}{\sqrt{2}} - \frac{y}{1.428}\right] y\,dy$

$0.121331\,\bar{y} = 0.0494834$

$\boxed{\therefore \bar{y} = 0.4078}$

Test Yourself 5.3

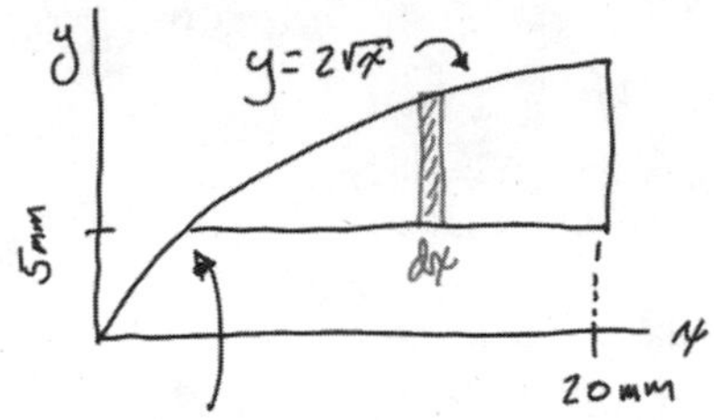

$y = 5 = 2\sqrt{x}$

$x = (5/2)^2 = 6.25\ \text{mm}$

① Find the Area

$dA = (y_{TOP} - y_{BOT})\,dx$

$(2\sqrt{x} - 5)\,dx$

$A = \int_{6.25}^{20} dA = \int_{6.25}^{20} (2\sqrt{x} - 5)\,dx$

$\boxed{A = 29.67\ \text{mm}^2}$

② Compute $\bar{x}$

$A\bar{x} = \int_{6.25}^{20} (2\sqrt{x} - 5)\,x\,dx$

$29.67\,\bar{x} = 450.62$

$\boxed{\bar{x} = 15.19\ \text{mm}}$

③

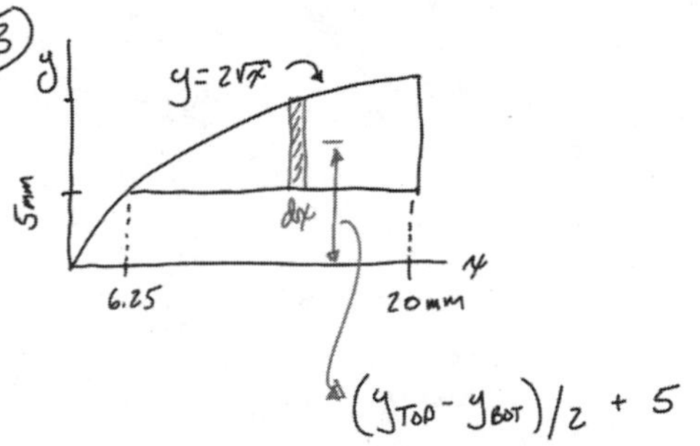

$A\bar{y} = \int_{6.25}^{20} (2\sqrt{x} - 5)\left[\frac{(\overset{=2\sqrt{x}}{y_{TOP}} - \overset{=5}{y_{BOT}})}{2} + 5\right] dx$

$29.67\,\bar{y} = \int_{6.25}^{20} (2\sqrt{x} - 5)\left[\frac{(2\sqrt{x} - 5)}{2} + 5\right] dx$

$\boxed{\bar{y} = \frac{189.06}{29.67} = 6.37\ \text{mm}}$

Test Yourself 5.4

(1) SKETCH THE PROFILE

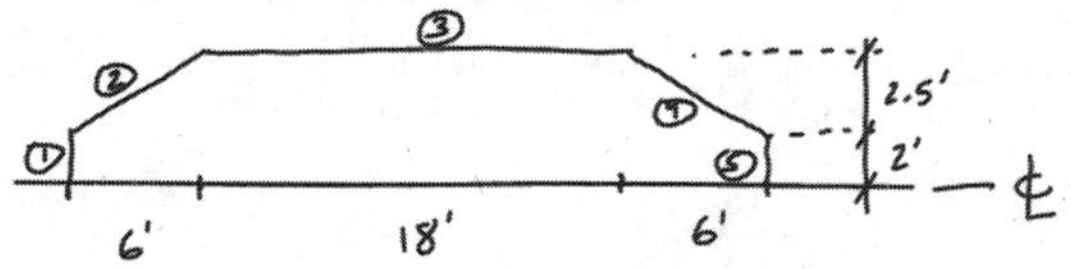

(2) BUILD THE TABLE FOR THE SURFACE AREA

	Length (L_i)	$\bar{y}_i$	$L_i y_i$
(1)	2	1	2
(2)	$\sqrt{6^2+2.5^2}=6.5$	3.25	21.125
(3)	18	4.5	81
(4)	6.5	3.25	21.125
(5)	2	1	2

Sum = 127.25

Surface Area = $2\pi(127.25)$

$= 799.54 \text{ ft}^2$

(3) SKETCH VOLUMES

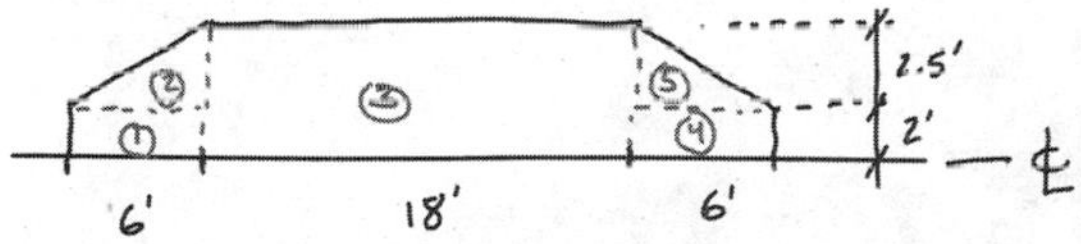

(4) BUILD TABLE

	Area (A_i)	$\bar{y}_i$	$A_i \bar{y}_i$
1	12	1	12
2	$\frac{1}{2}(6)(2.5)=7.5$	2.833	21.25
3	81	2.25	182.25
4	12	1	12
5	7.5	2.833	21.25

Sum = 248.75

$V = 2\pi(248.75)$

$= 1{,}563 \text{ ft}^3$

Test Yourself 5.5

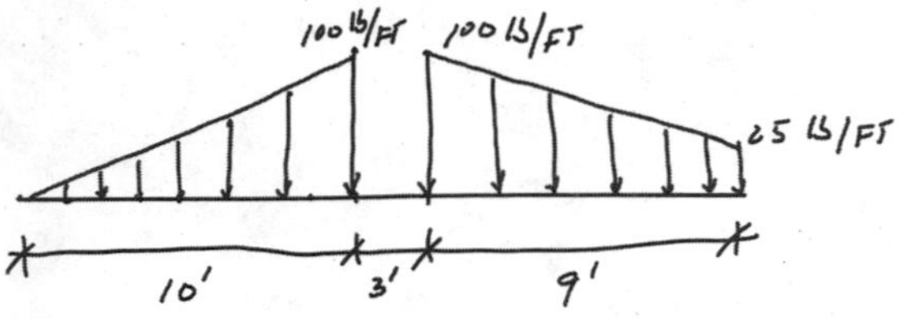

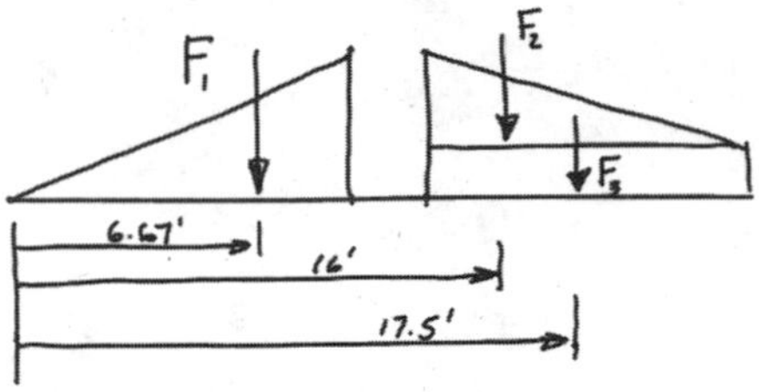

② Compute The MAGNITUDE & LOCATION

$F_1 = \frac{1}{2}(100)(10) = 500$ lb

$\bar{x}_1 = 2/3\,(10) = 6.67$ ft

$F_2 = \frac{1}{2}(9)(75) = 337.5$ lb

$\bar{x}_2 = 13 + \frac{1}{3}(9) = 16$ FT

$F_3 = (9)(25) = 225$ lb

$\bar{x}_3 = 10 + 3 + 9/2 = 17.5'$

FBD of Beam

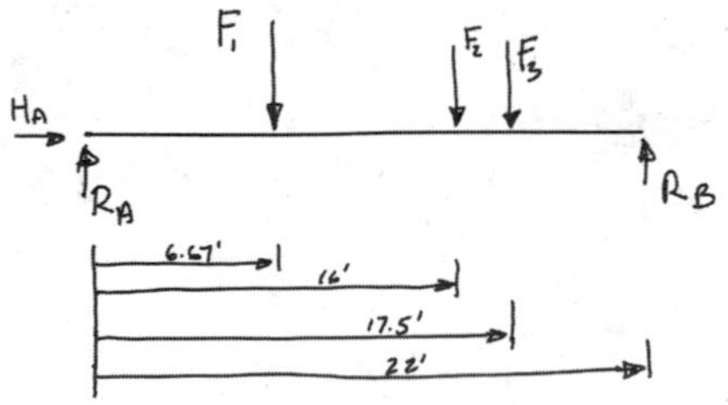

USE EQUILIBRIUM EQNS

$+\circlearrowleft \sum M_A = 0 \quad 500(6.67) + 337.5(16) + 225(17.5) - R_B 22 = 0$

$\therefore R_B = 576$ lb

$+\circlearrowleft \sum M_B = 0 \quad R_A(22) - 500(15.33) - 337.5(6) - 225(4.5) = 0$

$\therefore R_A = 486.5$ lb

Check: $\sum F_y = 0 \quad -500 - 337.5 - 225 + 576 + 486.5 = 0$ ✓

NOTE: $\sum F_x = 0 \quad \therefore H_A = 0$

Test Yourself 6.1

1. Machine
2. Frame
3. Truss
4. Truss (support structure)
5. Machine
6. Machine
7. Machine
8. Truss (in 3D!)
9. Could be argued to be a truss, frame, or a machine

Test Yourself 6.2

1. Tension
2. Compression
3. Compression
4. Tension

Test Yourself 6.3

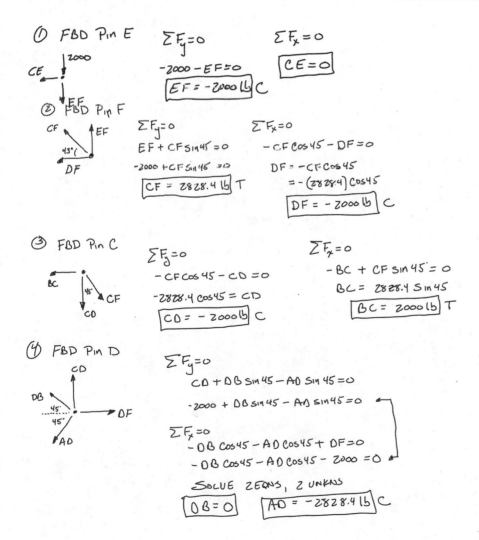

① FBD Pin E

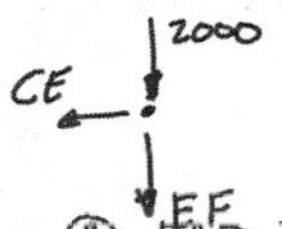

$\Sigma F_y = 0$

$-2000 - EF = 0$

$\boxed{EF = -2000\text{ lb}}$ C

$\Sigma F_x = 0$

$\boxed{CE = 0}$

② FBD Pin F

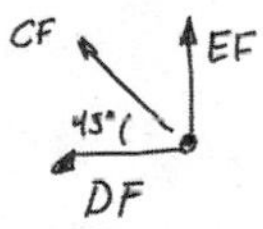

$\Sigma F_y = 0$

$EF + CF\sin 45 = 0$

$-2000 + CF\sin 45 = 0$

$\boxed{CF = 2828.4\text{ lb}}$ T

$\Sigma F_x = 0$

$-CF\cos 45 - DF = 0$

$DF = -CF\cos 45$

$= -(2828.4)\cos 45$

$\boxed{DF = -2000\text{ lb}}$ C

③ FBD Pin C

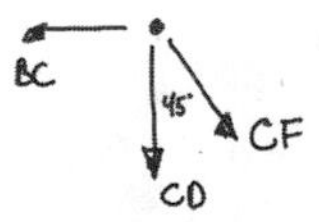

$\Sigma F_y = 0$

$-CF\cos 45 - CD = 0$

$-2828.4\cos 45 = CD$

$\boxed{CD = -2000\text{ lb}}$ C

$\Sigma F_x = 0$

$-BC + CF\sin 45 = 0$

$BC = 2828.4\sin 45$

$\boxed{BC = 2000\text{ lb}}$ T

④ FBD Pin D

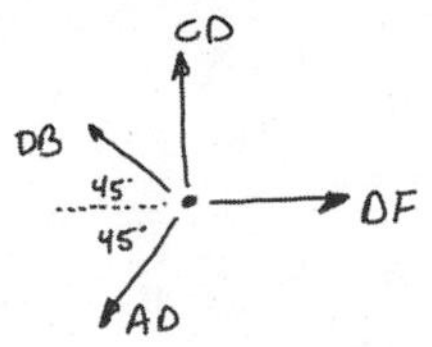

$\Sigma F_y = 0$

$CD + DB\sin 45 - AD\sin 45 = 0$

$-2000 + DB\sin 45 - AD\sin 45 = 0$

$\Sigma F_x = 0$

$-DB\cos 45 - AD\cos 45 + DF = 0$

$-DB\cos 45 - AD\cos 45 - 2000 = 0$

SOLVE 2 EQNS, 2 UNKNS

$\boxed{DB = 0}$ $\boxed{AD = -2828.4\text{ lb}}$ C

Test Yourself 6.4

FBD

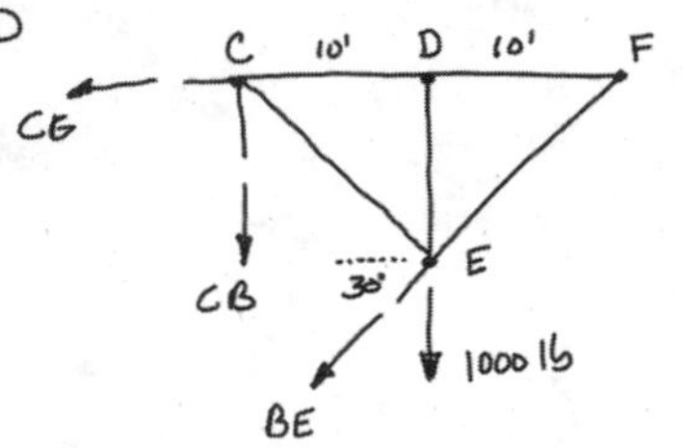

$\boxed{DF = 0}$
By observation

$\Sigma M_F = 0$

$1000(10) + CB(20) = 0$

$\therefore \boxed{CB = -500\text{ lb}}$ C

$\Sigma F_y = 0$

$-CB - 1000 - BE \sin 30 = 0$

$500 - 1000 - BE \sin 30 = 0$

$\boxed{BE = -1000\text{ lb}}$ C

$\Sigma F_x = 0$

$-CG - BE \cos 30 = 0$

$-CG - (-1000) \cos 30 = 0$

$CG = 1000 \cos 30$

$\boxed{CG = 866\text{ lb}}$ T

Test Yourself 6.5

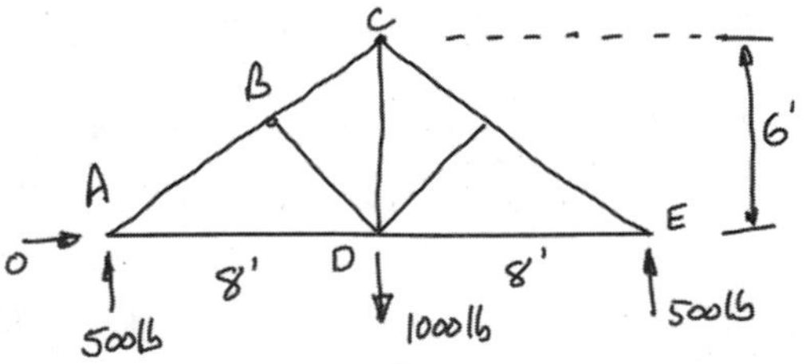

FBD Pin A

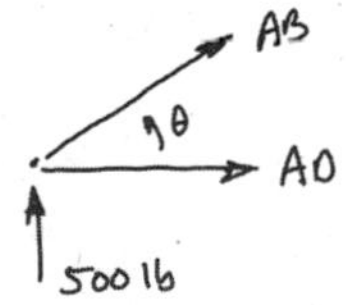

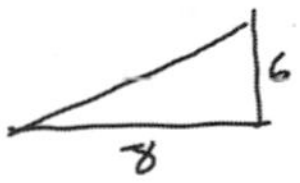

$\tan\theta = \frac{6}{8}$

$\theta = 36.87°$

$\Sigma F_y = 0 \quad AB \sin 36.87 + 500 = 0$

$\therefore \boxed{AB = -833.3\text{ lb}}$ C

$\Sigma F_x = 0$

$AB \cos 36.87 + AD = 0$

$AD = -AB \cos 36.87$

$\boxed{AD = 666.6\text{ lb}}$ T

Test Yourself 6.6

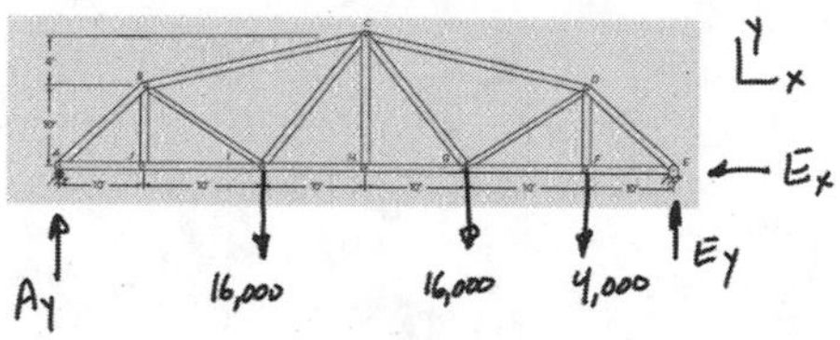

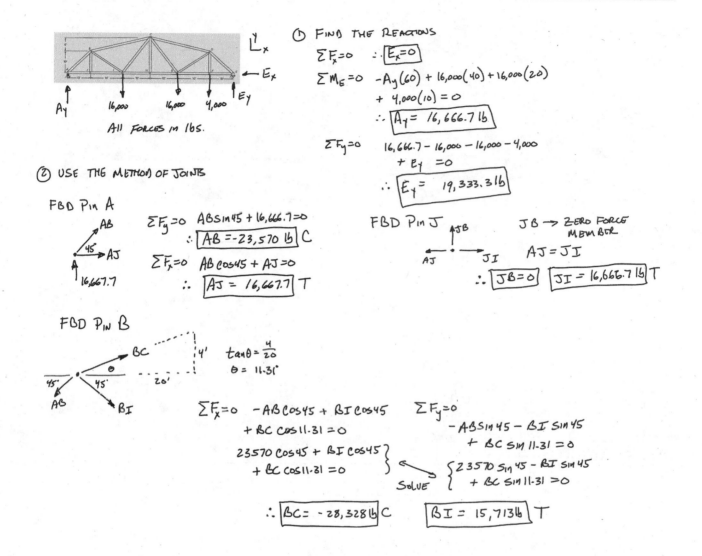

All forces in lbs.

① FIND THE REACTIONS

$\Sigma F_x = 0 \quad \therefore \boxed{E_x = 0}$

$\Sigma M_E = 0 \quad -A_y(60) + 16{,}000(40) + 16{,}000(20) + 4{,}000(10) = 0$

$\therefore \boxed{A_y = 16{,}666.7\text{ lb}}$

$\Sigma F_y = 0 \quad 16{,}666.7 - 16{,}000 - 16{,}000 - 4{,}000 + E_y = 0$

$\therefore \boxed{E_y = 19{,}333.3\text{ lb}}$

② USE THE METHOD OF JOINTS

FBD Pin A

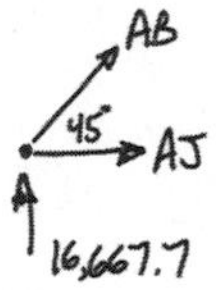

$\Sigma F_y = 0 \quad AB \sin 45 + 16{,}666.7 = 0$

$\therefore \boxed{AB = -23{,}570\text{ lb}}$ C

$\Sigma F_x = 0 \quad AB \cos 45 + AJ = 0$

$\therefore \boxed{AJ = 16{,}667.7}$ T

FBD Pin J

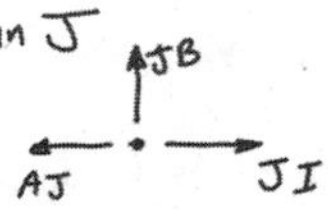

JB → ZERO FORCE MEMBER

$AJ = JI$

$\therefore \boxed{JB = 0} \quad \boxed{JI = 16{,}666.7\text{ lb}}$ T

FBD Pin B

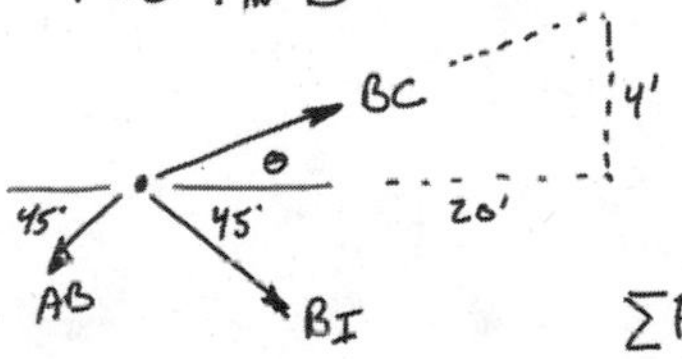

$\tan\theta = \frac{4}{20}$

$\theta = 11.31°$

$\Sigma F_x = 0 \quad -AB\cos 45 + BI\cos 45 + BC\cos 11.31 = 0$

$\Sigma F_y = 0 \quad -AB\sin 45 - BI\sin 45 + BC\sin 11.31 = 0$

$23570\cos 45 + BI\cos 45 + BC\cos 11.31 = 0$

$23570\sin 45 - BI\sin 45 + BC\sin 11.31 = 0$

SOLVE

$\therefore \boxed{BC = -28{,}328\text{ lb}}$ C $\quad \boxed{BI = 15{,}713\text{ lb}}$ T

FBD PIN I

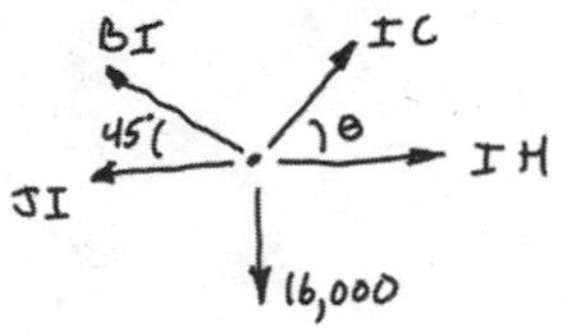

$\tan\theta = \frac{14}{10}$ $\therefore \theta = 54.46°$

$JI = 16{,}666.7$

$BI = 15{,}713$

$\Sigma F_y = 0$

$BI \sin 45 + IC \sin 54.46 - 16{,}000 = 0$

$IC \sin 54.46 = 16{,}000 - BI \sin 45$

$= 16{,}000 - 15713 \sin 45$

$IC \sin 54.46 = 4889.2$

$\therefore$ **IC = 6,008 lb T**

$\Sigma F_x = 0$

$-JI - BI\cos 45 + IC\cos 54.46 + IH = 0$

$-16{,}667.7 - 15713\cos 45 + 6008\cos 54.46 + IH = 0$

$\therefore$ **IH = 24,286 lb T**

FBD Pin H

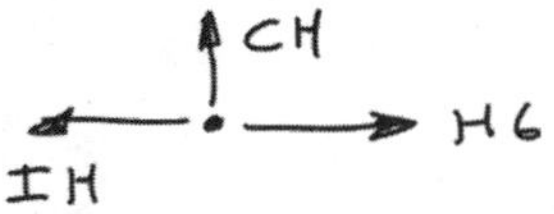

CH = 0

$\Sigma F_x = 0$ **HG = IH = 24,286 lb T**

FBD Pin E

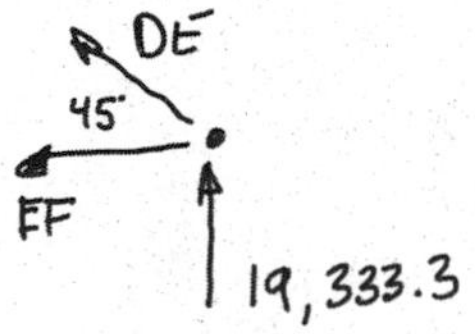

$\Sigma F_y = 0$

$DE \sin 45 + 19{,}333.3 = 0$

$\therefore$ **DE = −27,341 lb C**

$\Sigma F_x = 0$

$-DE\cos 45 - EF = 0$

$EF = -(-27{,}341)\cos 45$

$\therefore$ **EF = 19,333.3 lb T**

FBD Pin F

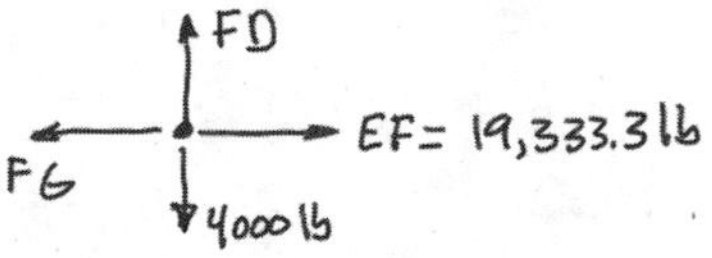

$\Sigma F_y = 0$ ∴ **FD = 4000 lb T**

$\Sigma F_x = 0$ **FG = 19,333.3 lb T**

FBD Pin D

DE = −27,341

FD = 4000

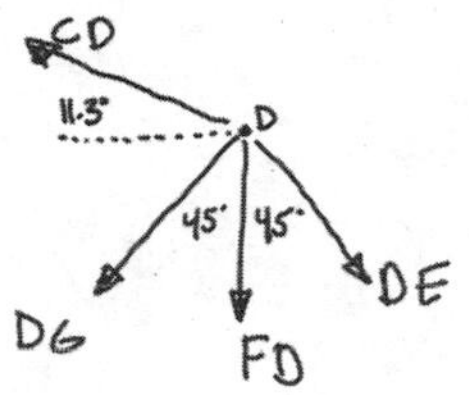

$\Sigma F_x = 0$

$DE \sin 45 - DG \sin 45 - CD \cos 11.3 = 0$

$\Sigma F_y = 0$

$CD \sin 11.3 - DG \cos 45 - DE \cos 45 - FD = 0$

$-27{,}341 \sin 45 - DG \sin 45 - CD \cos 11.3 = 0$

$CD \sin 11.3 - DG \cos 45 - (-27{,}341) \cos 45 - 4000 = 0$

∴ **CD = −29,464 lb C** **DG = 13,519 lb T**

FBD Pin G

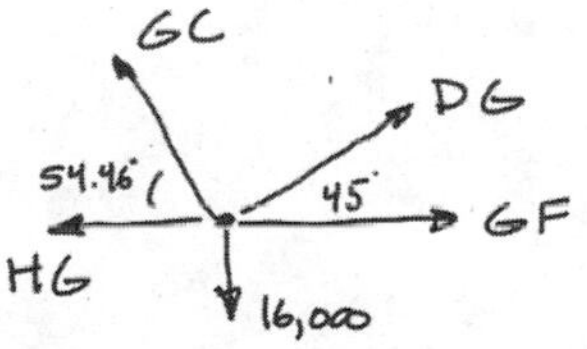

$\Sigma F_y = 0$

$DG \sin 45 + GC \sin 54.46 - 16{,}000 = 0$

$13519 \sin 45 + GC \sin 54.46 - 16{,}000 = 0$

∴ **GC = 7,915 lb T**

Test Yourself 6.7

FBD 1

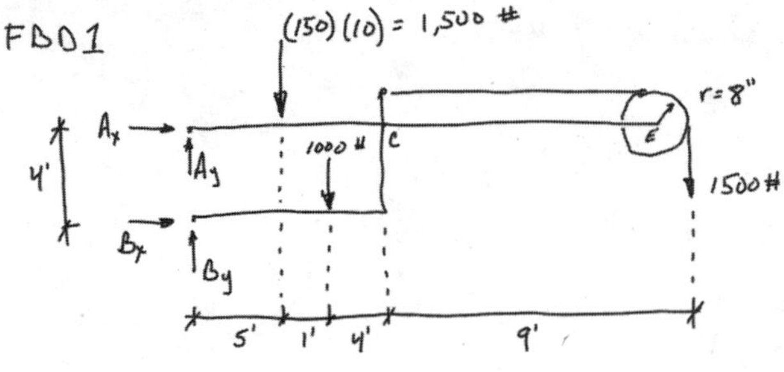

$\Sigma M_A = 0$

$B_x(4) - 1500(5) - 1000(6) - 1500(19) = 0$

$\therefore$ $\boxed{B_x = 10{,}500\#}$ To The RIGHT

$\Sigma F_x = 0$

$A_x + B_x = 0$

$\therefore$ $\boxed{A_x = -10{,}500\#}$ TO THE LEFT

FBD 2

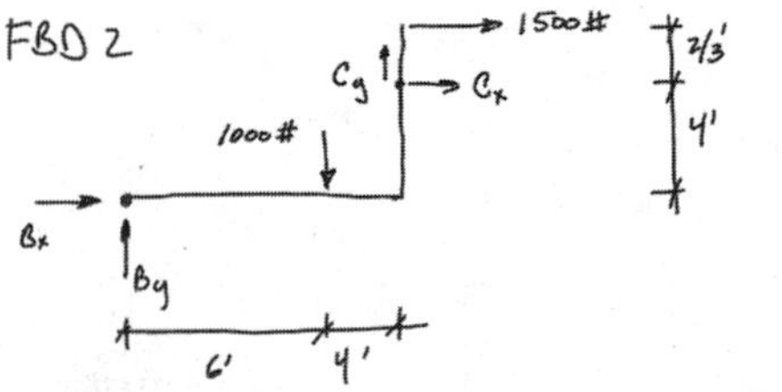

$\Sigma M_c = 0$

$-1500(2/3) + 1000(4) - B_y(10) + B_x(4) = 0$

$[B_x = 10{,}500]$ From Previous

$\therefore$ $\boxed{B_y = 4{,}500\#}$ UPWARD

FROM FBD 1

$\Sigma F_y = 0$

$B_y + A_y - 1500 - 1000 - 1500 = 0$

$[B_y = 4{,}500]$ From Previous

$\therefore$ $\boxed{A_y = -500\#}$ DOWNWARD

Test Yourself 6.8

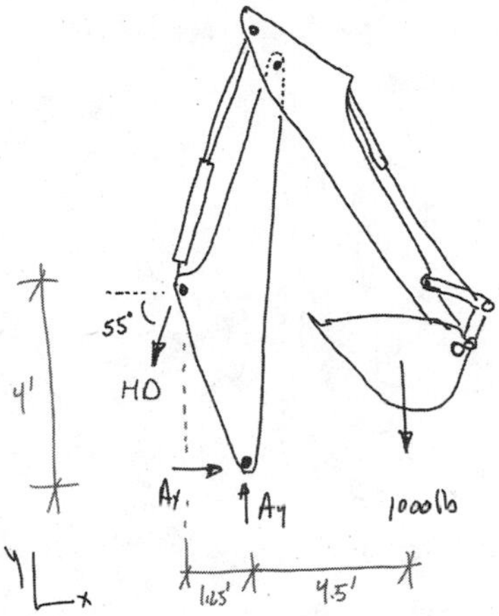

$$\Sigma M_A = 0 = -1000(4.5) + (HD\cos 55)4 + (HD\sin 55)(1.25)$$

$$HD = 1{,}356\ \text{lb}$$

$$\Sigma F_x = 0 \quad -HD\cos 55 + A_x = 0$$

$$A_x = 777.7\ \text{lb}$$

$$\Sigma F_y = 0 \quad -HD\sin 55 + A_y - 1000 = 0$$

$$A_y = 2{,}110\ \text{lb}$$

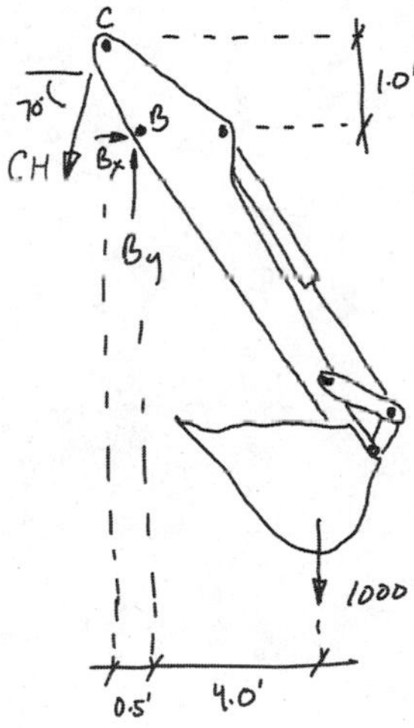

$$\Sigma M_B = 0$$

$$CH\cos 70(1) + CH\sin(70)(0.5) - 1000(4.0) = 0$$

$$CH[\cos 70 + 0.5\sin 70] = 4000$$

$$\therefore \boxed{CH = 4{,}927\ \#}$$

$$\Sigma F_x = 0$$

$$B_x - CH\cos 70 = 0$$

$$\therefore \boxed{B_x = 1{,}685\ \#}$$

$$\Sigma F_y = 0$$

$$B_y - 1000 - CH\sin 70^\circ = 0$$

$$\therefore \boxed{B_y = 5{,}630\ \#}$$

Test Yourself 6.9

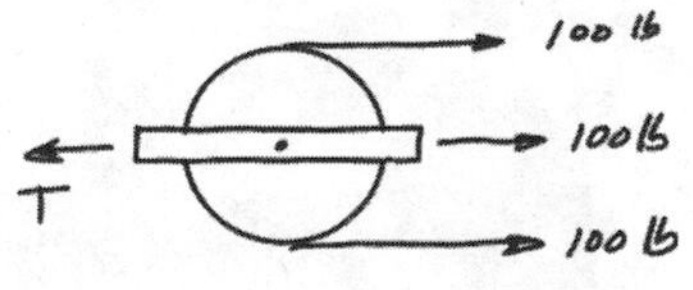

$$\Sigma F_x = 0 \quad T = 300\ \text{lb}$$

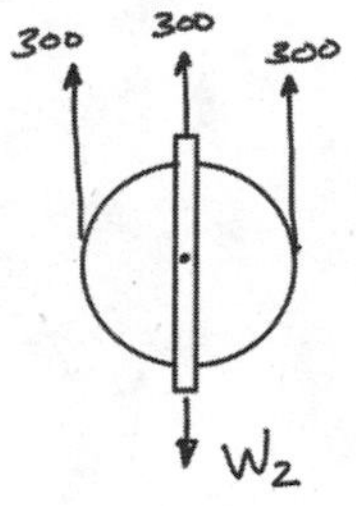

$$\Sigma F_y = 0$$

$$\boxed{W_2 = 900\ \text{lb}}$$

Test Yourself 7.1

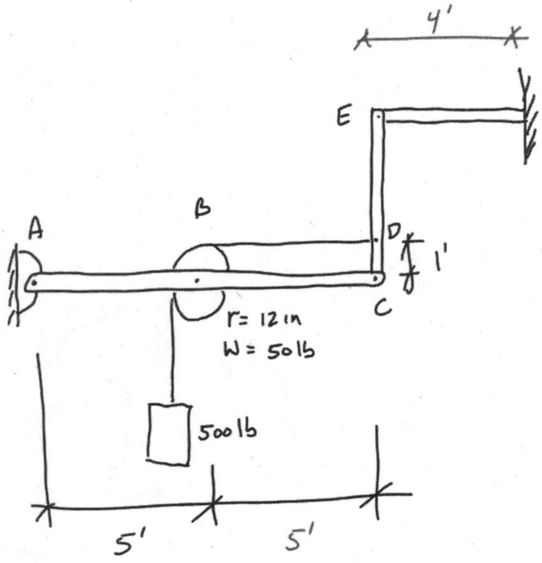

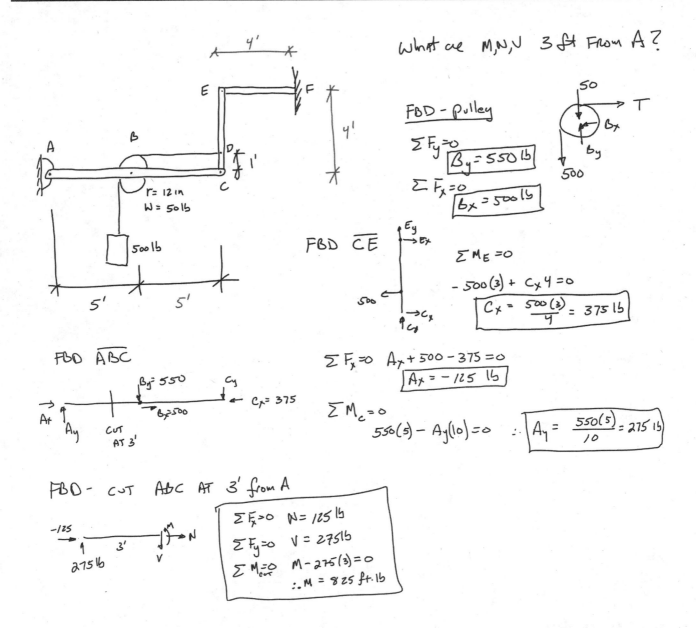

What are M, N, V 3 ft From A?

FBD - Pulley

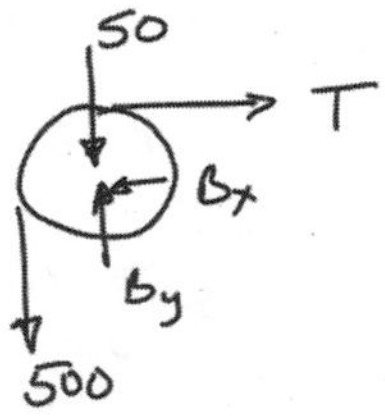

$\Sigma F_y = 0$

$\boxed{B_y = 550 \text{ lb}}$

$\Sigma F_x = 0$

$\boxed{B_x = 500 \text{ lb}}$

FBD $\overline{CE}$

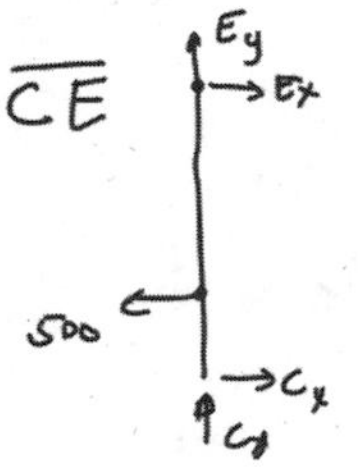

$\Sigma M_E = 0$

$-500(3) + C_x 4 = 0$

$\boxed{C_x = \frac{500(3)}{4} = 375 \text{ lb}}$

FBD $\overline{ABC}$

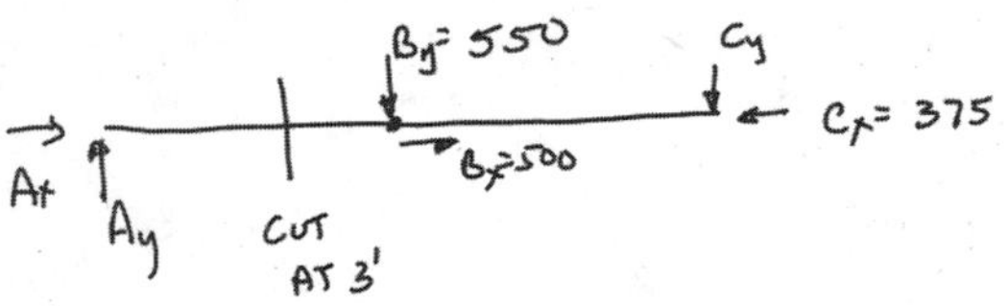

$\Sigma F_x = 0 \quad A_x + 500 - 375 = 0$

$\boxed{A_x = -125 \text{ lb}}$

$\Sigma M_c = 0$

$550(5) - A_y(10) = 0 \quad \therefore \boxed{A_y = \frac{550(5)}{10} = 275 \text{ lb}}$

FBD - CUT ABC AT 3' from A

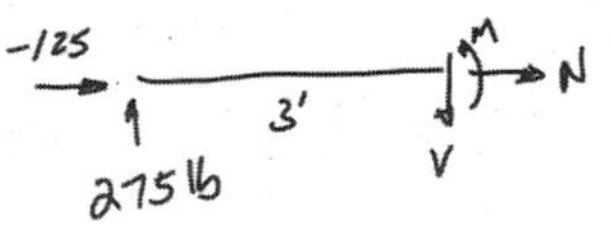

$\Sigma F_x = 0 \quad N = 125 \text{ lb}$

$\Sigma F_y = 0 \quad V = 275 \text{ lb}$

$\Sigma M_{cut} = 0 \quad M - 275(3) = 0$

$\therefore M = 825 \text{ ft} \cdot \text{lb}$

Test Yourself 7.2

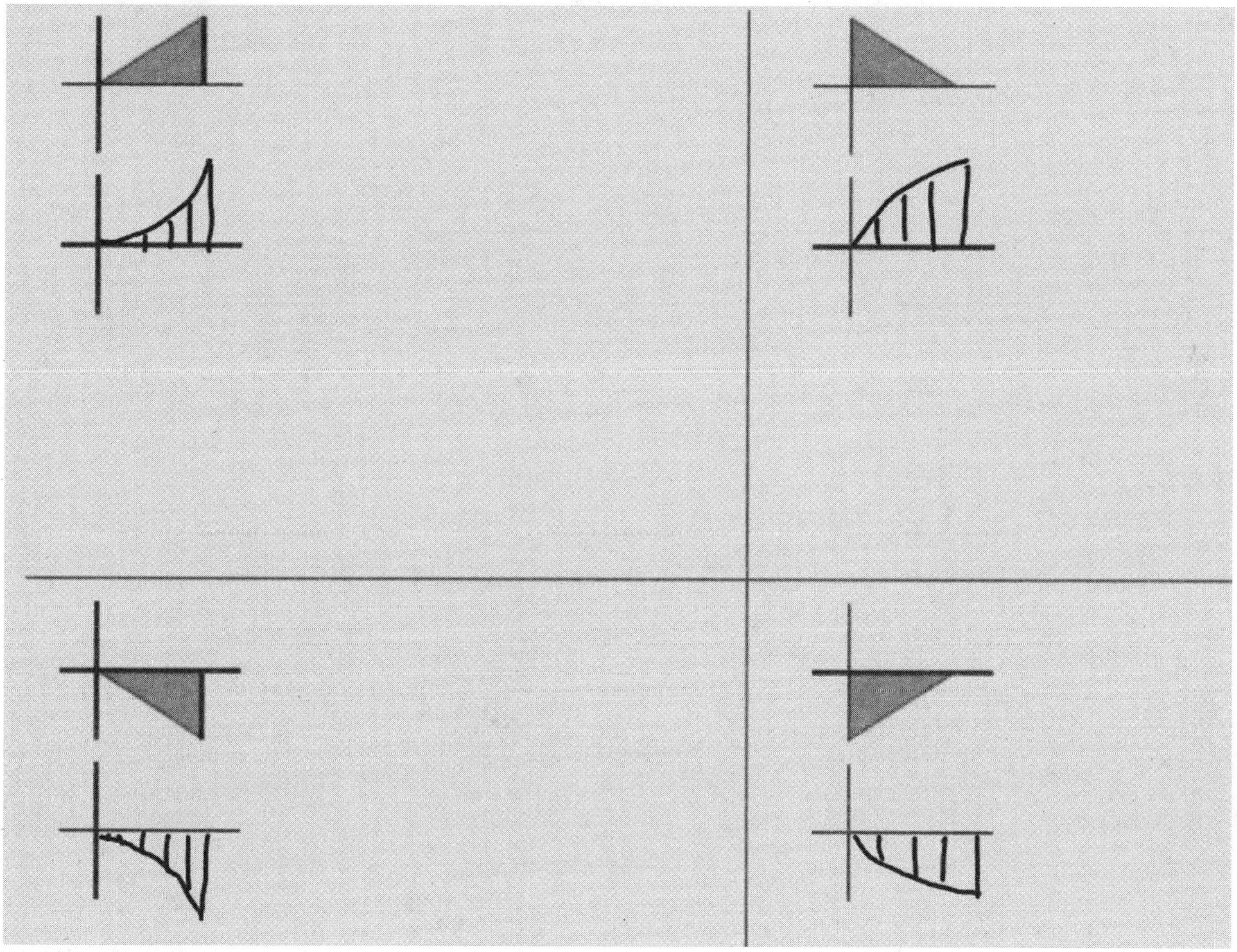

Test Yourself 7.3

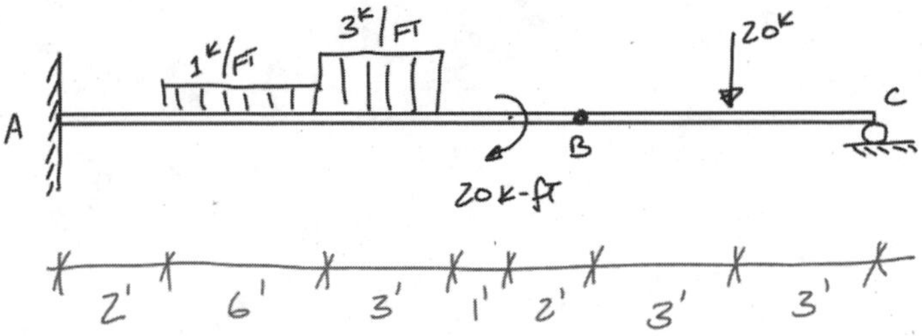

Global Equilibrium

$\Sigma F_y = 0 \quad A_y - 6 - 9 - 20 + 10 = 0$

$\boxed{A_y = 6 + 9 + 20 - 10 = 25^K}$

$\Sigma M_A = 0$

$M_A - 6(5) - 9(9.5) - 20 - 20(17) + 10(20) = 0$

$\boxed{M_A = 275.5 \text{ K}\cdot\text{ft}}$

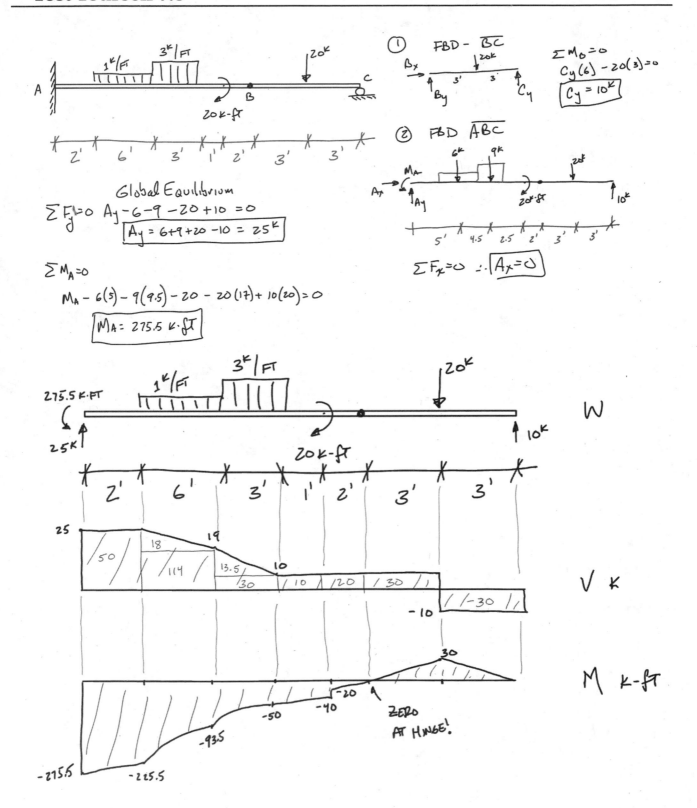

① FBD - $\overline{BC}$

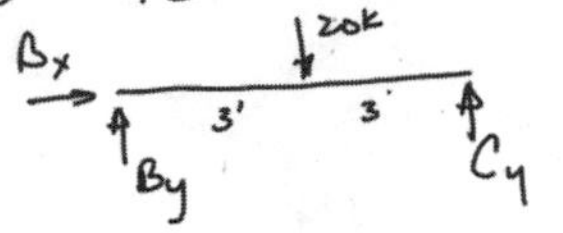

$\Sigma M_B = 0$

$C_y(6) - 20(3) = 0$

$\boxed{C_y = 10^K}$

② FBD $\overline{ABC}$

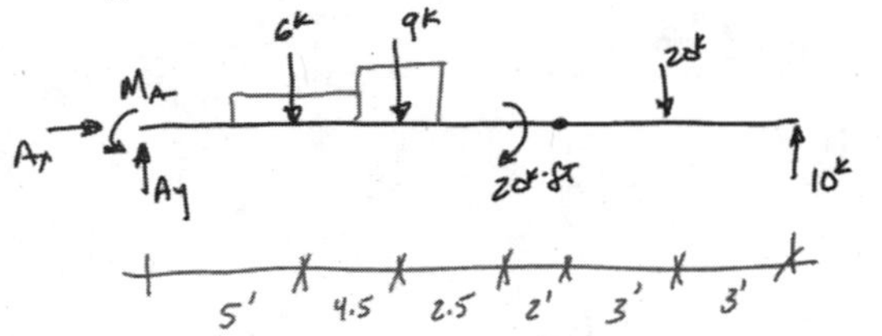

$\Sigma F_x = 0 \;\therefore\; \boxed{A_x = 0}$

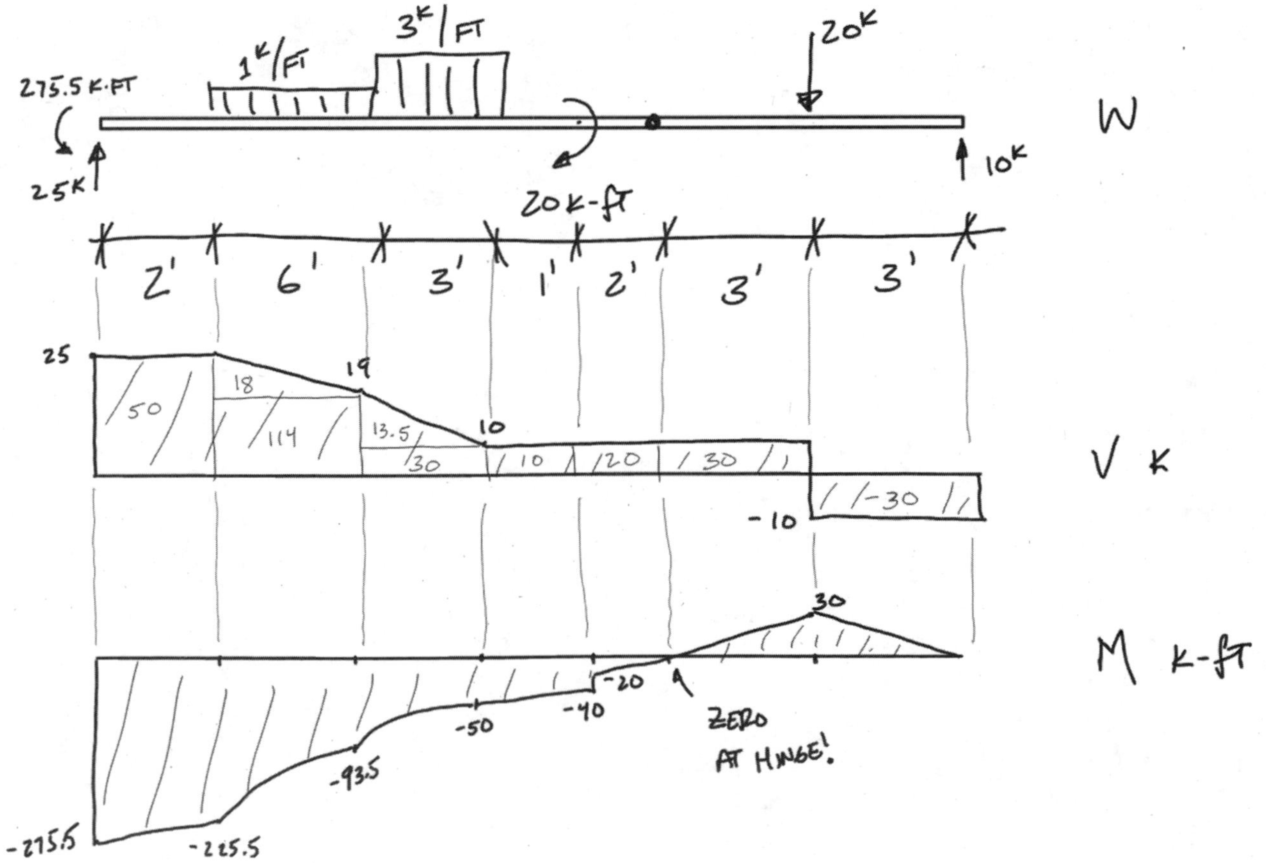

Test Yourself 7.4

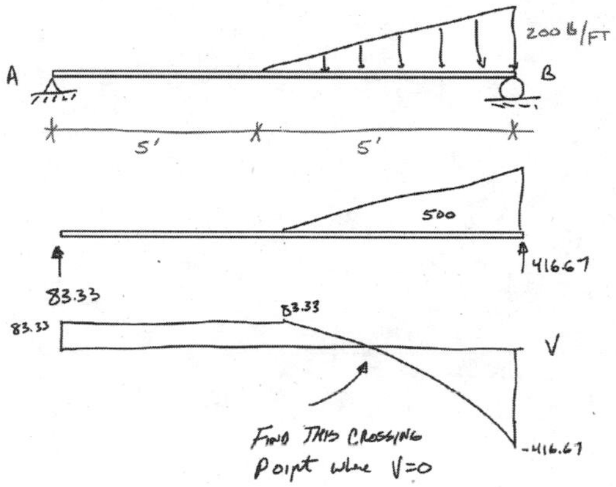

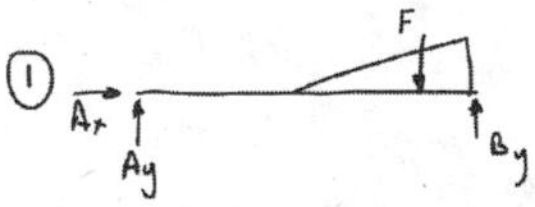

$F = \frac{1}{2}(200)5 = 500 \text{ lb}$

$\Sigma M_A = 0 \quad B_y(10) - 500(5 + \frac{2}{3}5) = 0$

$\therefore B_y = 416.67 \text{ lb}$

$\Sigma M_B = 0 \quad -A_y(10) + 500(\frac{1}{3}5) = 0$

$\therefore A_y = 83.33 \text{ lb}$

Check! $\Sigma F_y = 0 \quad 416.67 + 83.33 = 500$ ✓

for x: $5 \le x \le 10$

$w(x) = 40(x-5)$

$w(x) = 40(x-5)$ ($x=5$ $w=0$ ✓ $x=10$ $w=200$ ✓)

FBD $5 \le x \le 10$

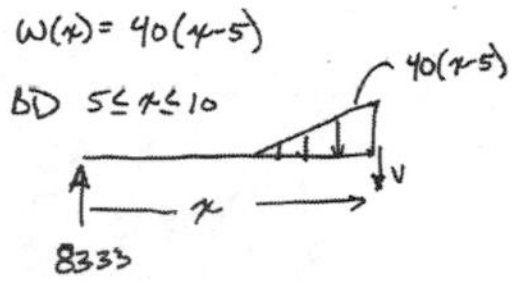

$\Sigma F_y = 0 \quad 83.33 - \left[\frac{1}{2}(x-5)\, 40(x-5)\right] - V = 0$

$V(x) = 83.33 - 20(x-5)^2$

2.96
7.04

When does $V(x) = 0$?

$83.33 - 20(x-5)^2 = 0$

} Two Solutions $x = 2.96$ and $x = 7.04$ 〉 Note: this Equation is only valid $5 \le x \le 10$ So $x = 7.04$ is our solution

$V = 0$ when $x = 7.04$ ft

Also: when $V = 0$, we need to compute M_{MAX}

FBD $5 \le x \le 10$

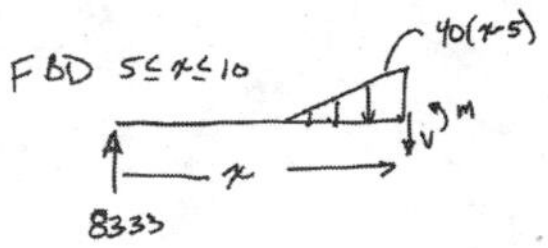

$\Sigma M_{cut} = 0$

$-83.33x + \left[\frac{1}{2}(x-5)\, 40(x-5) \cdot \frac{1}{3}(x-5)\right] + M = 0$

$\therefore M(x) = 83.33x - 6.67(x-5)^3$

$$\boxed{M_{max} = M(7.04) = 530 \text{ ft lb}}$$

FBD $5 \le x \le 10$

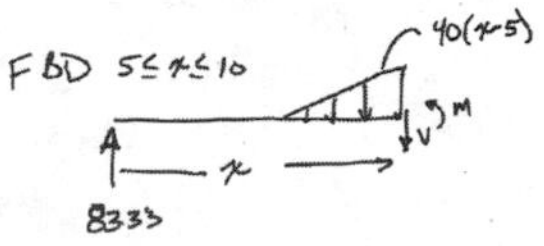

$\Sigma M_{cut} = 0$

$-83.33x + \left[\frac{1}{2}(x-5)\, 40(x-5) \cdot \frac{1}{3}(x-5)\right] + M = 0$

$\therefore M(x) = 83.33x - 6.67(x-5)^3$

$$\boxed{M_{max} = M(7.04) = 530 \text{ ft lb}}$$

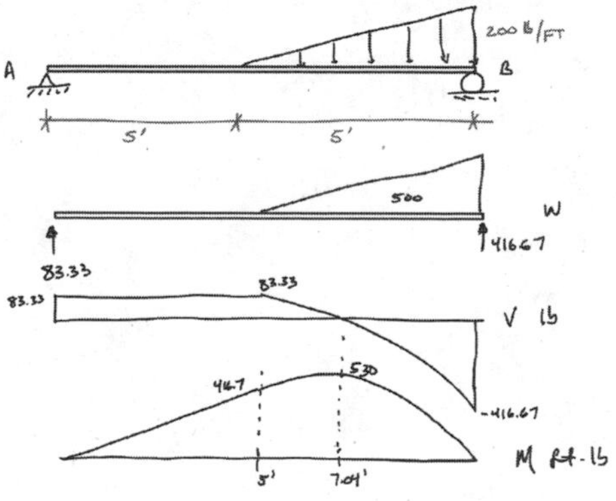

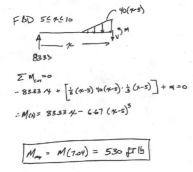

Test Yourself 8.1

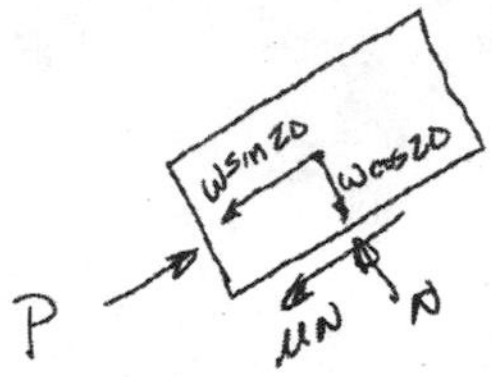

AT Slip $f = \mu N$

$\Sigma F_y = 0$

$\therefore N = W \cos 20$

$= 100(\cos 20)$

$N = 93.97 \text{ lb}$

$\Sigma F_x = 0$

$P - W \sin 20 - \mu N = 0$

$P = W \sin 20 + \mu N$

$= 100(\sin 20) + 0.4(93.97)$

$\therefore \boxed{P = 71.79 \text{ lb}}$

Test Yourself 8.2

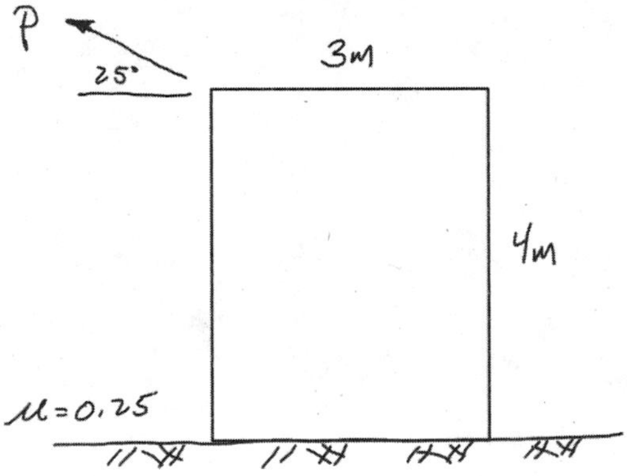

FBD

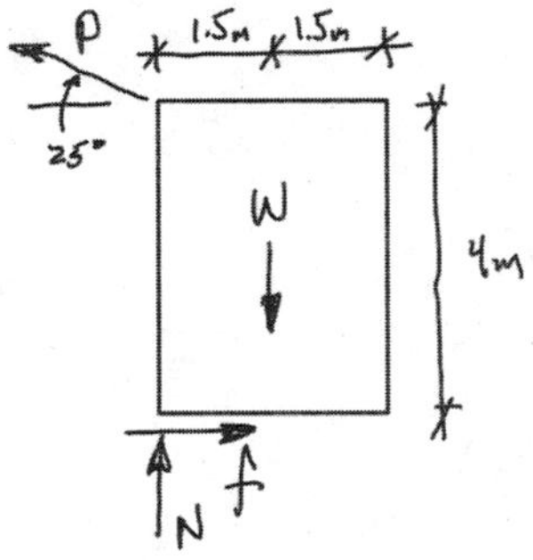

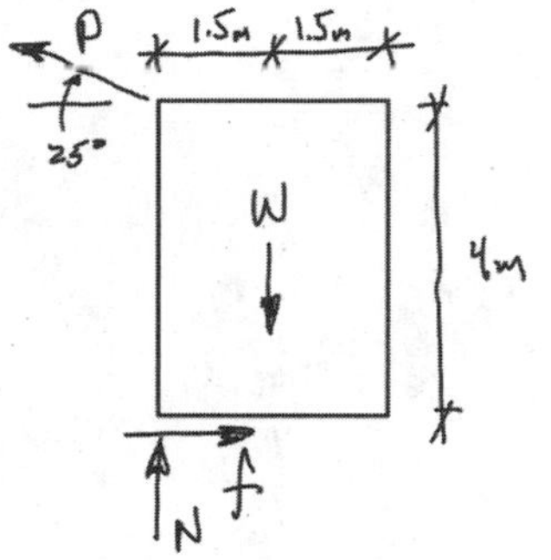

SLIP

$\Sigma F_y = 0$

$P\sin 25 - W + N = 0$

$N = W - P\sin 25$

At slip: $f = \mu N$

$\mu = 0.25$

$\Sigma F_x = 0$

$P\cos 25 - f = 0$

At slip, $f = \mu N$

$P\cos 25 - \mu[W - P\sin 25] = 0$

$P\cos 25 - \mu W + \mu P\sin 25 = 0$

$P[\cos 25 + \mu\sin 25] = \mu W$

$$P = \frac{\mu W}{(\cos 25 + \mu\sin 25)}$$

$$= \frac{(0.25)\,W}{\cos 25^\circ + 0.25\sin 25^\circ}$$

$\therefore$ $\boxed{P = 0.247W}$

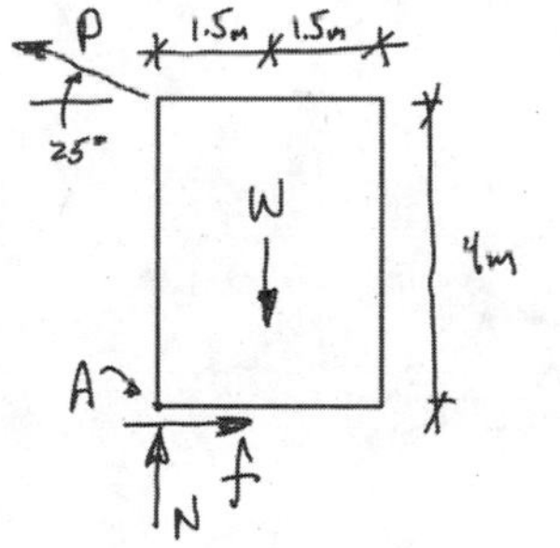

TIP

$\Sigma M_A = 0$

$P\cos 25\,(4) - W\,(1.5) = 0$

$$P = \frac{W\,(1.5)}{4\cos 25} = 0.414W$$

$\boxed{P_{Tip} = 0.414W}$

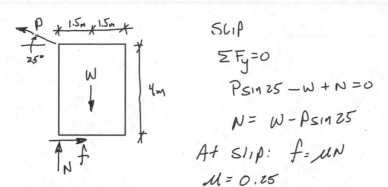

$\therefore$ $\boxed{\text{BOX SLIPS W/O TIPPING}}$

Test Yourself 8.3

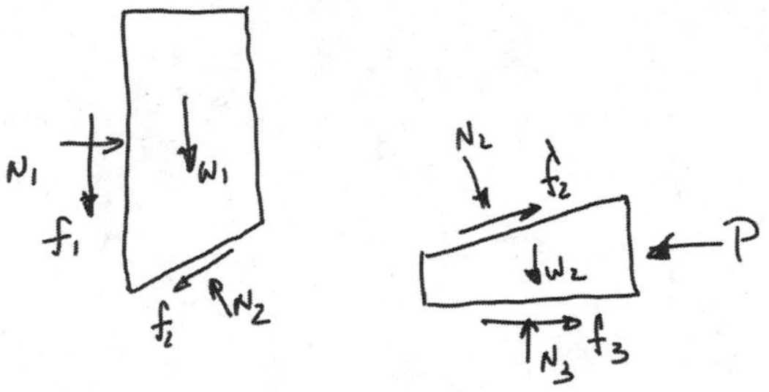

1.

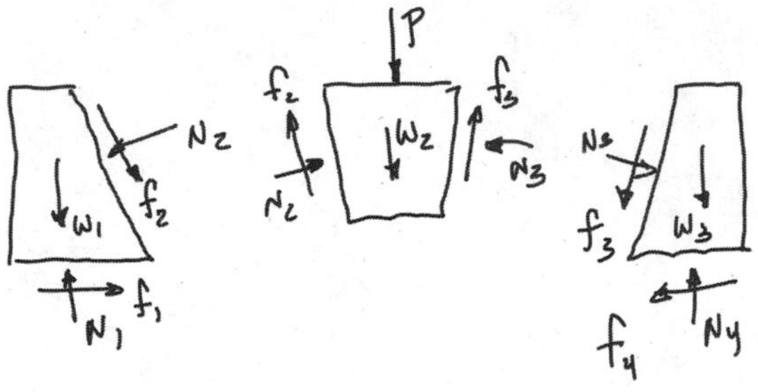

2.

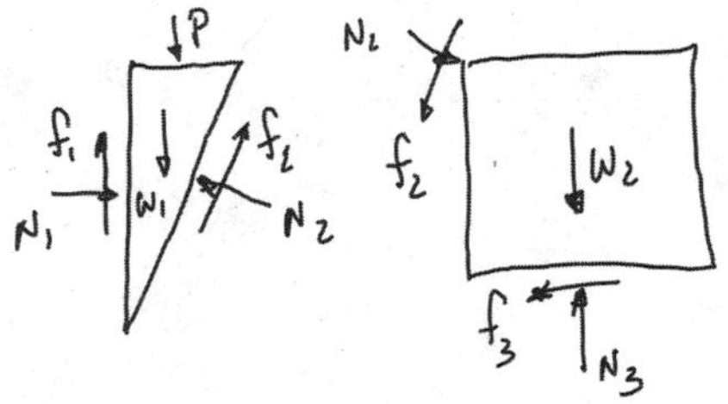

3.

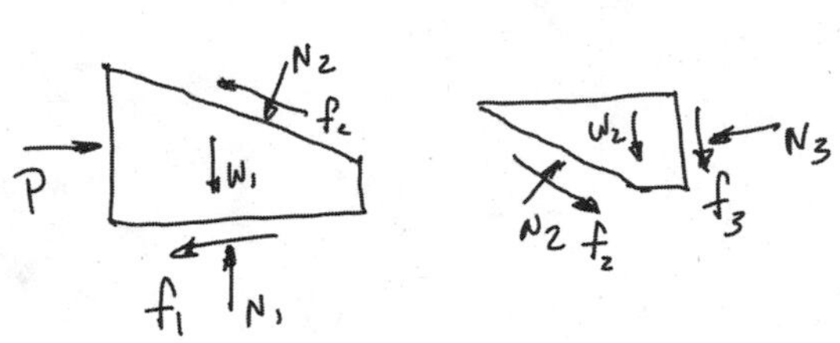

4.

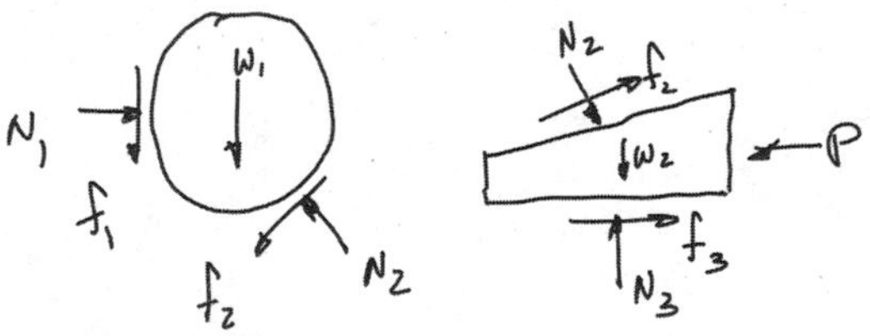

5.

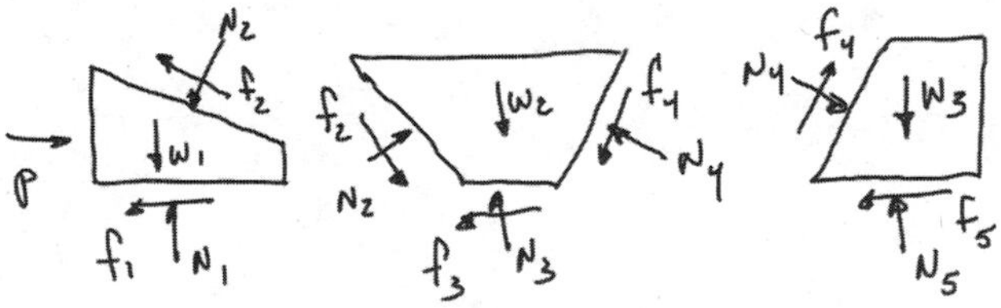

6.

Test Yourself 8.4

FBD BLOCK B

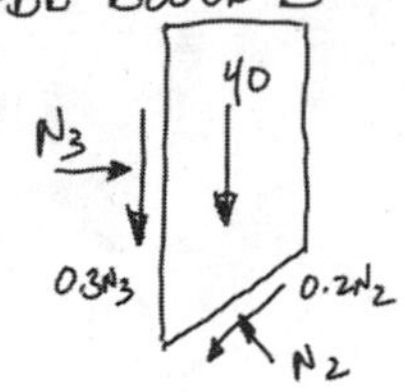

$\Sigma F_y = 0$

$$-40 - 0.3N_3 - 0.2N_2 \sin 45 + N_2 \cos 45 = 0$$

$\Sigma F_x = 0$

$$N_3 - 0.2N_2 \cos 45 - N_2 \sin 45 = 0$$

SOLVE

$N_2 = 128.6$ lb

$N_3 = 109.1$ lb

FBD BLOCK A

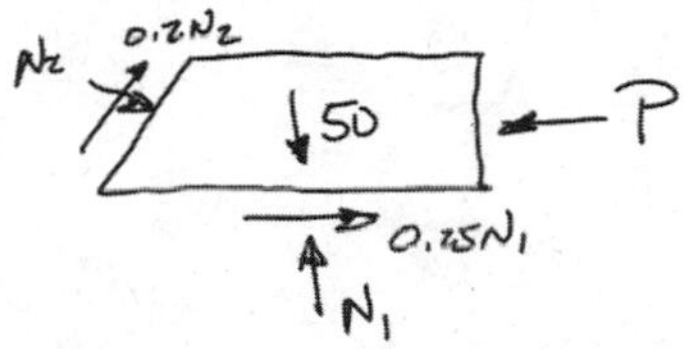

$\Sigma F_x = 0$

$$0.2N_2 \cos 45 + N_2 \sin 45 + 0.25N_1 - P = 0$$

$\Sigma F_y = 0$

$$N_1 - 50 + 0.2N_2 \sin 45 - N_2 \cos 45 = 0$$

$\therefore$ $N_1 = 122.8$ lb $\quad P = 139.8$ lb

Test Yourself 8.5

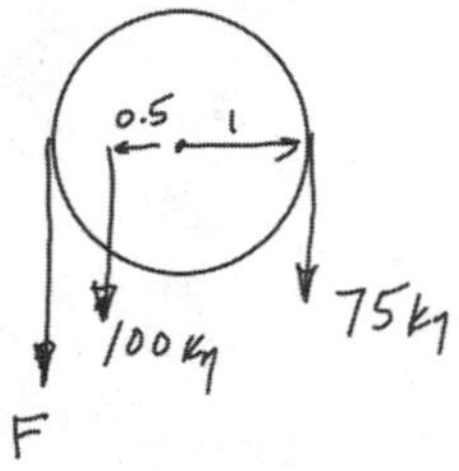

ASSUME IMPENDING ROTATION IS TO THE LEFT.

$$\Sigma M_{center} = F(1) + 100(0.5) - 75(1) = 0$$

$$\therefore F = 75 - 50 = 25 \text{ kg}$$

FOR BELT FRICTION ON DRUM

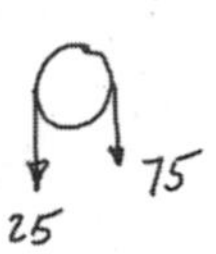

$$\frac{T_1}{T_2} = e^{\mu\beta} \qquad T_1 = 75, \quad T_2 = 25, \quad \beta = 180° = \pi$$

$$\frac{75}{25} = e^{\mu\pi} \rightarrow 3 = e^{\mu\pi}$$

$$\ln(3) = \mu\pi \quad \therefore \boxed{\mu = 0.35}$$

Test Yourself 9.1

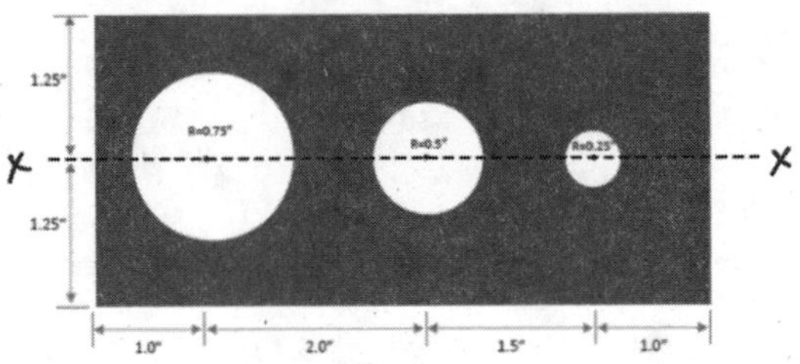

$$I_x = \frac{1}{12}(5.5)(2.5)^3 - \frac{1}{4}\pi\left[(0.75)^4 + (0.5)^4 + (0.25)^4\right]$$

$$\boxed{I_x = 6.86 \text{ in}^4}$$

Test Yourself 9.2

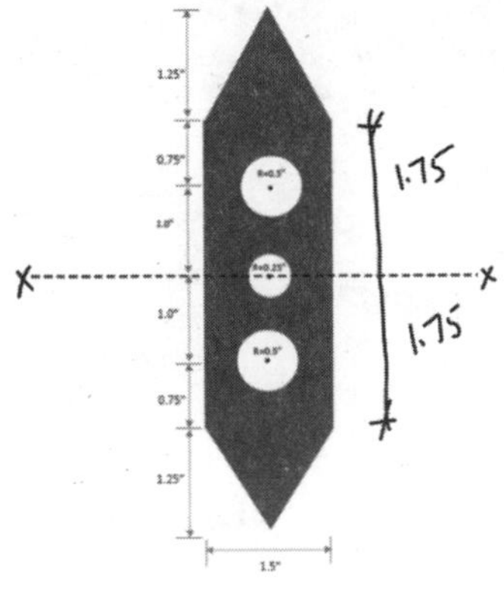

By Symmetry, The Centroid is Located As Shown By the Dashed Line.

$$I_x = \frac{1}{12}(1.5)(3.5)^3 - 2\left[\frac{\pi}{4}(0.5)^4 + (\pi(0.5)^2)(1)^2\right] - \frac{\pi}{4}(0.25)^4 + 2\left[\frac{1}{36}(1.5)(1.25)^3 + \left(\frac{1}{2}(1.5)(1.25)\right)\left(1.75 + \frac{1.25}{3}\right)^2\right]$$

$$\boxed{I_x = 12.65 \text{ in}^4}$$

Test Yourself 9.3

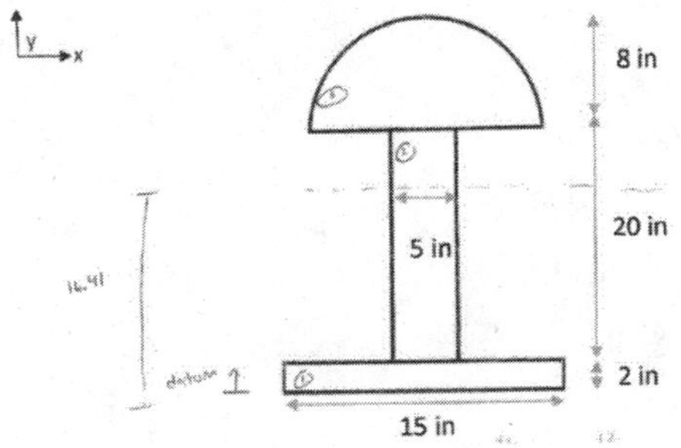

① FIND THE CENTROIDAL X-AXIS

- SET THE DATUM AT THE BASE
- USE THE Three BASIC SHAPES SHOWN

$$\bar{y} = \frac{\Sigma A_i \bar{y}_i}{\Sigma A_i}$$

#	AREA	$\bar{y}_i$	$A\bar{y}_i$
1	30	1	30
2	100	12	1200
3	100.5	25.4	2552.7
	230.5 in²		3782.7 in³

$$\bar{y} = \frac{3782.7}{230.5} = 16.41 \text{ in}$$

② Compute I_x About The Centroid

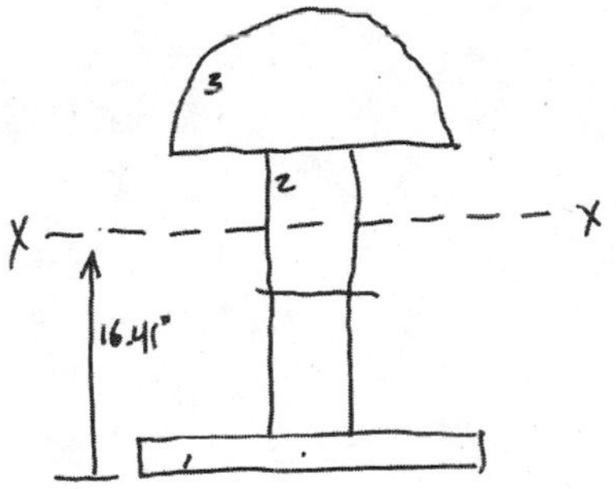

$$I_x = \frac{1}{12}(15)(2)^3 + (30)(15.41)^2$$
$$+ \frac{1}{12}(5)(20)^3 + (100)(4.41)^2$$
$$+ I_c + (100.5)(25.4 - 16.41)^2$$

$$I_c = \frac{1}{8}\pi r^4 - \frac{8r^4}{9\pi}$$
$$= \frac{1}{8}\pi(8)^4 - \frac{8(8)^4}{9\pi} = 449.6$$

$$\therefore \; I_x = 20{,}984 \text{ in}^4$$

Test Yourself 9.4

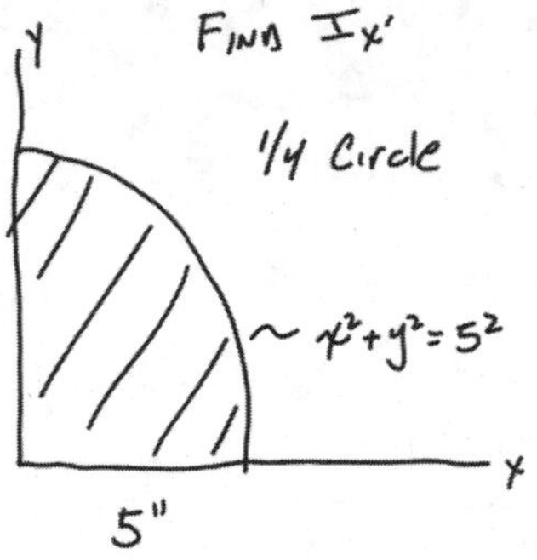

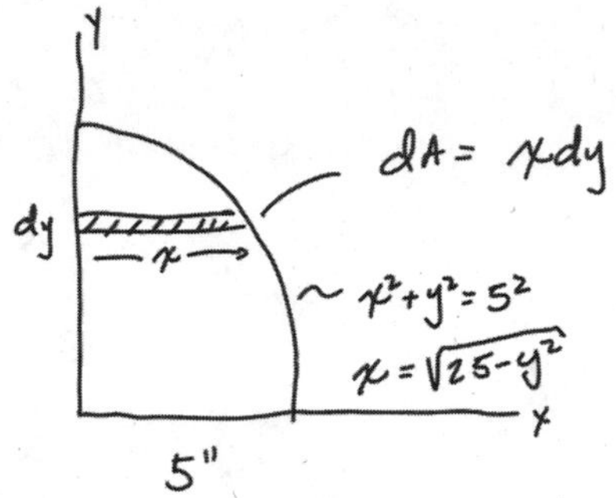

$$I_{x'} = \int y^2 dA$$

$$= \int y^2 (x\,dy)$$

$$= \int_0^5 y^2 \left(\sqrt{25-y^2}\right) dy$$

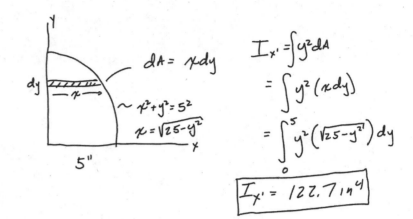

$$\boxed{I_{x'} = 122.7\ in^4}$$

Exam 1—Statics, Practice Set 1

Department of Mechanical Engineering

My Favorite University

Good Luck!

120 MINUTES

This class: Sec. 001 – Dr. Hanson

(This exam covers Level 1 through part of Level 4.)

RULES:

1. The solutions to this exam are video solutions which you can access at the QR code; however, you are not permitted to look at them until after your practice exam!
2. All your work must be done on the papers provided including this cover sheet.
3. Closed book and closed notes.
4. You may have a FE approved calculator, pens, pencils, erasers.
5. Use of a phone of any kind is considered cheating.
6. Treat this as your real exam and time yourself in a nice quiet place at a desk, as if you are taking this in your own classroom.

HONOR CODE—YOUR ME DEPARTMENT

I hereby certify that I will follow the Code of Student Conduct as defined by the University and the Department, that I will not cheat nor will I condone cheating.

Name ______________________________ (Print legibly)

Name ______________________________ (Signature)

PROBLEM 1 (20 POINTS)

Determine the magnitude and coordinate direction angles of the resultant of F_1, F_2, and F_3.

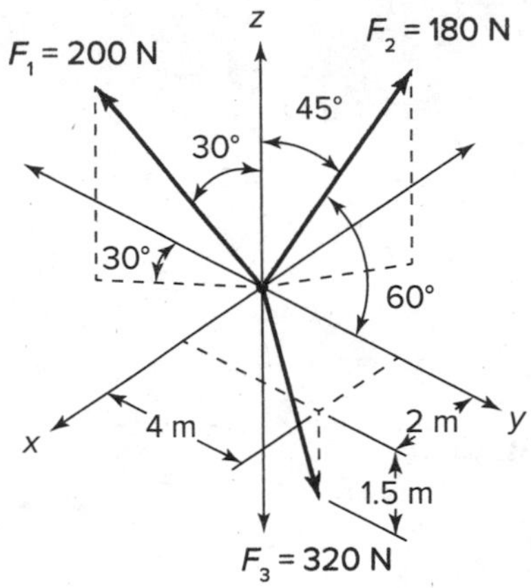

$|\vec{F}_R| =$ ________________

$\theta_x(\alpha) =$ ________________

$\theta_y(\beta) =$ ________________

$\theta_z(\gamma) =$ ________________

PROBLEM 2 (20 POINTS)

Bears have invaded your campsite, and you're forced to hang your backpack from a tree. Find the max weight of the backpack if cable *AE* and *AD* have a breaking strength of 200 lbs, and cable *ABC* has a breaking strength of only 100 lbs.

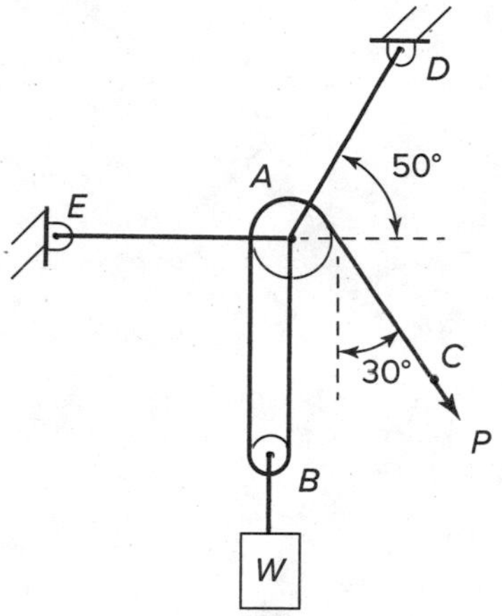

Weight of the backpack = ____________

PROBLEM 3 (20 POINTS)

If the suspended weight is 800 lbs, find the tension in cables *AC*, *AE*, and the single cable *DABA*. (This cable goes from point *D* through a ring at point *A*, around a pulley at *B*, and then back to point *A* where it's tied off.)

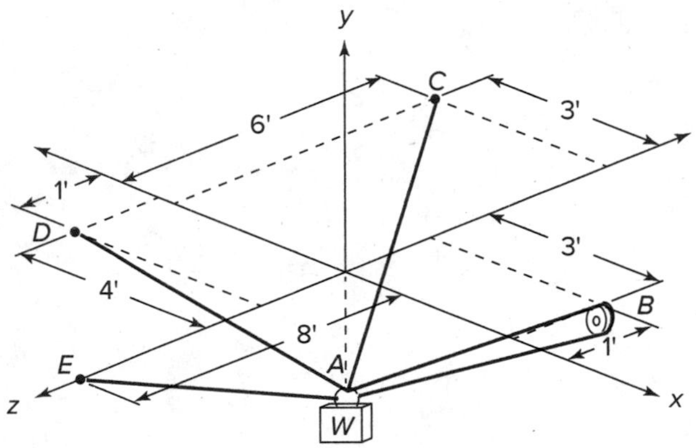

Tension in *AC* = ____________________

Tension in *AE* = ____________________

Tension in *DABA* = ____________________

PROBLEM 4 (20 POINTS)

If the resultant moment is to be zero on the system, find force P.

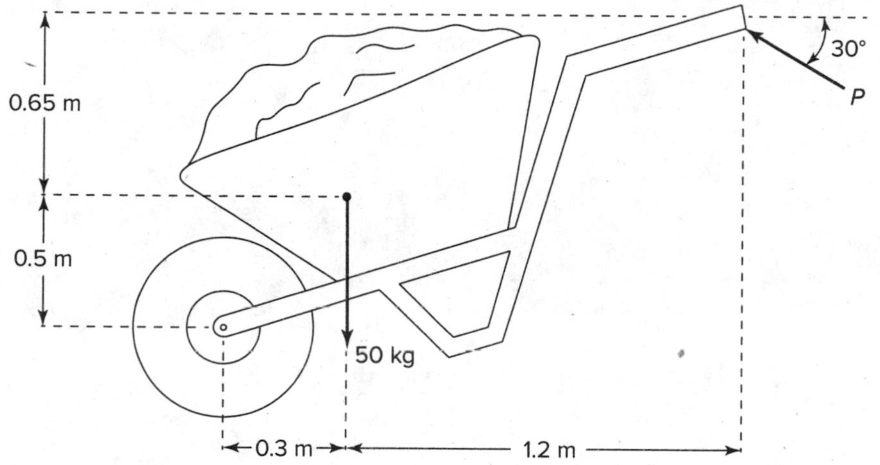

Force $P =$ ______________________

PROBLEM 5 (20 POINTS)

A board supports an awning over a door. The force in the board is 57 lbs. What is the moment at point *A* produced by board *ED*?

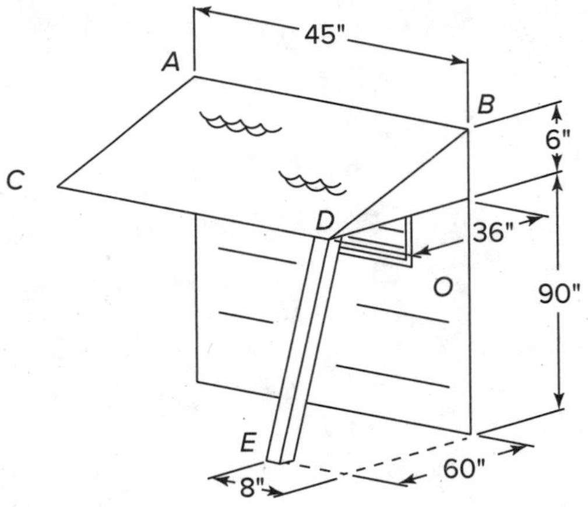

Moment at point *A* = ______________________

Exam 1—Statics, Practice Set 2

Department of Mechanical Engineering

My Favorite University

Good Luck!

120 MINUTES

This class: Sec. 001 – Dr. Hanson

(This exam covers Level 1 through part of Level 4.)

RULES:

1. The solutions to this exam are in the back of the book; however, you are not permitted to look at them until after your practice exam!
2. All your work must be done on the papers provided including this cover sheet.
3. Closed book and closed notes.
4. You may have a FE approved calculator, pens, pencils, erasers.
5. Use of a phone of any kind is considered cheating.
6. Treat this as your real exam and time yourself in a nice quiet place at a desk, as if you are taking this in your own classroom.

HONOR CODE—YOUR ME DEPARTMENT

I hereby certify that I will follow the Code of Student Conduct as defined by the University and the Department, that I will not cheat nor will I condone cheating.

Name ________________________________ (Print legibly)

Name ________________________________ (Signature)

PROBLEM 1 (20 POINTS)

For the given system, find the resultant of the four vectors and find the directional cosine angles of the resultant vector.

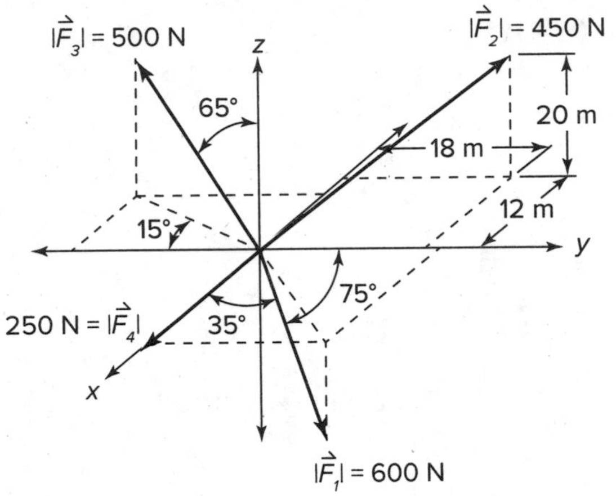

$|\vec{F}_R| =$ ________________

$\theta_x(\alpha) =$ ________________

$\theta_y(\beta) =$ ________________

$\theta_z(\gamma) =$ ________________

PROBLEM 2 (20 POINTS)

Three ropes are tied to a ring at *E*. If the max tension in any rope can be no more than 8 kN, find the maximum weight of the suspended box.

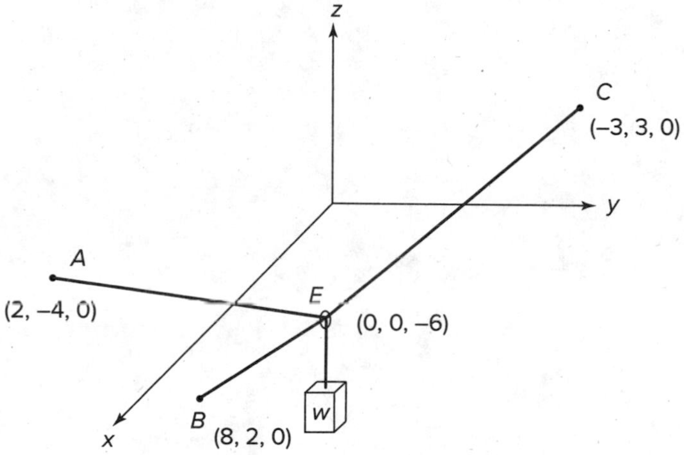

Weight of the box = ____________________

PROBLEM 3 (20 POINTS)

If the flowerpot weighs 80 lbs and the weight of the cylinder (W) is 100 lbs, find the tension in cable BC, the angle θ, and the stretch in the spring (x).

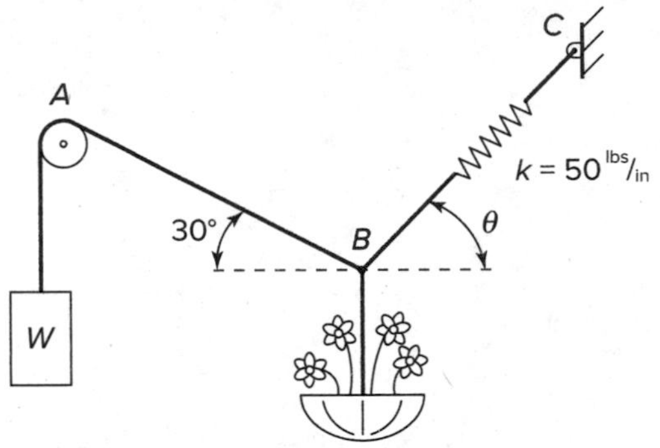

Tension in $BC =$ ____________________

Angle $\theta =$ ____________________

Stretch in spring $=$ ____________________

PROBLEM 4 (20 POINTS)

For the given system, find the moment about point E if $\theta = 35°$. If θ is variable, also calculate the maximum and the minimum possible clockwise moments about point E.

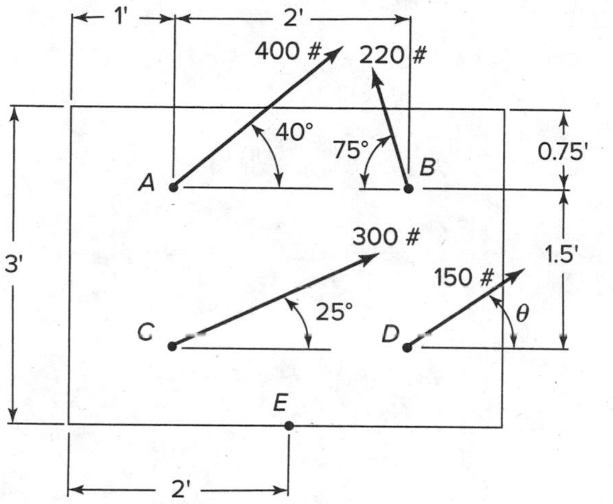

Moment about E = ____________________

Max moment = ____________________

Min moment = ____________________

PROBLEM 5 (20 POINTS)

Find the moment about point *A* produced by the force vector applied at point *B*.

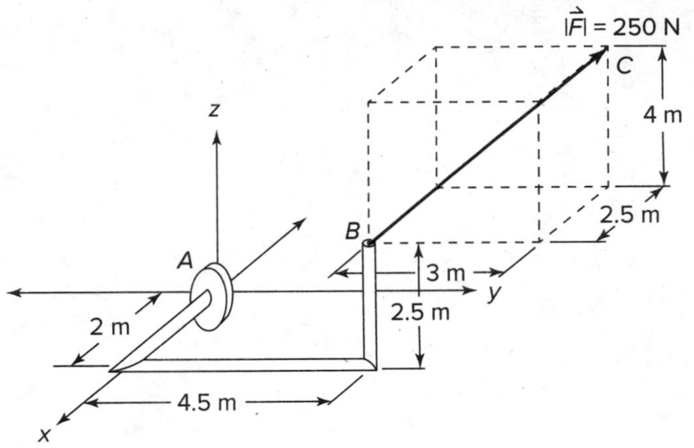

Moment about $A =$ ________________________

Exam 2—Statics, Practice Set 1

Department of Mechanical Engineering

My Favorite University

Good Luck!

120 MINUTES

This class: Sec. 001 – Dr. Hanson

(This exam covers the end of Level 4 through Level 6.)

RULES:

1. The solutions to this exam are video solutions which you can access at the QR code; however, you are not permitted to look at them until after your practice exam!
2. All your work must be done on the papers provided including this cover sheet.
3. Closed book and closed notes;
4. You may have a FE approved calculator, pens, pencils, erasers.
5. Use of a phone of any kind is considered cheating.
6. Treat this as your real exam and time yourself in a nice quiet place at a desk, as if you are taking this in your own classroom.

HONOR CODE—YOUR ME DEPARTMENT

I hereby certify that I will follow the Code of Student Conduct as defined by the University and the Department, that I will not cheat nor will I condone cheating.

Name ______________________________ (Print legibly)

Name ______________________________ (Signature)

PROBLEM 1 (16.7 POINTS)

Replace the given system with a single force and determine where that force intersects line *OA* measured from point *O*.

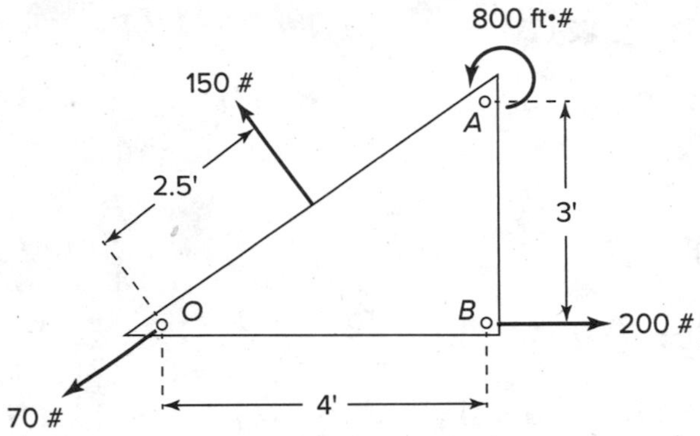

$|\vec{F}_R| =$ ____________________

Distance = ____________________

PROBLEM 2 (16.7 POINTS)

Find the reactions at points B and D.

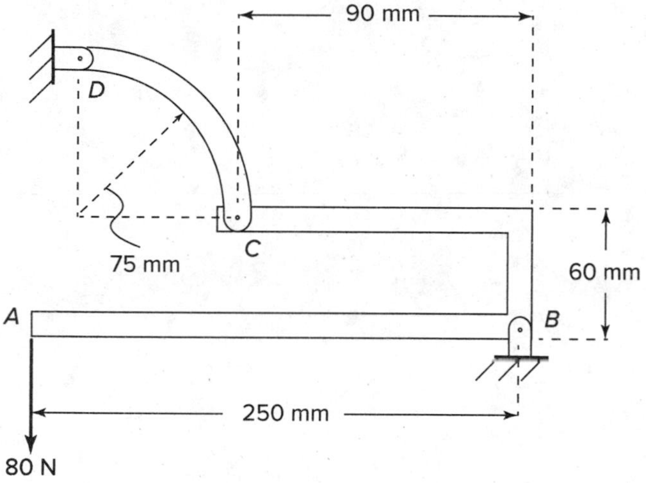

$B_x =$ ________________

$B_y =$ ________________

$D_x =$ ________________

$D_y =$ ________________

PROBLEM 3 (16.7 POINTS)

The bracket is supported by ball and socket joints at A and B. Find the tension in cable CD.

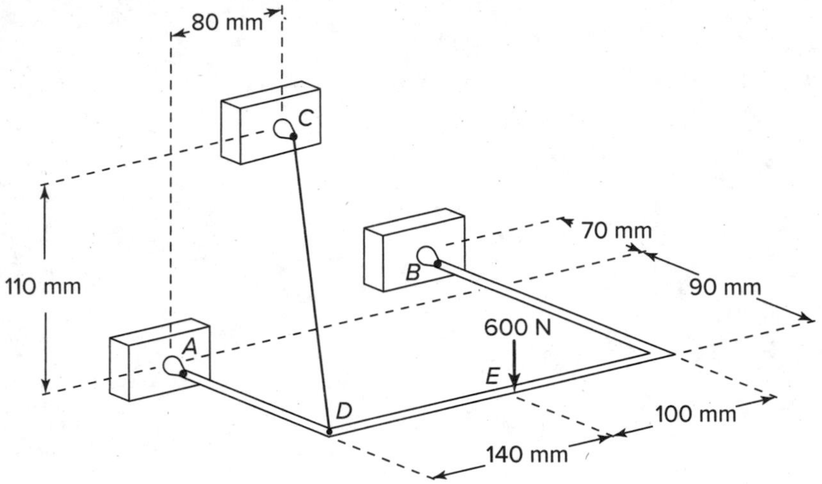

Tension in CD = ______________

PROBLEM 4 (16.7 POINTS)

Find the centroid $(\bar{x}, \bar{y}, \bar{z})$ for the thin part given.

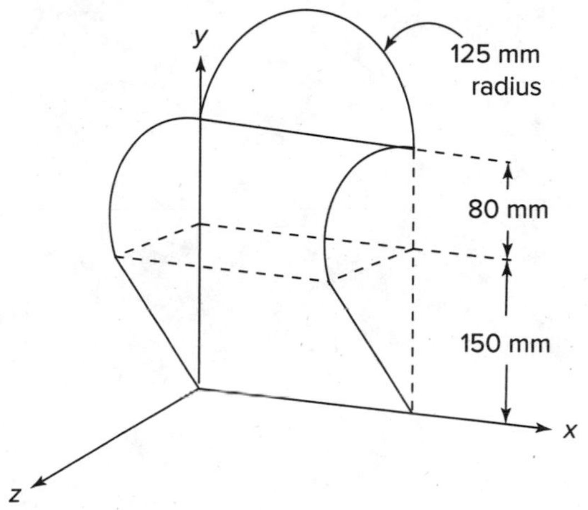

$\bar{x} =$ ______________________________

$\bar{y} =$ ______________________________

$\bar{z} =$ ______________________________

PROBLEM 5 (16.7 POINTS)

Find $\bar{x}$ and $\bar{y}$ for the shape using the centroid by calculus method.

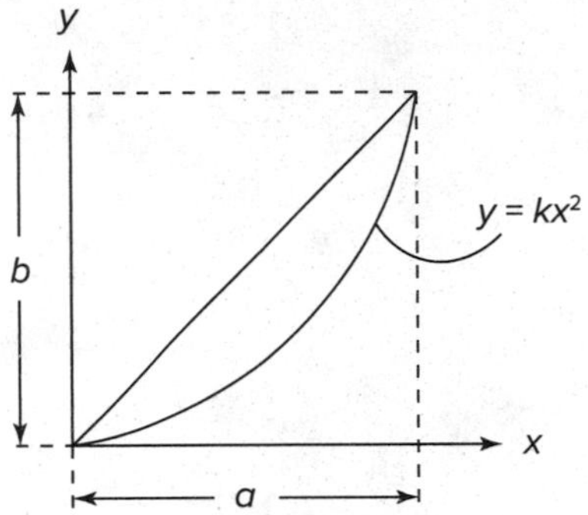

$\bar{x}$ = ______________________

$\bar{y}$ = ______________________

PROBLEM 6 (16.7 POINTS)

Find the surface area and volume of the given shape. The groove is semicircular with a radius of 20 mm.

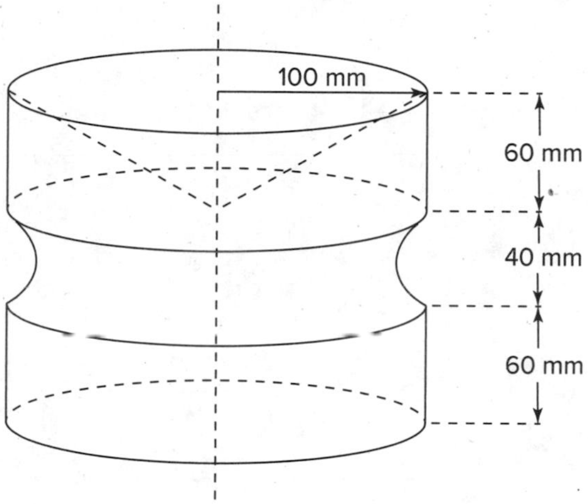

Surface area = ____________________

Volume = ____________________

Exam 2—Statics, Practice Set 2

Department of Mechanical Engineering

My Favorite University

Good Luck!

120 MINUTES

This class: Sec. 001 – Dr. Hanson

(This exam covers the end of Level 4 through Level 6.)

RULES:

1. The solutions to this exam are in the back of the book; however, you are not permitted to look at them until after your practice exam!
2. All your work must be done on the papers provided including this cover sheet.
3. Closed book and closed notes.
4. You may have a FE approved calculator, pens, pencils, erasers.
5. Use of a phone of any kind is considered cheating.
6. Treat this as your real exam and time yourself in a nice quiet place at a desk, as if you are taking this in your own classroom.

HONOR CODE—YOUR ME DEPARTMENT

I hereby certify that I will follow the Code of Student Conduct as defined by the University and the Department, that I will not cheat nor will I condone cheating.

Name ____________________________________ (Print legibly)

Name ____________________________________ (Signature)

PROBLEM 1 (20 POINTS)

Four forces and a couple are applied to a frame as shown in the figure below. Determine (a) the magnitude and direction of the resultant and (b) the perpendicular distance d_A from point A to the line of action of the resultant along a horizontal line passing through A and G.

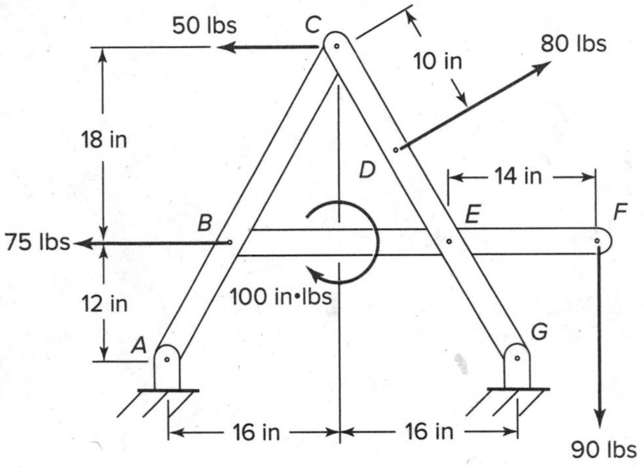

$|\vec{F}_R| =$ ________________

Distance from $A =$ ________________

PROBLEM 2 (20 POINTS)

A bar is loaded and supported as shown in the figure below. The bar has a uniform cross section and weighs 100 lbs. Determine the reaction at supports *A*, *B*, and *C*.

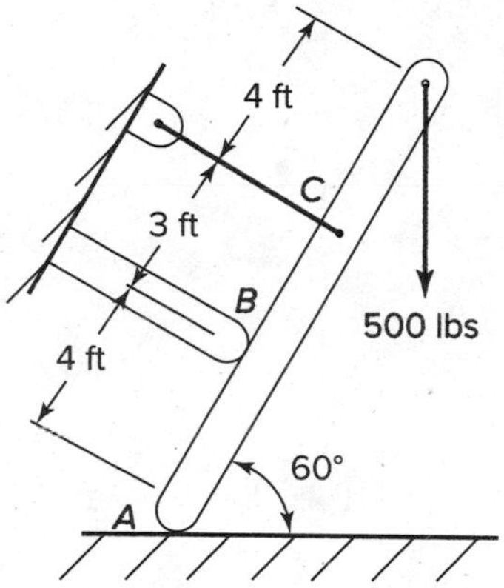

Reaction *A* = ____________________

Reaction *B* = ____________________

Reaction *C* = ____________________

PROBLEM 3 (20 POINTS)

A post and bracket is used to support a pulley as shown in the figure below. A cable passing over the pulley transmits a 500-lb load as shown in the figure. Determine the reaction at support *A* of the post.

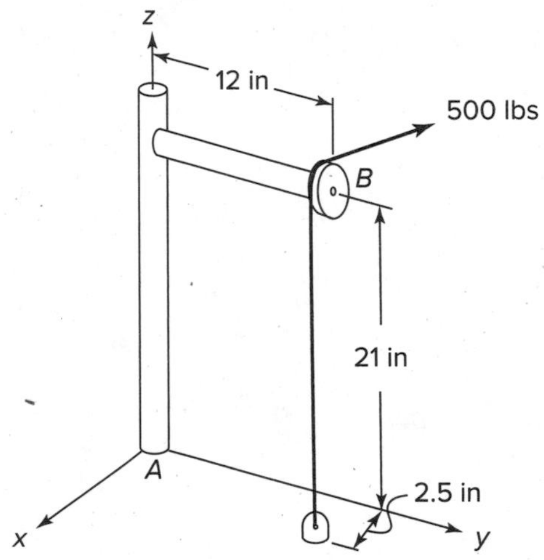

$A_x =$ ______________________ $MA_x =$ ______________________

$A_y =$ ______________________ $MA_y =$ ______________________

$A_z =$ ______________________ $MA_z =$ ______________________

PROBLEM 4 (20 POINTS)

Rotate the shaded area shown in the figure below 360° around the x-axis. Find $\bar{x}$ of the shaded area and the volume of that area swept around the x-axis.

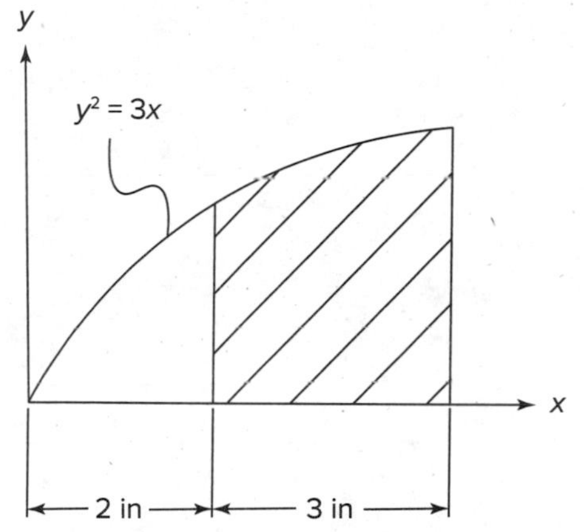

$\bar{x}$ = ______________________

Volume = ______________________

PROBLEM 5 (20 POINTS)

Locate the centroid $(\bar{x}, \bar{y})$ of the area given in the figure below. The figure is a quarter circle on top of a narrow rectangle with a square hole removed.

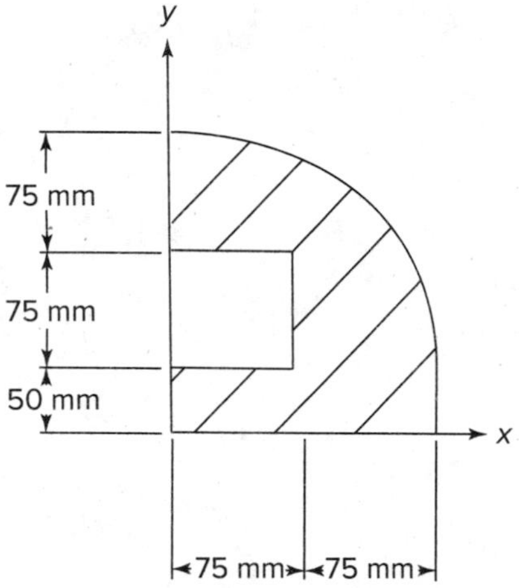

$\bar{x} =$ ________________________

$\bar{y} =$ ________________________

Exam 3—Statics, Practice Set 1

Department of Mechanical Engineering

My Favorite University

Good Luck!

120 MINUTES

This class: Sec. 001 – Dr. Hanson

(This exam covers Level 7 through Level 9.)

RULES:

1. The solutions to this exam are video solutions which you can access at the QR code; however, you are not permitted to look at them until after your practice exam!
2. All your work must be done on the papers provided including this cover sheet.
3. Closed book and closed notes.
4. You may have a FE approved calculator, pens, pencils, erasers.
5. Use of a phone of any kind is considered cheating.
6. Treat this as your real exam and time yourself in a nice quiet place at a desk, as if you are taking this in your own classroom.

HONOR CODE—YOUR ME DEPARTMENT

I hereby certify that I will follow the Code of Student Conduct as defined by the University and the Department, that I will not cheat nor will I condone cheating.

Name ______________________________ (Print legibly)

Name ______________________________ (Signature)

PROBLEM 1 (14.3 POINTS)

Find the force in members *AB, AD, AE,* and *BD*. State whether each member is in tension or compression.

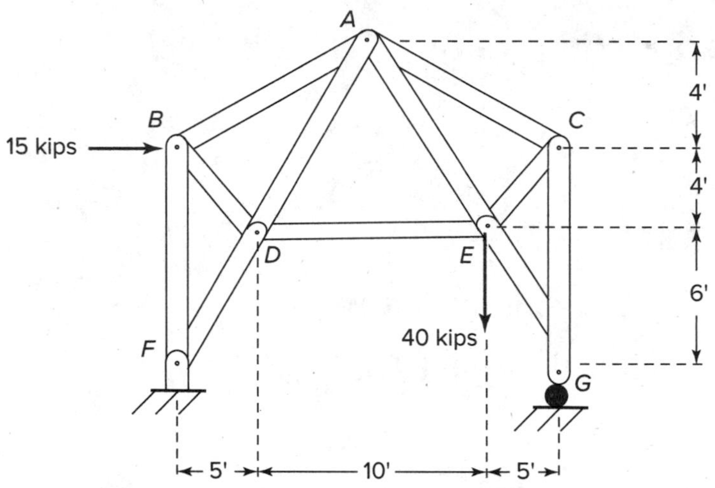

AB = ______________________

AD = ______________________

AE = ______________________

BD = ______________________

PROBLEM 2 (14.3 POINTS)

Find all forces on member *DABC*.

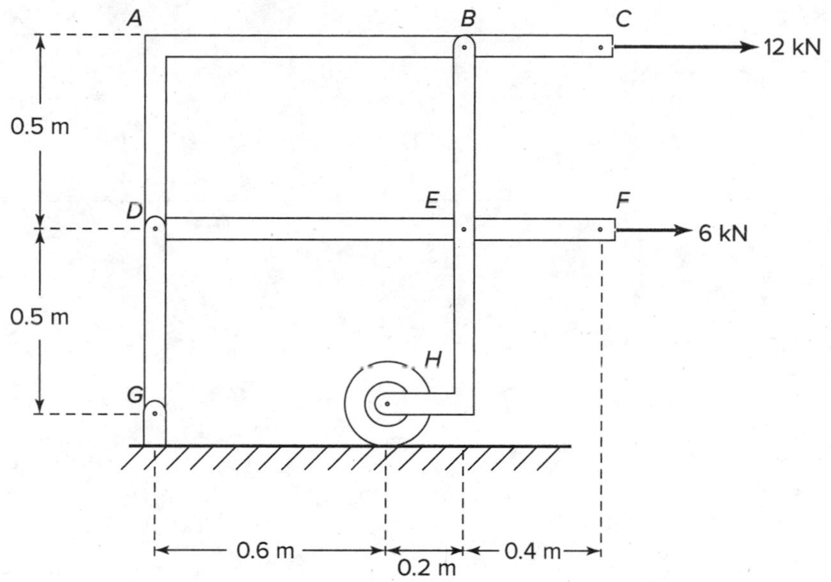

$B_x =$ ____________________

$B_y =$ ____________________

$D_x =$ ____________________

$D_y =$ ____________________

PROBLEM 3 (14.3 POINTS)

If the radius of the pulleys is 150 mm, find the internal forces at J and K.

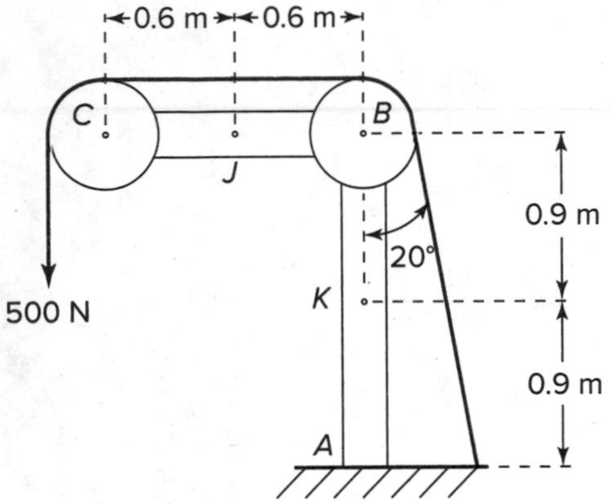

$M_J =$ __________ $M_K =$ __________

$N_J =$ __________ $N_K =$ __________

$V_J =$ __________ $V_K =$ __________

PROBLEM 4 (14.3 POINTS)

Draw the shear/moment diagram for the given loaded compound beam.

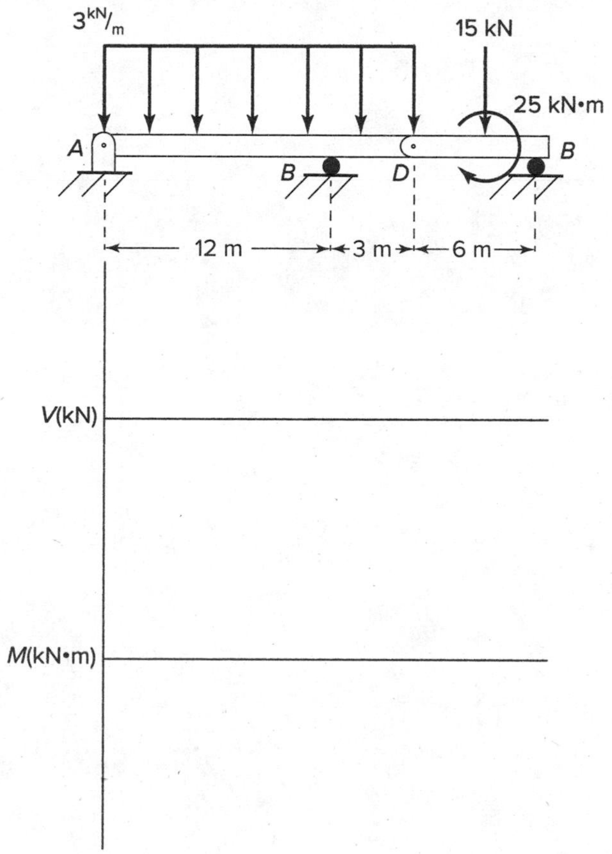

PROBLEM 5 (14.3 POINTS)

Find the angle θ where motion occurs. The static coefficient of friction between all surfaces is 0.15.

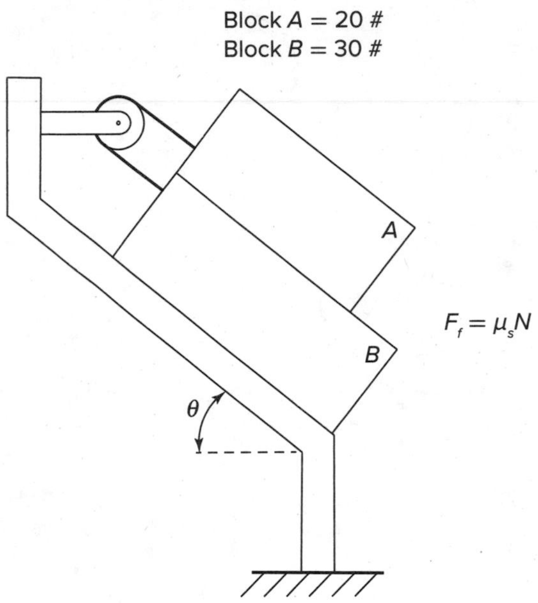

Angle $\theta =$ ______________________

PROBLEM 6 (14.3 POINTS)

Find force P where motion in the system occurs. The uniform crate has a mass of 60 kg, and the uniform cart has a mass of 10 kg. The static coefficient of friction between the cart and the floor is 0.35, and the static coefficient of friction between the cart and the crate is 0.50.

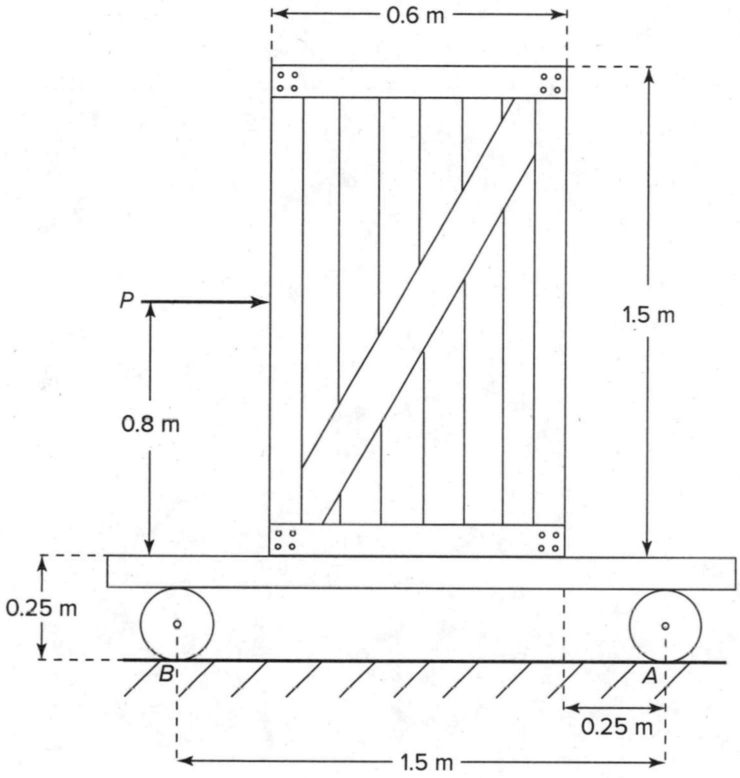

Force $P =$ ____________________

PROBLEM 7 (14.3 POINTS)

Find force P required to raise the 3 kN weight on block A. The weight of block A and block B is negligible. The static coefficient of friction is 0.25 between all surfaces.

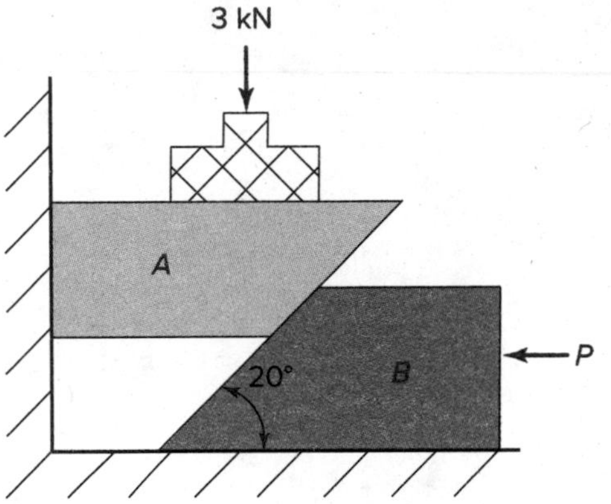

Force $P =$ ______________________

Exam 3—Statics, Practice Set 2

Department of Mechanical Engineering

My Favorite University

Good Luck!

120 MINUTES

This class: Sec. 001 – Dr. Hanson

(This exam covers Level 7 through Level 9.)

RULES:

1. The solutions to this exam are in the back of the book; however, you are not permitted to look at them until after your practice exam!
2. All your work must be done on the papers provided including this cover sheet.
3. Closed book and closed notes.
4. You may have a FE approved calculator, pens, pencils, erasers.
5. Use of a phone of any kind is considered cheating.
6. Treat this as your real exam and time yourself in a nice quiet place at a desk, as if you are taking this in your own classroom.

HONOR CODE—YOUR ME DEPARTMENT

I hereby certify that I will follow the Code of Student Conduct as defined by the University and the Department, that I will not cheat nor will I condone cheating.

Name ______________________________ (Print legibly)

Name ______________________________ (Signature)

PROBLEM 1 (20 POINTS)

The homogeneous block has a mass of m kg, and the coefficient of friction between the block and the plane is 0.40. The incline is 20 degrees. If the force P increases gradually until motion ensues, will the block slide or tip and for what value of P? Solve in terms of m (mass) and P (force).

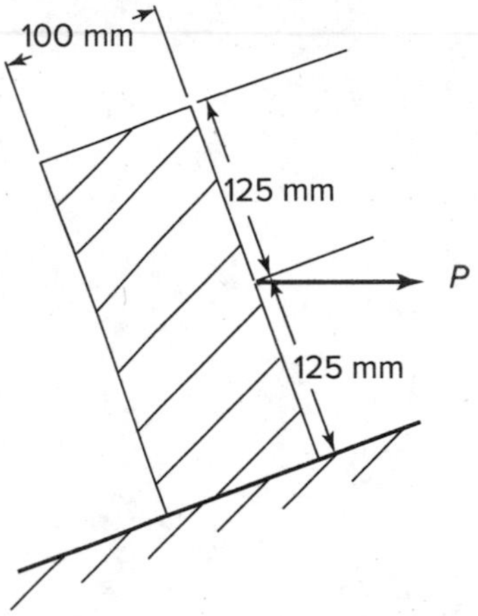

P for slipping = ____________________

P for tipping = ____________________

Will it slip or tip? ____________________

PROBLEM 2 (20 POINTS)

Find the tension in cable *FH*. Disc *G* has a weight of 500 lbs to be applied at point *C*.

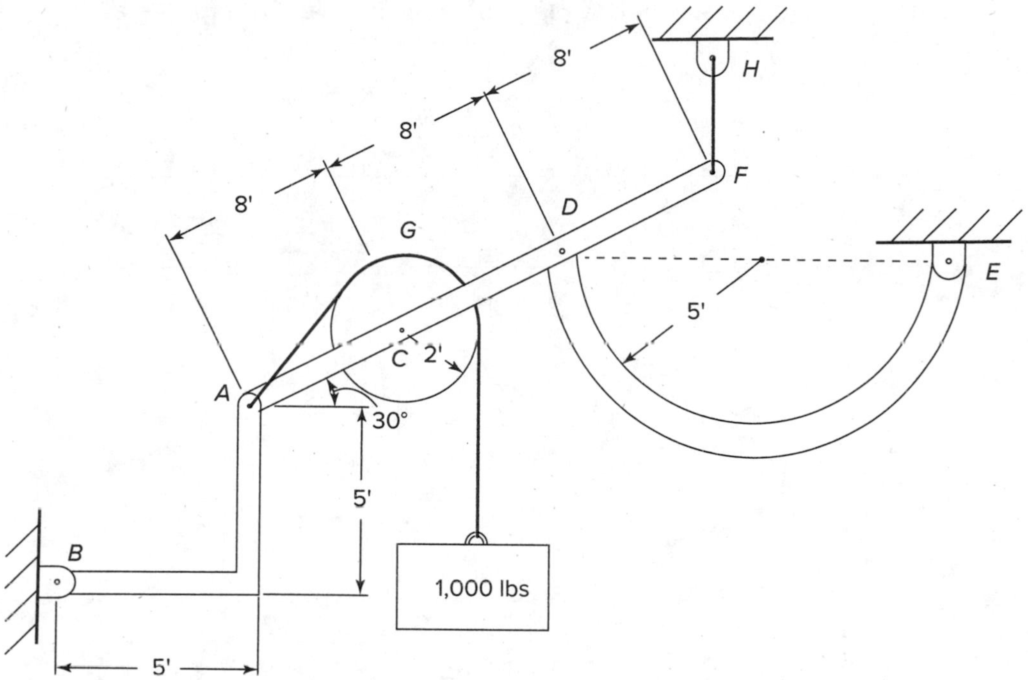

Tension in *FH* = ____________________

PROBLEM 3 (20 POINTS)

The light bar is used to support the 50 kg block in its vertical guides. If the coefficient of static friction is 0.30 at the upper end of the bar and 0.40 at the lower end of the bar, find the friction force acting at each end for $x = 75$ mm. Also find the maximum value of x for which the bar will not slip.

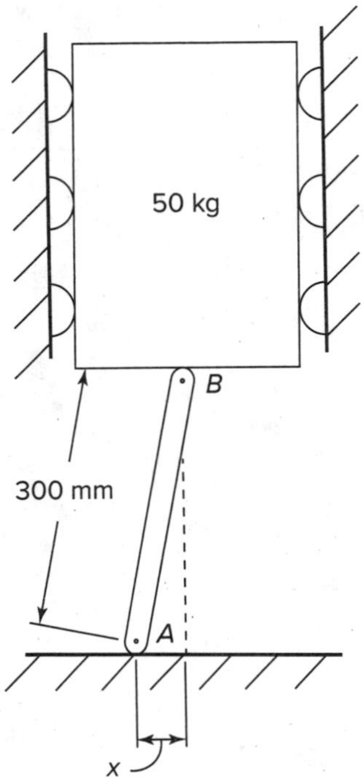

At 75 mm:

$F_A =$ ____________________

$F_B =$ ____________________

Maximum $x =$ ____________________

PROBLEM 4 (20 POINTS)

Find the forces in members *CD*, *DG*, *FG*, and *CG* and state whether each member is in tension or compression.

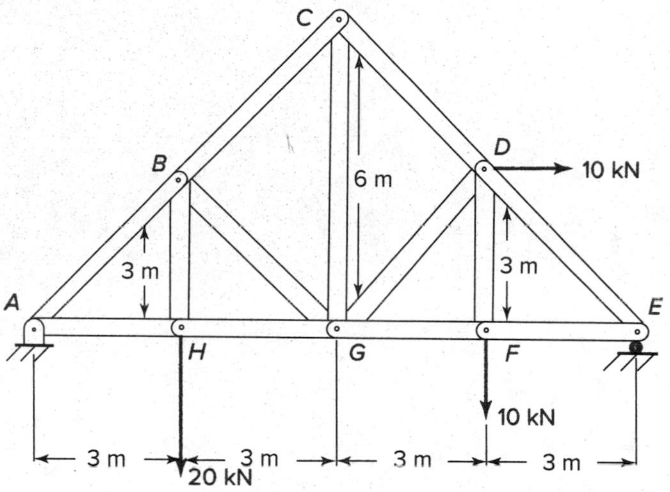

$CD =$ __________

$DG =$ __________

$FG =$ __________

$CG =$ __________

PROBLEM 5 (20 POINTS)

Draw the shear/moment diagram for the following load.

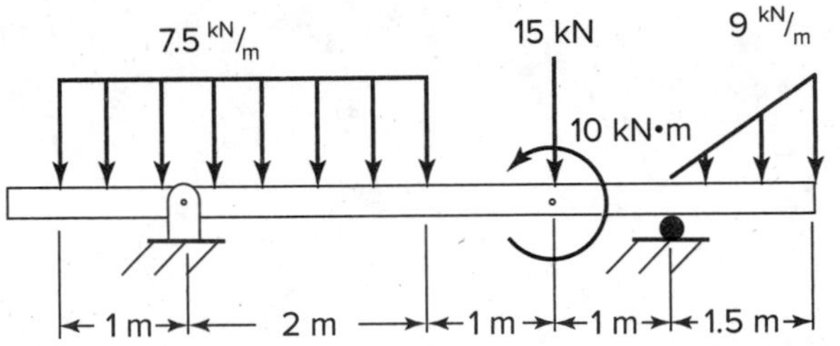

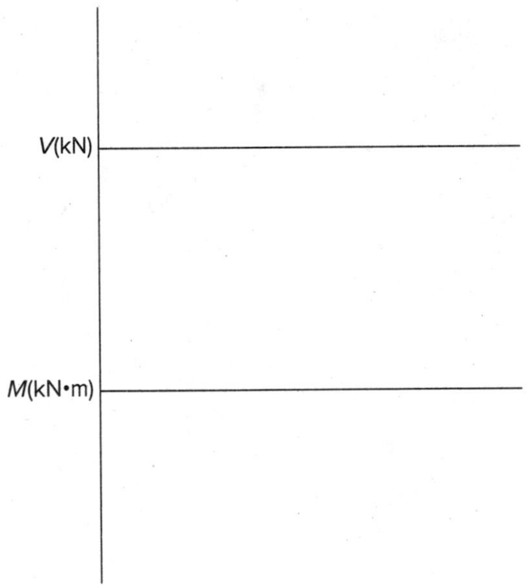

Exam Solutions

EXAM 1: PROBLEM 1

$\vec{F}_1 \quad \theta_x = 35^\circ \quad \theta_y = 75^\circ \quad \theta_z = ?$

From the Figure $\theta_z > 90^\circ$

$\cos^2 35^\circ + \cos^2 75^\circ + \cos^2\theta_z = 1$

$0.671 + 0.0670 + \cos^2\theta_z = 1$

$\therefore \cos^2\theta_z = 0.262$

$\cos\theta_z = \pm 0.512$

$\therefore \theta_z = 59.2^\circ; \boxed{120.8^\circ}$

$F_x = 600(\cos 35) = 491.5\,N$

$F_y = 600(\cos 75) = 155.3\,N$

$F = 600(\cos 120.8) = -307.2\,N$

$\vec{F}_2$ Find Unit Vector

$$\hat{\lambda} = \frac{-12\hat{i} + 18\hat{j} + 20\hat{k}}{\sqrt{12^2 + 18^2 + 20^2}}$$

$$= \frac{-12\hat{i} + 18\hat{j} + 20\hat{k}}{29.46}$$

$$= -0.407\hat{i} + 0.611\hat{j} + 0.679\hat{k}$$

$$\therefore \vec{F}_2 = 450(-0.407\hat{i} + 0.611\hat{j} + 0.679\hat{k})\,N$$

$$= -183.2\hat{i} + 275.0\hat{j} + 305.6\hat{k}\quad N$$

$\vec{F}_3$ - use "blue triangles" Method

$F_x = 500[\sin(65)\cos(-105)] = -117.3\,N$

$F_y = 500[\sin(65)\sin(-105)] = -437.7\,N$

$F_z = 500\cos 65 = 211.3\,N$

$\vec{F}_4 = 250\hat{i}\,N$

$\vec{F}_1 = 491.5\hat{i} + 155.3\hat{j} - 307.2\hat{k}\ N$

$\vec{F}_2 = -183.2\hat{i} + 275.0\hat{j} + 305.6\hat{k}\ N$

$\vec{F}_3 = -117.3\hat{i} - 437.7\hat{j} + 211.3\hat{k}\ N$

$\vec{F}_4 = 250\hat{i}\ N$

$\boxed{\vec{F}_R = 441.0\hat{i} - 7.400\hat{j} + 209.7\hat{k}\ N}$

$|\vec{F}_R| = \sqrt{441.0^2 + 7.4^2 + 209.7^2} = 488.4\,N$

Direction Cosines:

$\alpha = \cos^{-1}[441/488.4] = 25.5^\circ$

$\beta = \cos^{-1}[-7.4/488.4] = 90.9^\circ$

$\gamma = \cos^{-1}[209.7/488.4] = 64.6^\circ$

EXAM 1: PROBLEM 2

FBD

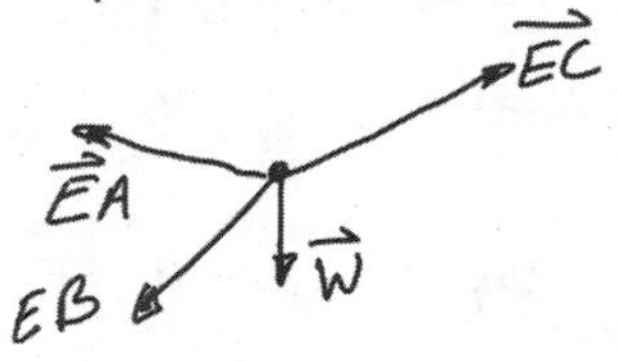

Express As Vectors

EA: $(2,-4,0)-(0,0,-6)=(2,-4,6)$

$$\hat{\lambda}_{EA}=\frac{2\hat{i}-4\hat{j}+6\hat{k}}{\sqrt{2^2+4^2+6^2}}$$

$$=0.267\hat{i}-0.535\hat{j}+0.802\hat{k}$$

① $\vec{F}_{EA}=F_{EA}(0.267\hat{i}-0.535\hat{j}+0.802\hat{k})$

EB: $(8,2,0)-(0,0,-6)=(8,2,6)$

$$\hat{\lambda}_{EB}=\frac{8\hat{i}+2\hat{j}+6\hat{k}}{\sqrt{8^2+2^2+6^2}}$$

$$=0.784\hat{i}+0.196\hat{j}+0.588\hat{k}$$

② $F_{EB}(0.784\hat{i}+0.196\hat{j}+0.588\hat{k})$

EC: $(-3,3,0)-(0,0,-6)=(-3,3,6)$

$$\hat{\lambda}_{EC}=\frac{-3\hat{i}+3\hat{j}+6\hat{k}}{\sqrt{3^2+3^2+6^2}}$$

$$=-0.408\hat{i}+0.408\hat{j}+0.816\hat{k}$$

③ $\vec{F}_{EC}=F_{EC}(-0.408\hat{i}+0.408\hat{j}+0.816\hat{k})$

$$\vec{W}=W\hat{k}$$

$$\Sigma\vec{F}=\vec{0}$$

$\Sigma F_x=0=0.267F_{EA}+0.784F_{EB}-0.408F_{EC}$

$\Sigma F_y=0=-0.535F_{EA}+0.196F_{EB}+0.408F_{EC}$

$\Sigma F_z=0=0.802F_{EA}+0.588F_{EB}+0.816F_{EC}-W$

Let 8 kN be the max in a rope

Set $F_{EA}=8$ kN ∴ $F_{EB}=2.188$ kN ($F_{EC}=9.439$ kN)
$W=15.4$ kN (No Good!)

Set $F_{EC}=8$ kN $F_{EA}=6.780$ kN $F_{EB}=1.854$ kN
$W=13.1$ kN (OK)

Set $F_{EB}=8$ kN ($F_{EA}=29.25$ kN) ($F_{EC}=34.5$ kN)
$W=56.3$ kN (No Good!)

∴ F_{EC} is limited to 8 kN

$$\boxed{W_{MAX}=13.1\text{ kN}}$$

EXAM 1: PROBLEM 3

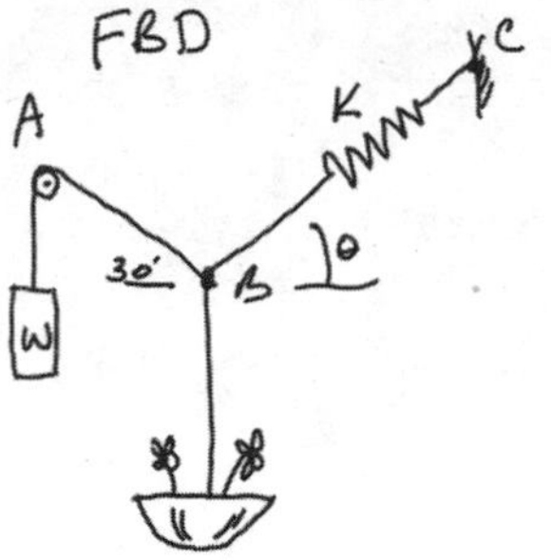

$K = 50\text{ lb/in}$ Basket $= 80\text{ lb}$

$W = 100\text{ lb}$

FBD Point B

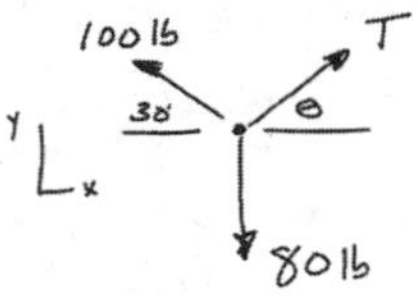

$\Sigma F_x = 0$

① $-100\cos 30 + T\cos\theta = 0$

$\Sigma F_y = 0$

② $100\sin 30 + T\sin\theta - 80 = 0$

From ①

$T = \dfrac{100\cos 30}{\cos\theta} = \dfrac{86.6}{\cos\theta}$

Sub into ②

$100\sin 30 + \dfrac{86.6}{\cos\theta}\cdot\sin\theta - 80 = 0$

$50 + 86.6\tan\theta - 80 = 0$

$\tan\theta = 30/86.6$

$\boxed{\theta = 19.1°}$

From ① $T = 86.6/\cos\theta =$

$86.6/\cos 19.1 = \boxed{91.7\text{ lb} = T}$

Since $K = 50\dfrac{\text{lb}}{\text{in}} =$

$\text{Force} = K\cdot\Delta$

$91.7\text{ lb} = 50\dfrac{\text{lb}}{\text{in}}\Delta$

$\therefore \boxed{\Delta = 1.83''}$

EXAM 1: PROBLEM 4

Assume Points A & C As well As B & D Are Alingned Vertically.

$\Sigma M_E = 400[-2.25\cos 40 - \sin 40]$

$+ 220[2.25\cos 75 + \sin 75]$

$+ 300[-0.75\cos 25 - \sin 25]$

$+ 150[-0.75\cos 35 + \sin 35]$

$M_E = -942.76\text{ ft lb}$

Max Moment From D is when the Force is ⊥ To Point E.

Moment Arm

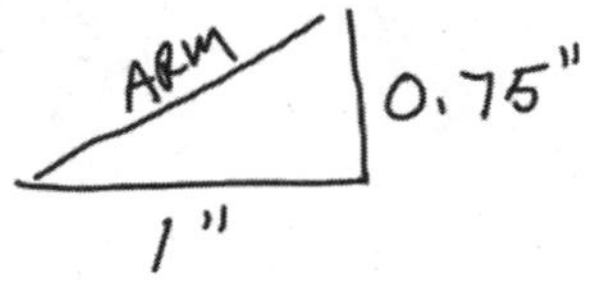

$\text{Arm} = \sqrt{1^2 + 0.75^2} = 1.25''$

$M_D = 150(1.25) = \pm 187.5$

$M_{MAX} = -1,124.14\text{ ft lb}$

$M_{MIN} = -749.10\text{ ft lb}$

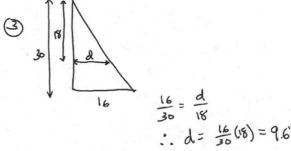

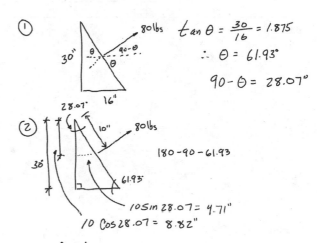

EXAM 1: PROBLEM 5

① $\hat{\lambda}_{BC} = \dfrac{-2.5\hat{\imath} + 3\hat{\jmath} + 4\hat{k}}{\sqrt{2.5^2 + 3^2 + 4^2}}$

$= -0.447\hat{\imath} + 0.537\hat{\jmath} + 0.716\hat{k}$

$\therefore \vec{F} = 250(-0.447\hat{\imath} + 0.537\hat{\jmath} + 0.716\hat{k})$

$= -111.75\hat{\imath} + 134.25\hat{\jmath} + 179.00\hat{k}$ N

② $\vec{M}_A = \vec{r}_{AB} \times \vec{F}$

$= (2\hat{\imath} + 4.5\hat{\jmath} + 2.5\hat{k}) \times$

$(-111.75\hat{\imath} + 134.25\hat{\jmath} + 179.00\hat{k})$

③ $\begin{array}{c|ccc|} & \hat{\imath} & \hat{\jmath} & \hat{k} \\ \vec{r}_{AB} & 2 & 4.5 & 2.5 \\ \vec{F} & -111.75 & 134.25 & 179 \end{array}$

$[(4.5)(179) - (2.5)(134.25)]\hat{\imath}$

$- [(2)(179) - (2.5)(-111.75)]\hat{\jmath}$

$+ [(2)(134.25) - (4.5)(-111.75)]\hat{k}$

$\boxed{\therefore \vec{M}_A = 469.88\hat{\imath} - 637.38\hat{\jmath} + 771.38\hat{k} \text{ N·m}}$

EXAM 2: PROBLEM 1

①

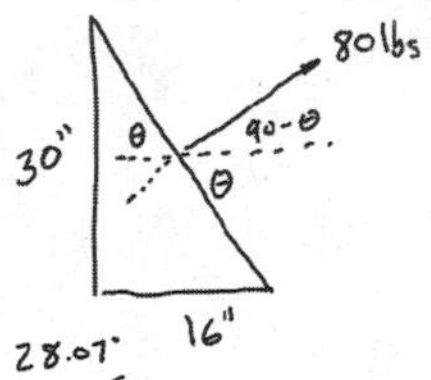

$\tan\theta = \dfrac{30}{16} = 1.875$

$\therefore \theta = 61.93°$

$90 - \theta = 28.07°$

②

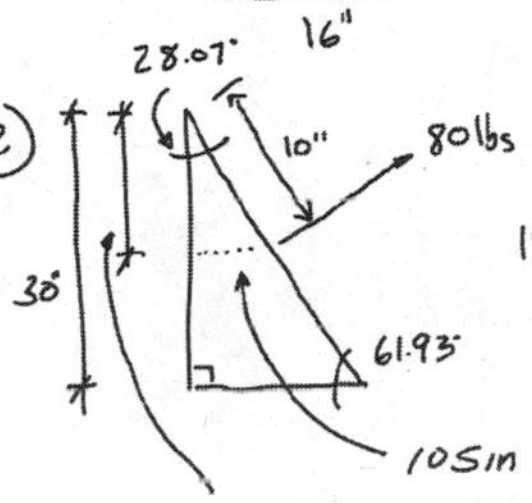

$180 - 90 - 61.93$

$10 \sin 28.07 = 4.71"$

$10 \cos 28.07 = 8.82"$

③

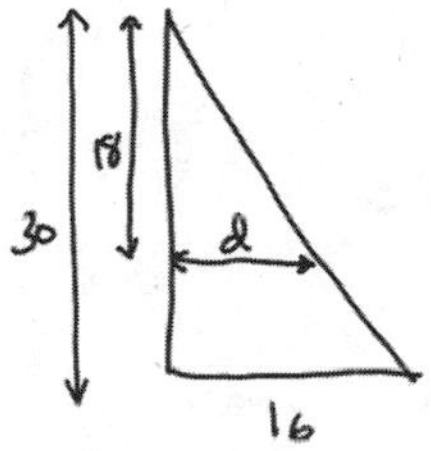

$\dfrac{16}{30} = \dfrac{d}{18}$

$\therefore d = \dfrac{16}{30}(18) = 9.6"$

④ $\Sigma M_A = 75(12) + 50(30) - 90(16 + 9.6 + 14) - 100$

$-80\cos 28.07(12 + 18 - 8.82) + 80\sin 28.07(16 + 4.71)$

$= -1{,}979.5$ lb·in

$\Sigma F_x = -75 - 50 + 80\cos 28.07 = -54.41$ lb (Left)

$\Sigma F_y = -90 + 80\sin 28.07 = -52.36$ lb (Down)

$\vec{F}_R = -54.41\hat{\imath} - 52.36\hat{\jmath}$ lb

⑤ $-1{,}979.5 = d_A(-52.36)$

$\boxed{d_A = 37.8 \text{ in}}$

EXAM 2: PROBLEM 2

FBD

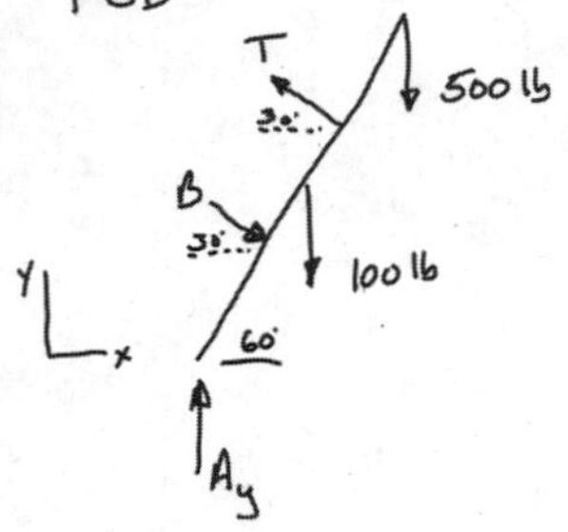

$\sum F_x = 0$

$B\cos 30 - T\cos 30 = 0$

$\therefore\ B = T$

$\sum F_y = 0$

$T\sin 30 - B\sin 30 - 100 + A_y - 500 = 0$

$\boxed{A_y = 600\text{ lb}}$

$\sum M_A = 0$

$-500(11\cos 60) - 100(5.5\cos 60) + T(7) - B(4) = 0$

$\therefore\ \boxed{T = 1{,}008.3\text{ lb} \quad B = 1{,}008.3\text{ lb}}$

EXAM 2: PROBLEM 3

FBD

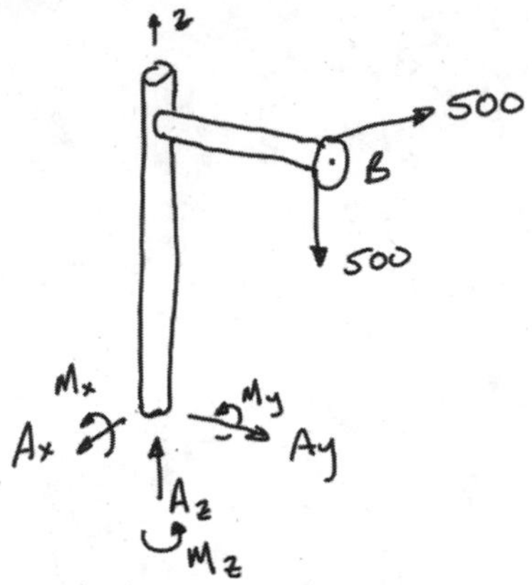

$\sum F_x = A_x - 500 = 0$

$\therefore\ \boxed{A_x = 500\text{ lb}}$

$\sum F_y = \boxed{A_y = 0}$

$\sum F_z = A_z - 500 = 0$

$\therefore\ \boxed{A_z = 500\text{ lb}}$

$\sum M_x = M_x - 500(12) = 0$

$\therefore\ \boxed{M_x = 6{,}000\text{ in}\cdot\text{lb}}$

$\sum M_y = M_y - 500(21 + 2.5) + 500(2.5) = 0$

$\therefore\ \boxed{M_y = 10{,}500\text{ in}\cdot\text{lb}}$

$\sum M_z = M_z + 500(12) = 0$

$\therefore\ \boxed{M_z = -6{,}000\text{ in}\cdot\text{lb}}$

EXAM 2: PROBLEM 4

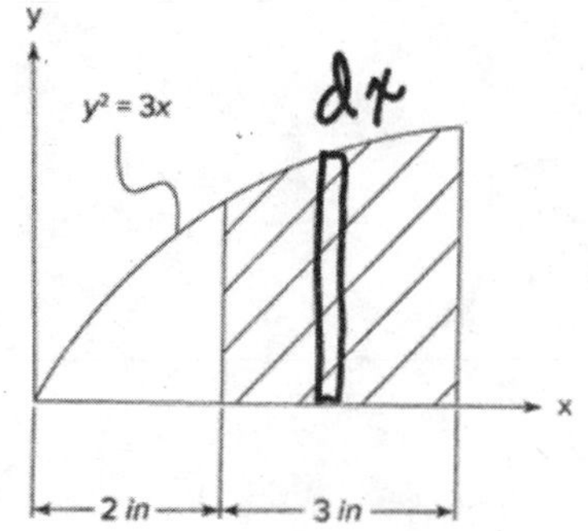

① $dA = y\,dx = \sqrt{3}\,x^{1/2}\,dx$

$$A = \int_2^5 \sqrt{3}\,x^{1/2}\,dx = \frac{2\sqrt{3}}{3}\left[x^{3/2}\right]_2^5 = \frac{2}{3}\sqrt{3}\left(5^{3/2} - 2^{3/2}\right)$$

$\therefore A = 9.644 \text{ in}^2$

②
$$A\bar{x} = \int x\,dA = \int_2^5 x\left(\sqrt{3}\,x^{1/2}\right)dx$$
$$= \sqrt{3}\int_2^5 x^{3/2}\,dx = \frac{2}{5}\sqrt{3}\left[x^{5/2}\right]_2^5$$
$$= \frac{2}{5}\sqrt{3}\left(5^{5/2} - 2^{5/2}\right)$$

$A\bar{x} = 34.811$

$V = 2\pi A\bar{x}$

$= 2\pi(34.811)$

$V = 218.7 \text{ in}^3$

$\therefore \bar{x} = 34.811/9.644 = 3.61 \text{ in}$

EXAM 2: PROBLEM 5

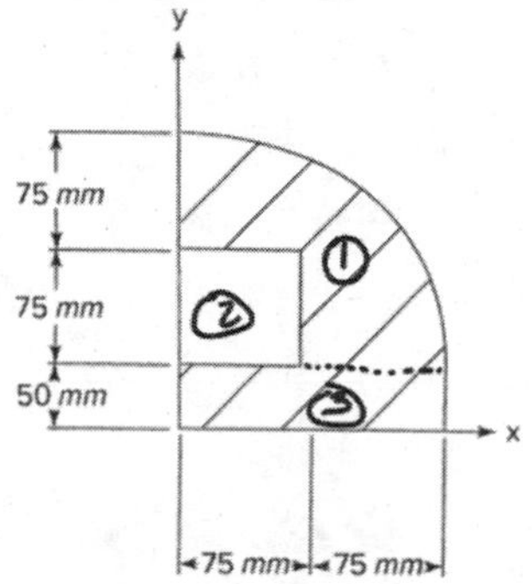

	Area (mm²)	$\bar{x}$	$A\bar{x}$	$\bar{y}$	$A\bar{y}$
1	$\frac{\pi(150)^2}{4} = 17{,}671.5$	63.662	1,125,003.	113.662	2,008,578
2	$-(75)^2 = -5625$	37.5	-210,937.5	87.5	-492,187.5
3	7,500	75	562,500	25	187,500
	19,546.5		1,476,565.5		1,703,890.5

$$\therefore \bar{x} = \frac{1{,}476{,}565.5}{19{,}546.5} = 75.54 \text{ mm}$$

$$\bar{y} = \frac{1{,}703{,}890.5}{19{,}546.5} = 87.17 \text{ mm}$$

EXAM 3: PROBLEM 1

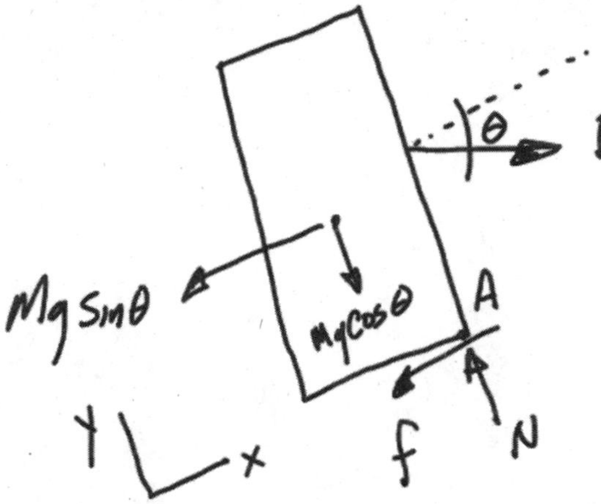

① $\Sigma F_y = 0 \quad N - Mg\cos 20 - P\sin 20 = 0$

② $\Sigma F_x = 0 \quad P\cos 20 - Mg\sin 20 - f = 0$

For slip $f = \mu N$

① → $N = 9.218M + 0.342P$

$\therefore f = 0.4(9.218M + 0.342P)$

sub into ②

$0.940P - 3.355M - 0.4(9.218M + 0.342P) = 0$

$0.940P - 3.355M - 3.687M - 0.137P = 0$

$0.803P - 7.042M = 0$

$\therefore$ $\boxed{P = 8.77M}$ ONSET OF SLIPPING

$\Sigma M_A = 0 \quad -(P\cos 20)\,125 + (Mg\cos 20)\,50 + (Mg\sin 20)(125) = 0$

$-117.46P + 460.9M + 419.4M = 0$

$\therefore$ $\boxed{P = \frac{880.3}{117.46}M = 7.49M}$ TIP

Since P for tipping is smaller than P for slipping ....

$\therefore$ Box Tips ✓

EXAM 3: PROBLEM 2

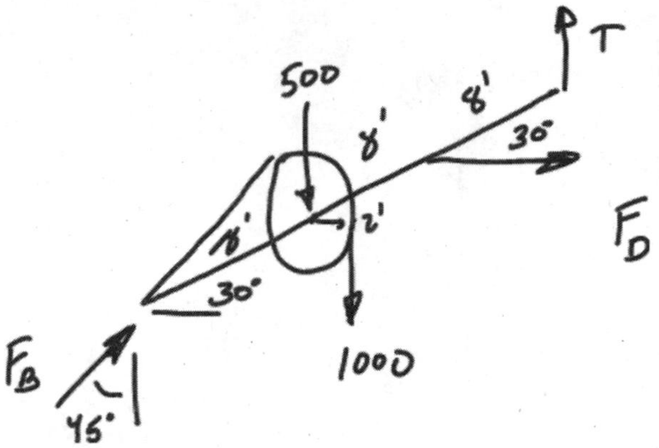

$\Sigma M_A = 0$

$-500(8\cos 30) - 1000(2 + 8\cos 30) - F_D(16\sin 30) + T(24\cos 30) = 0$

$\Sigma F_x = 0$

$F_D + F_B\sin 45 = 0$

$\Sigma F_y = 0$

$F_B\cos 45 - 500 - 1000 + T = 0$

$-12{,}392.3 - 8F_D + 20.7846T = 0$

$0.7071F_B + F_D = 0$

$\therefore F_D = -0.7071F_B$

$-1500 + 0.7071F_B + T = 0$

$\therefore T = 1500 - 0.7071F_B$

$$-13,392.3 - 8\left[-0.7071 F_B\right] +$$
$$20.7846\left[1500 - 0.7071 F_B\right] = 0$$

$$-12,392.3 + 5.657 F_B + 31,176.9$$
$$- 14.697 F_B = 0$$

$$18,784.6 - 9.040 F_B = 0$$

$$\therefore \boxed{F_B = 2077.9 \text{ lb}}$$

$$T = 1500 - 0.7071 F_B$$

$$\therefore \boxed{T = 30.72 \text{ lb}}$$

$$F_D = -0.7071 F_B$$

$$\therefore \boxed{F_D = -1,469.3 \text{ lb}}$$

EXAM 3: PROBLEM 3

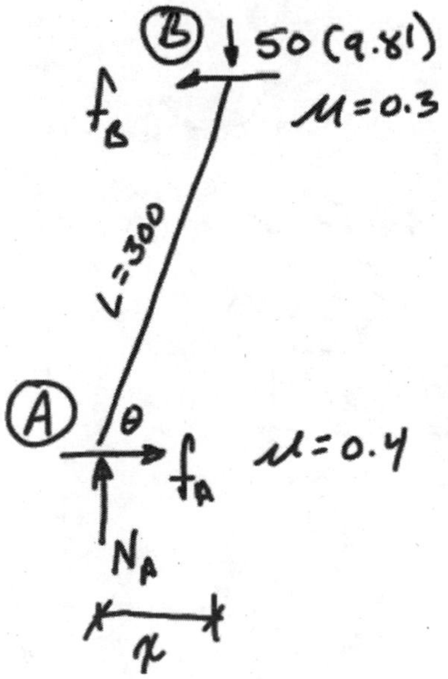

$$\Sigma F_y = 0$$
$$N_A = 50(9.81) = 490.5 \text{ N}$$
$$(f_A)_{MAX} = 0.4(490.5) = 196.2 \text{ N}$$
$$(f_B)_{MAX} = 0.3(490.5) = 147.2 \text{ N}$$

$$\Sigma M_A = 0$$
$$f_B\left(\sqrt{300^2 - 75^2}\right) - 490.5(75) = 0$$
$$\therefore f_b = 126.65 \text{ N} < 147.2 \text{ N}$$

∴ No slip at B

$$\Sigma F_x = 0$$
$$f_A = f_B = 126.65 \text{ N}$$

∴ No slip at A

$$\boxed{f_A = f_B = 126.65 \text{ N}}$$ ✓

$$\Sigma M_A = 0$$
$$-490.5(x) + 147.2\left(\sqrt{300^2 - x^2}\right) = 0$$
$$3.332x = (300^2 - x^2)^{1/2}$$
$$(3.332x)^2 = 300^2 - x^2$$
$$11.102x^2 = 300^2 - x^2$$
$$12.102x^2 = 300^2$$
$$x^2 = 7,436.79$$
$$\therefore \boxed{x = 86.2 \text{ mm}}$$

EXAM 3: PROBLEM 4

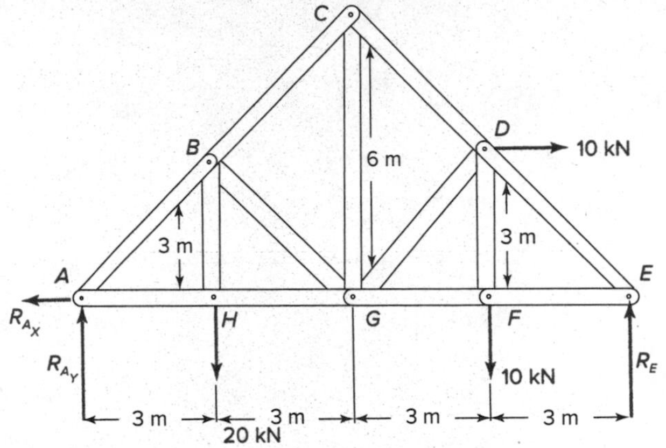

$$\Sigma M_A = 0$$

$$-(20)(3) - (10)(9) - (10)(3) + R_E(12) = 0$$

$$\therefore \boxed{R_E = 15 \text{ kN}}$$

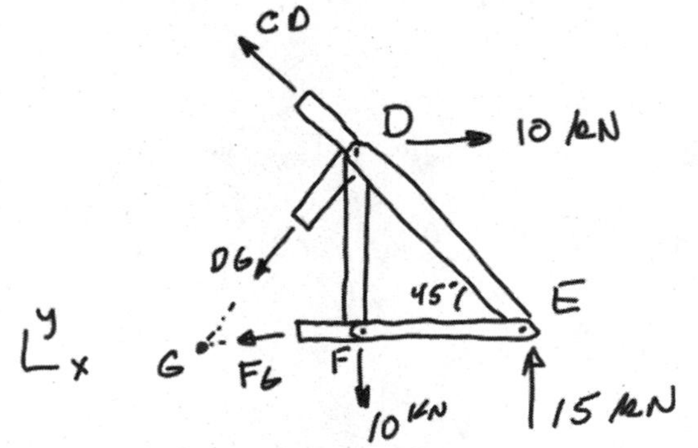

$$\Sigma M_G = 0$$

$$-(10)(3) + (15)(6) - (10)(3) + CD\left(\sqrt{3^2 + 3^2}\right) = 0$$

$$\boxed{CD = -7.07 \text{ kN}} \ C$$

$$\Sigma M_D = 0$$

$$15(3) - FG(3) = 0$$

$$\therefore \boxed{FG = 15 \text{ kN}} \ T$$

$$\Sigma F_y = 0$$

$$15 - 10 + CD\cos 45 - DG\cos 45 = 0$$

$$CD = -7.07$$

$$5 - 7.07\cos 45 = DG\cos 45$$

$$0 = DG\cos 45$$

$$\therefore \boxed{DG = 0}$$

FBD Pin C

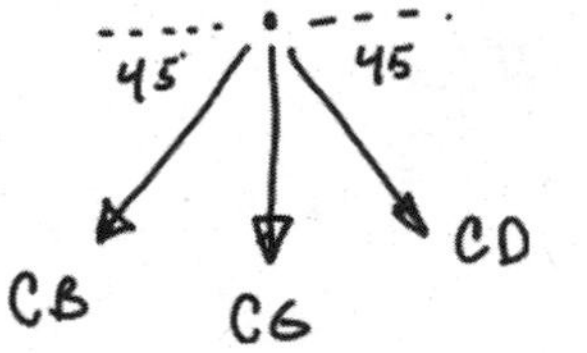

$$\Sigma F_x = 0$$

$$CD\cos 45 - CB\cos 45 = 0$$

$$CD = CB = -7.07 \text{ kN}$$

$$\Sigma F_y = 0$$

$$-CB\sin 45 - CD\sin 45 - CG = 0$$

$$-(-7.07)\sin 45 - (-7.07)\sin 45 = CG$$

$$2(7.07)\sin 45 = CG$$

$$\boxed{CG = 10 \text{ kN}}$$

EXAM 3: PROBLEM 5

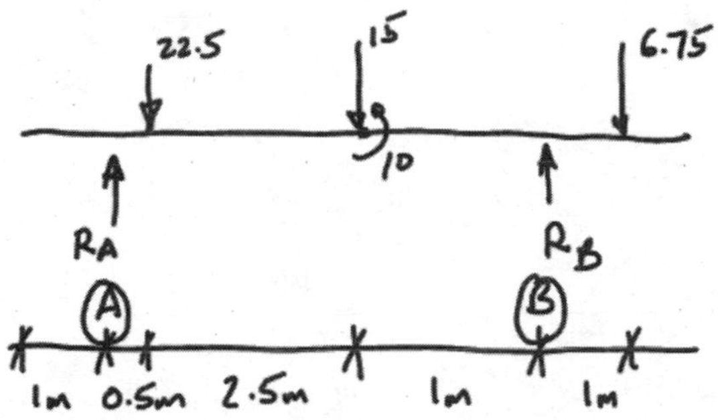

$\Sigma M_A = 0$

$-22.5(0.5) - 15(3) + 10 + R_B(4) - 6.75(5) = 0$

$\therefore \boxed{R_B = 20.0 \text{ kN}}$

$\Sigma F_y = 0$

$-22.5 + 20.0 - 15 + R_A - 6.75 = 0$

$\therefore \boxed{R_A = 24.25 \text{ kN}}$

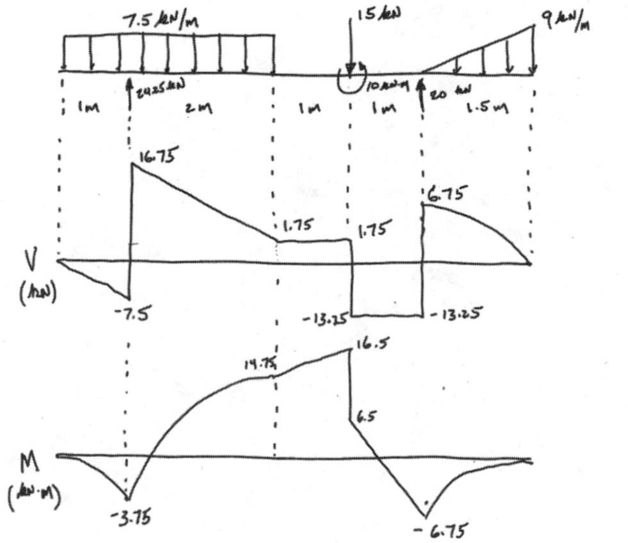